Advanced Research in Shipping Informatics and Communications

Advanced Research in Shipping Informatics and Communications

Guest Editor

Nikitas Nikitakos

MDPI

Basel • Beijing • Wuhan • Barcelona • Belgrade • Novi Sad • Cluj • Manchester

Guest Editor
Nikitas Nikitakos
Department of Shipping,
Trade and Transport
University of the Aegean
Chios
Greece

Editorial Office
MDPI AG
Grosspeteranlage 5
4052 Basel, Switzerland

This is a reprint of the Special Issue, published open access by the journal *Journal of Marine Science and Engineering* (ISSN 2077-1312), freely accessible at: https://www.mdpi.com/journal/jmse/special_issues/vl_shipping_informatics_communications.

For citation purposes, cite each article independently as indicated on the article page online and as indicated below:

Lastname, A.A.; Lastname, B.B. Article Title. *Journal Name* **Year**, *Volume Number*, Page Range.

ISBN 978-3-7258-5543-8 (Hbk)
ISBN 978-3-7258-5544-5 (PDF)
https://doi.org/10.3390/books978-3-7258-5544-5

Contents

About the Editor

Nikitas Nikitakos

Nikitas Nikitakos is a graduate of the Hellenic Naval Academy (1980) and holds a B.Sc. in Economics (University of Piraeus 1986) and two M.Sc. degrees from the Naval Postgraduate School, Monterey, CA, USA (M.Sc. Electrical Engineering and M.Sc. in Appl. Mathematics). He spent 25 years as Naval Officer (Captain H.N. ret.) when he participated in several NATO and EU committees. He received a Ph.D. in Electrical and Computer Engineering from the National Technical University of Athens (1996). He was Professor of Shipping Informatics and New Technologies (2001–2023) and Head of the Dept. of Shipping Trade and Transport at the University of the Aegean (2005–2009). He participated in more than 50 European research projects and studies (as coordinator and principal researcher) on ports and shipping. He holds three international patents on renewable energies at sea and he was awarded by Lloyd's List on Maritime Technological Innovation in 2006. From Aug. 2023 to Aug. 2024 he was Head of Dept. of Maritime Logistics and Supply Chain management at Sharjah Maritime Academy, UAE. He has published five books and many articles in internationally referenced journals and conferences. He is a Visiting Professor at Shanghai Maritime University, the World Maritime University (WMU) Sweden, Hangzhou Dianzi University, China (Honorary Doctorate), and Kaliningrad State Technical University (Honorary Doctorate). He holds ISPS, PMP, and PMI-RPM, Prince2 certifications.

Preface

The subject of this Special Issue and its reprint is "Advanced Research in Shipping Informatics and Communications". The scope of this Special Issue is to present research covering the incorporation of information and communication technologies in shipping and the maritime domain. The implementation of new technologies from additive manufacturing to e-navigation technologies and Artificial Intelligence, all part of the digitalization of the shipping industry, is examined through published research. The multi-faced cyber security threat landscape in maritime security is covered as well, showcasing the operational and technological vulnerabilities of the shipping sector for OT and IT systems onshore in ports and facilities and onboard maritime assets. This Special Issue aims to fill a gap in the research of shipping informatics and communication, extending to the technological revolution experienced through Shipping 4.0 in the maritime domain.

The Guest Editor, Prof. Dr. Nikitas Nikitakos, would like to acknowledge the support provided by Dr. Iosif Progoulakis (University of the Aegean) in the coordination of the manuscript submittals and reviews for this Special Issue as well as the editing, review and co-writing of the Editorial.

Nikitas Nikitakos
Guest Editor

Journal of
Marine Science
and Engineering

MDPI

Editorial

Advanced Research in Shipping Informatics and Communications

Nikitas Nikitakos * and Iosif Progoulakis

Department of Shipping Trade and Transport, University of the Aegean, Korai 2a, GR82132 Chios, Chios, Greece; i.progoulakis@aegean.gr
* Correspondence: nnik@aegean.gr

The shipping industry, a vital part of global trade and the maritime transport system (MTS) and a critical infrastructure for all nations, is undergoing a technological evolution through the integration of advanced information and communication technologies. This Special Issue has explored innovations in informatics and communications in the maritime domain, highlighting their pivotal role in shaping the sector's future.

At the forefront of this technological shift are maritime cyber physical systems [1], which are merging the physical components of maritime operations with digital innovations to create smart, interconnected networks. These systems facilitate live tracking and command and information analysis, enabling informed decision-making and the optimal use of technical resources. However, the increased connectivity of maritime cyber physical systems leads to new cybersecurity weaknesses, creating the need for robust protective measures to secure vital maritime infrastructure onboard vessels and at shore in port facilities.

The rise of autonomous vessels [2] marks a significant change in the maritime transport system. These vessels create an expectation of increased efficiency and a decrease in human errors through sophisticated communication networks, advanced sensors, and complex algorithms. The legal and regulatory aspects underlying the adaptation of unmanned vessels are a major concern, requiring global industry cooperation and the cultivation of a solid regulatory framework to guarantee their safe and accountable use.

Maritime satellite communication technologies [3] are essential for maintaining contact in the vast and often isolated marine environment. These technologies enable seamless data transfer, voice exchange, and internet access, supporting a wide array of maritime applications, including navigation, weather prediction, and emergency responses. They act as the essential link for reliable communication in the offshore maritime environment, guaranteeing the flow of crucial data.

The incorporation of IoT, AI, and data science [4] is radically altering shipping operations. AI methods, such as those used to predict ship performance metrics, enable real-time tracking of vessel operations, preventive maintenance, and streamlined route planning. Data fusion and analysis support data-driven decisions, improving operational effectiveness and reducing asset operational expenditure. These tools offer unique insights into operational patterns and optimize the distribution of technical resources.

Offshore technologies [5], including those used in oil and gas exploration and renewable energy production, rely heavily on advanced communication and information networks. These systems assist in the remote monitoring, control, and maintenance of offshore platforms, ensuring safe and productive operations in a challenging environment. The ability to remotely manage and monitor these assets is crucial for safety, productivity, and environmental protection.

E-navigation technologies [6], including AIS, ECDIS, VTS, GMDSS, and SAR, are crucial in enhancing maritime safety and efficiency. The IMO's e-navigation plan aims to merge these tools into a comprehensive and user-friendly navigation system. These systems facilitate real-time information distribution, collision prevention, and search and rescue operations, contributing to a safer maritime environment and more efficient shipping operations.

Maritime cybersecurity [7–9] is an increasing concern in today's digital age. Cyber threats pose substantial risk to maritime operations, in the form of data breaches, system failures, and piracy. Strong cybersecurity procedures are essential to protect vital asset infrastructure and ensure the safety of personnel and cargo. Understanding interdependencies and risk factors is crucial for implementing efficient security measures both on board vessels and at the shore, in port infrastructure and facilities.

Additive manufacturing [10], or 3D printing, is transforming the procurement and handling of maritime spare parts. This technology allows for on-demand, Just-In-Time (JIT) production and stock management of parts, cutting down on overall inventory costs and downtime. This capability is especially valuable in remote locations, where traditional supply chains may be slow or unreliable.

Smart port technologies [11] are improving the efficiency, resiliency, and sustainability of port operations. These tools include automated cargo handling, intelligent traffic control, and real-time data fusion and analysis. These advancements are vital for managing the increasing volume of global trade and minimizing the environmental impact of port activities.

The Human–Machine Interface (HMI) [12] in maritime systems is vital for guaranteeing safe and effective interaction between operators and technology. User-friendly interfaces and intuitive controls are essential for reducing human error and maximizing operational efficiency in increasingly complex systems.

Maritime digitalization [13–16] is driving a major shift in the shipping industry. Digital platforms, data analysis, and automation are enabling greater transparency, efficiency, resiliency, and sustainability. This digital transformation is not just about adopting new technologies, but about altering the ways in which maritime operations are conducted.

Upon concluding this Special Issue, it becomes evident that the integration of these technologies is not merely a technological advancement, but a strategic imperative for the maritime sector. By accepting this technological shift and its tools as force multipliers in the industry and embracing digitalization, the maritime sector can enhance safety, efficiency, resiliency, and sustainability, ultimately ensuring its continued crucial role in global trade.

Author Contributions: Conceptualization, N.N. and I.P.; resources, N.N. and I.P.; writing—original draft preparation, N.N. and I.P.; writing—review and editing, N.N. and I.P.; supervision, N.N.; project administration, N.N. and I.P. All authors have read and agreed to the published version of the manuscript.

Conflicts of Interest: The authors declare no conflicts of interest.

References

1. Progoulakis, I.; Rohmeyer, P.; Nikitakos, N. Cyber physical systems security for maritime assets. *J. Mar. Sci. Eng.* **2021**, *9*, 1384. [CrossRef]
2. Johansson, T.M.; Fernández, J.E.; Dalaklis, D.; Pastra, A.; Skinner, J.A. (Eds.) *Autonomous Vessels in Maritime Affairs: Law and Governance Implications*; Springer Nature: Berlin/Heidelberg, Germany, 2023.
3. Chen, W.; Li, C.; Yu, J.; Zhang, J.; Chang, F. A survey of maritime communications: From the wireless channel measurements and modeling perspective. *Reg. Stud. Mar. Sci.* **2021**, *48*, 102031. [CrossRef]
4. Alexiou, K.; Pariotis, E.G.; Zannis, T.C.; Leligou, H.C. Prediction of a ship's operational parameters using artificial intelligence techniques. *J. Mar. Sci. Eng.* **2021**, *9*, 681. [CrossRef]

5. Lee, P.T.W.; Cullinane, K. (Eds.) *Dynamic Shipping and Port Development in the Globalized Economy*; Palgrave Macmillan: London, UK, 2016.
6. International Maritime Organization (IMO). *E-Navigation Strategy Implementation Plan, Update 1*; (MSC.1/Circ.1595); IMO: London, UK, 2018.
7. Mohsendokht, M.; Li, H.; Kontovas, C.; Chang, C.-H.; Qu, Z.; Yang, Z. Decoding dependencies among the risk factors influencing maritime cybersecurity: Lessons learned from historical incidents in the past two decades. *Ocean Eng.* **2024**, *312*, 119078. [CrossRef]
8. Katsikas, S.K.; Kavallieratos, G.; Amro, A. *Future Trends in Maritime Cybersecurity*; Elsevier eBooks; Elsevier: Amsterdam, The Netherlands, 2024; pp. 1663–1678. [CrossRef]
9. Peng, P.; Xie, X.; Claramunt, C.; Lu, F.; Gong, F.; Yan, R. Bibliometric analysis of maritime cybersecurity: Research status, focus, and perspectives. *Transp. Res. Part E Logist. Transp. Rev.* **2025**, *195*, 103971. [CrossRef]
10. Kostidi, E.; Nikitakos, N.; Progoulakis, I. Additive manufacturing and maritime spare parts: Benefits and obstacles for the end-users. *J. Mar. Sci. Eng.* **2021**, *9*, 895. [CrossRef]
11. Zarzuelo, I.d.l.P.; Soeane, M.J.F.; Bermúdez, B.L. Industry 4.0 in the port and maritime industry: A literature review. *J. Ind. Inf. Integr.* **2020**, *20*, 100173. [CrossRef]
12. Kitada, M.; Baldauf, M.; Mannov, A.; Svendsen, P.A.; Baumler, R.; Schröder-Hinrichs, J.-U.; Dalaklis, D.; Fonseca, T.; Shi, X.; Lagdami, K. Command of vessels in the era of digitalization. In *Advances in Human Factors, Business Management and Society, Proceedings of the AHFE 2018 Advances in Intelligent Systems and Computing, Orlando, FL, USA, 21–25 July 2018*; Kantola, J.I., Nazir, S., Barath, T., Eds.; Springer: Cham, Switzerland, 2019; Volume 783, pp. 385–394. [CrossRef]
13. Raza, Z.; Woxenius, J.; Vural, C.A.; Lind, M. Digital transformation of maritime logistics: Exploring trends in the liner shipping segment. *Comput. Ind.* **2023**, *145*, 103811. [CrossRef]
14. Aiello, G.; Giallanza, A.; Mascarella, G. Towards Shipping 4.0. A preliminary gap analysis. *Procedia Manuf.* **2020**, *42*, 24–29. [CrossRef]
15. Lind, M.; Michaelides, M.; Ward, R.; Watson, R.T. (Eds.) *Maritime Informatics*; Springer: Berlin/Heidelberg, Germany, 2021. [CrossRef]
16. Artikis, A.; Zissis, D. (Eds.) *Guide to Maritime Informatics*; Springer: Berlin/Heidelberg, Germany, 2021.

Journal of
Marine Science and Engineering

MDPI

Article

A Vulnerability Centric System of Systems Analysis on the Maritime Transportation Sector Most Valuable Assets: Recommendations for Port Facilities and Ships

Chronis Kapalidis [1,*], Stavros Karamperidis [2], Tim Watson [3] and Georgios Koligiannis [4]

1 Cyber Security Centre, WMG, University of Warwick, Coventry CV4 7AL, UK
2 Plymouth Business School, University of Plymouth, Plymouth PL4 8AA, UK
3 Allan Turing Institute, London NW1 2DB, UK
4 Hellenic Navy, 15561 Cholargos, Greece
* Correspondence: chronis.kapalidis.1@warwick.ac.uk

Abstract: The maritime sector is a vital component of the global economy. Its international nature supersedes state boundaries and any disruption in its operations could have consequent and collateral global effects, affecting the socio-economic wellbeing of regions, states and peoples. The increasing adoption of digitalisation in the sector, primarily increases efficiency, minimize cost and maximise benefit, while improving safety, simultaneously introduces a new cyber threat landscape. The attack surface has broadened further due to the COVID-19 pandemic, as recent, high-profile cyber attacks to shipping companies have indicated. Cybersecurity is not limited to technology but involves people and business processes. Hence, to mitigate the security risk introduced by cyber threat actors, the industry, like any other, should initially focus on identifying its most critical assets and then adopt risk mitigation measures, spreading from legislative initiatives to company-specific technology solutions. Industry-led initiatives should promote the adoption of cyber-related policies and mechanisms that focus on business continuity. It should be the role of international bodies, classification societies and national authorities to ensure compliance and full implementation of these measures. This paper adopts a System of Systems Analysis to carry out a vulnerability assessment of port and ship ecosystem, while providing insights on the role of the aforementioned entities. Our analysis decomposes the industry's major assets; ports and ships, to specific subcomponents which are used as the basis of the vulnerability assessment. According to our findings, this approach highlights that the majority of these subcomponents; ports and ships, are increasingly vulnerable to cyber attacks.

Keywords: maritime sector; maritime security; cybersecurity; ports; ships

1. Introduction

Over the last decade, cybersecurity attracted a significant attention in the maritime domain. Reports for cyberattacks on ports and ships have increased substantially from 2017 to 2022 [1,2]. The impact of cyber-related incidents on the maritime economy is emerging dramatically, especially during the COVID-19 pandemic, as illustrated in Appendix A, where office staff are, in their majority working from home, and ship crew demand increased connectivity to keep in touch with people ashore. Maritime transport and all related activities, holistically defined as the Maritime Transportation System (MTS), are conducted by technology-intensive platforms, which today rely heavily on information systems [3].

This evolution of the maritime sector, based on innovation and digitalization, is helping the sector flourish. With the introduction of digital services apart from the reduction of burdens and the administration costs, time required to load and locate cargo, control the performance of critical onboard systems, and manage ship traffic has significantly reduced [4]. As a result, this has introduced a widened set of access points to maritime Industrial Control Systems (ICS). Considering that cyber attacks are becoming ever more sophisticated and commonplace [5], there is growing pressure from governments, regulatory bodies and international institutions for immediate action in prioritizing cybersecurity in the maritime sector.

The maritime domain consists of multiple and diverse components both on the infrastructure level and the services provided. Specifically, the MTS infrastructure consists of two key components: ports and ships. This argument does not intent to limit the components of a MTS but only to highlight the two predominant ones. According to the EU NIS Directive States can determine which essential service providers are to be considered as part of their CNI cybersecurity landscape. In January 2019, the Danish Maritime Authority released Order 46 which argues that service providers like VTS or AIS providers are essential services. Therefore, MTS infrastructure consists of ships, ports and other essential services. Any disruption to these systems can cause major disruption not only to the directly affected business, but even to the global supply chain. Due to increased maritime cyber threats policy should be targeted not only at the specific threat but to each component too, and provide a holistic security posture.

In June 2017 A.P. Moller Maersk's terminal in Ukraine was affected by the notPetya ransomware, in one of the most notable maritime-related cyber incident presented in Appendix A [6]. According to Gronholt-Pedersen [7] all business units at Maersk were affected including those in Rotterdam, Los Angeles, Mumbai and Auckland. Operations were forced either to completely shut down or temporarily slow down, resulting in a loss of revenue for the shipping giant initially estimated around $300 m [8–10]. Among other lessons learned from this incident, what is also highlighted is the importance of testing the adopted cybersecurity policies and measures in order to identify vulnerabilities and back doors to corporate systems, since the specific malware code was a known vulnerability that could have been spotted if an efficient assessment was conducted [9]. As the chairman of the company stated, A.P. Moller Maersk was also forced to change all its affected IT equipment and applications and restructure its global network, cyber policy and procedures to prevent and minimise the effect of any potential future attack [11].

This paper looks at the unique nature of cyber threats to the MTS by presenting the findings of new research conducted by the authors. These findings were based on a combination of desk-based research and stakeholder engagement. Specifically, semi-structured interviews and a roundtable discussion were conducted with subject matter experts, academics and practitioners (representatives from the public and private sectors along with international experts). Through a System-of-Systems Analysis (SOSA) the paper provides an overview of the cyber vulnerabilities that exist in ports and ships, the potential consequences of cyber attacks against them and the consequent affected fields. To illustrate the extent of such consequences on a socio-economic level, the paper uses the Persian Gulf and the Strait of Hormuz, one of the world's most important maritime chokepoint as a case study. Within that scope, the aim of the paper is to raise awareness on the importance of cybersecurity for the maritime sector, identify gaps in existing policies and consider how these should be addressed, while looking briefly at potential solutions. Section 2 demonstrates the current issues and highlights contemporary research directions of cybersecurity in the maritime sector. Section 3 demonstrates the methodology adopted for data collection and analysis. Section 4 demonstrates a thorough analysis of the vulnerability of the most critical assets in MTS. Section 5 demonstrates the findings of a case study of an MTS related cyber incident at the Persian Gulf through a socio-economic lens. Finally, Section 6 demonstrates the conclusion and recommendations.

Having reviewed the existing literature for the topic, the academic contribution of the paper focuses on providing initially a research-based identification of port and ship subcomponents through the lens of technology, and secondly a qualitative vulnerability assessment of cyber risks. Our paper provides useful insights to academics, researchers, students but also to industry practitioners.

2. Overview of the Current Landscape

The security of the maritime sector has been in the focus of the IMO for several years. In the aftermath of the 9/11 terrorist attacks in New York the organisation assembled expert groups to work on securing the sector from similar threats. This led to the introduction of the International Ship and Port Facility Security (ISPS) Code, which is an amendment to the Safety of Life at Sea (SOLAS) Convention (1974/1988) on minimum security arrangements for ships, ports and government agencies. Similarly, for cybersecurity, the IMO published the Interim Guidelines on Maritime Cyber Risk Management in 2016 and the consequent MSC-FAL.1/Circ.3 in 2017 (Guidelines on Maritime Cyber Risk Management) [12]. This document suggests that as of 1 January 2021 all organisations in the shipping industry must demonstrate cyber capability. Judging from the IMO's reactive approach to security, like the publication of the ISP code indicates, the fact that the IMO has already published an official documentation for cybersecurity indicates the importance of protecting the MTS against cyber threats.

Hence, risks related to cybersecurity are one of the main threats that the MTS faces with disruptive consequences that could lead even to loss of life, as elaborated later in the paper. Apart from the direct effects of a cyber attack on the system itself, the potential consequences of such an attack to areas intertwined with the industry, namely the economic and social sphere, should not be neglected. In order to examine the broader picture, we should look at: (1) The impact to the economy of an attack on a major port, (2) The consequences to people's everyday life, (3) The disruption to shipping routes if a ship is mishandled due to GPS spoofing and is run aground or causes a collision at a chokepoint.

In an effort to minimize the likelihood of these disruptive incidents occurring, we should, initially, identify the vulnerabilities of the systems used in the sector. Second, we should analyze the consequences of a cyber breach to these systems and third, the fields that may be affected by such an incident, since, as it will be highlighted throughout this paper the cyber-physical nature of these systems should not be neglected. To achieve the aforementioned, we have adopted SOSA. Such methodology brings together isolated systems to form a new more complex system, which offers more functionality and performance than the sum of its parts [13]. For the purposes of this paper, the port and ship ecosystems are regarded as system of systems and thus analyzed through a SOSA lens. Nowadays, as the adoption of digital technologies and services in the maritime domain is increasing, cybersecurity is becoming rather relevant in most maritime related activities. Ships, shipping companies' offices, port facilities, and any other company or organization, directly or indirectly, related to the shipping industry are increasingly incorporating digitalization into their daily operations. Computers and cyber-enabled technologies are being increasingly used for navigation, communications, cargo and ballast management, engineering, emergency systems, safety and security, environmental control, and many other purposes [14–16]. Digitization acts as an enabler to further improve the impressive record of efficiency and reliability for the MTS [17]. Building on the existent safety and security culture developed in the sector over the years, everyday operations, conducted manually in the past, are beginning to be facilitated using digital systems.

Great benefits come at great risk. The maritime sector could be described as a high value target for those engaged in unlawful acts, like smuggling, drug and human trafficking, piracy, terrorism and any other malicious activities due to large amounts of revenue involved in daily operations and the unpreparedness of the sector with regards its cyber protection [9,18]. The results of a cyber attack, depending on the threat actors' motivators, may vary from data loss or manipulation to disruption of trade activity, physical damage, environmental disaster or even injury or loss of life.

The predominant concern in the MTS is safety at sea. Maritime traffic control centres depend heavily on ICT equipment, such as navigation aid systems (GPS, GALILLEO, GLONASS, and AIS), safety communication systems (GMDSS, ship-to-ship and ship-to-shore, UHF/VHF/HF), satellite and GSM communications, in order to remain connected and monitor ships, while coordinating with port authorities, national shipping authorities, pilot agencies and other maritime organisations.

As an example, the Automatic Identification System (AIS), an automatic tracking system that uses transponders on ships to facilitate the safety of navigation at sea is vulnerable to jamming and spoofing attacks affecting the confidentiality and integrity of the system. Consequently, if a ship's navigation is tampered, it could lead to the ship running aground or cause collision and, in the worst-case scenario, an extended oil spill or even a gas explosion. The potential of a similar attack in offshore drilling units can have devastating effects [19]. Similarly, as it is analysed in the next section, ports, usually constituting part of a state's Critical National Infrastructure have several systems and services prone to cyber attacks.

3. Methodology

3.1. Justification of the Research Approach

It would be reasonable to approach this research in a rigorous way by constructing a simulation or theoretical environment, and then exploring how it might be attacked by developing a proof-of-concept attack, based on MITRE's ATT&CK framework [20] together with estimates of the likely resources and capabilities of attackers. However, this would highlight particular risks, but would be in danger of missing the more important aspects of the emergent risk in real, very large-scale systems of systems. As a consequence, the most appropriate approach to this research is based on a qualitative analysis of the component systems, and how they combine in modern port and maritime facilities. Identifying the key security controls, and how they are changing in the latest port and maritime systems of systems, provides insights into the areas of focus that ought to be highlighted to enable effective cybersecurity.

This approach, combined with sense-checking analysis by industry subject matter experts, balances appropriate academic rigour. It also provides a focus on the right scale of system of systems, to guard against an oversimplified test environment not capturing some of the most important aspects of cyber risk.

3.2. Data Collection and Analysis

Our research followed a qualitative approach. Research had three stages: (1) Desk-based research, (2) Stakeholder engagement, (3) Findings validation.

For the 1st stage (desk-based research) we conducted a semi systematic literature review, to identify main research on maritime cybersecurity. The literature review was conducted using an exploratory design and a thorough qualitative analysis of primary and secondary literature. Preliminary research identified a total of 295 documents using the following search terms:

"maritime cybersecurity"; "maritime cybersecurity"; "maritime" AND "cybersecurity"; "maritime" AND "cybersecurity"; "maritime" AND (cyber* or security*); "marine cybersecurity"; "marine cybersecurity"; "port cybersecurity"; "port cybersecurity".

These documents were then filtered using the following inclusion/exclusion criteria:

- The documentation should be in English
- The documents should be academic journal papers, reports, white papers, non-journal conference papers, books and international or national publications. No opinion articles, news agencies articles, blogs or any other online resource was assessed
- The document should be cyber-specific

2nd stage (Stakeholder engagement): Primary data was collected during the stakeholder engagement. Fifteen semi-structured interviews were conducted with a variety of stakeholders including: (1) shipowners and port operators, (2) Chief Information Security Officers (CISO), (3) Information security and technology providers, (4) Marine insurers, (5) Brokers, (6) Academics and (7) Policy makers. Saturation was achieved thus we stopped our data collection at 15 interviews. The interview questionnaire is presented in Appendix B. The years of experience of all interviewees is presented in Figure 1. Data collected was analysed to identify key challenges and it was then presented during three targeted workshops (3rd stage). Please see Figure 2 below, where a schematic representation of the three stages is demonstrated.

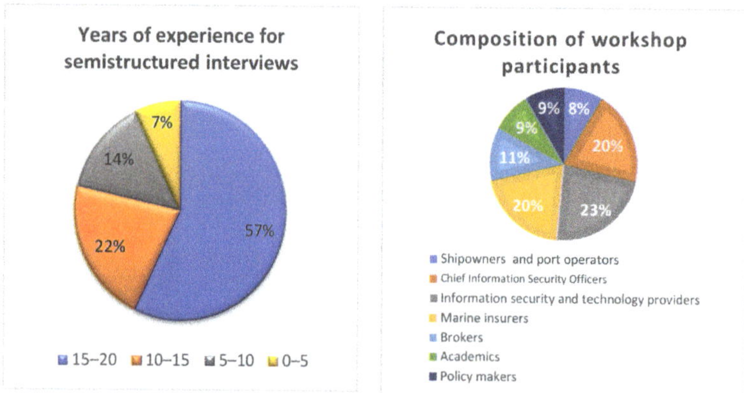

Figure 1. Participants' data. Source: Authors.

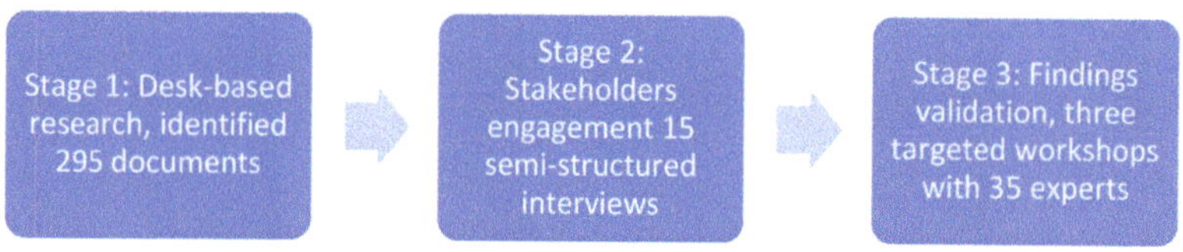

Figure 2. Research design. Source: Authors.

These three industry focused workshops were designed as validation activities of primary research findings. All the participants were senior representatives of the broader maritime sector as illustrated in Figure 1, with more than 15 years of experience. The total number of participants was 35. Both workshops were conducted under the Chatham House Rule facilitating participants in sharing sensitive information and real-life case studies. The outcomes of these workshops informed the findings presented in this paper.

A visual illustration of the research findings is demonstrated at Figures 3 and 4, in Sections 4.1 and 4.2 below, providing a detailed overview of vulnerabilities and consequences of a cyber incident to the industry's most valuable assets; the port and the ship.

3.3. *Limitations of the Methodology*

The authors applied a qualitative approach, focusing on understanding the vital cyber challenges in the MTS. Consequently, all the primary data collected could not be quantified and analysed through statistical procedures as in quantitative research. Compared to the quantitative methods, a significant constraint was encouraging more experts to participate in our research project, and thus we did not use a large sample that could numerically be considered a representative population. Mixed method research could further triangulate the findings and minimise the drawbacks of a traditional approach; however, the paper's purpose is not to present numerical parameters. Another limitation of the research is that the systematic literature review (i.e., 1st stage (desk-based research)) focused on scientific articles and reports published only in English.

4. The Vulnerabilities of the Most Critical Assets in the MTS: An Analysis

This part of the paper analyses the two main components of the MTS as identified initially and tries to examine how vulnerable these are to potential malicious cyber related activities. The paper will try to untangle the complexity of the sub-components of ports and ships, highlight the consequences of a cyber-related disruption to these components, while categorising the affected fields from such an incident. Consequent findings are the result of research conducted by the authors, consisting of desk-based research and qualitative, semi-structured interviews with industry practitioners, government officials, and academics, and were informed by the discussions instigated during a workshop with industry experts.

4.1. Port's Cyber Ecosystem

A port, as a cyber ecosystem, is a complex set of land and waterside systems and procedures, where the human factor still retains a predominant role. Over the last few years this ecosystem has rapidly become more digitalized, allowing the sector to thrive [21]. Naturally, this increased digitalization is expanding the attack landscape for cyber criminals and other threat actors, but also increases the likelihood of unintentional human error by, unfamiliar to the new technologies and cyber-hygiene practices and standards, company rotating crew staff. According to Allianz Global Corporate & Specialty [22], more than 75% of the marine casualties and accidents are attributable to human errors. The adoption of software-enabled systems and services offer a wide range of access points from where malicious software may infiltrate one or more of the port's systems. A breach to any of these systems or services may cause a wide range of disruptions to the port environment, varying from tampering with timely, efficient, and safe port operations, jeopardising health and safety of port and third-party staff, causing financial losses, environmental pollution and damaging an entity's reputation to facilitating smuggling or trafficking [23,24].

The cyber ecosystem or "cyber environment", as defined in the UK Department for Transport (DfT) Code of Practice, comprises the interconnected networks of both information and cyber-physical systems that use electronic, computer-based, and wireless systems, including information, services and social and business functions that exist only in cyberspace [14]. Applying this definition to ports, four main components can be identified: buildings, linear infrastructure, plant and machinery, and information and communication systems [14]. These four main components consist of 18 sub-components as illustrated in Table 1.

Table 1. Port ecosystem components.

Buildings (IT Based Systems)	Linear Infrastructure	Plant, Machinery and Operational Facilities (OT Based and SCADA Systems)	Information and Communication Systems
Port Facility offices	Roads	Cargo handling vehicles, cranes and conveyer assets	Financial Management and Monitoring
Port Facility accommodation	Access control points	Refineries	Container and vehicle movement systems
Warehouse and storage (chemical, gas, bulk and dry cargo) facilities	Rail systems	Power plants	Asset management systems (cargo, vehicle)
Vessel Traffic Control Tower	Utilities	Terminal Facilities (Container Terminal, Oil tank terminal)	Access control management system
Maintenance areas	Cargo handling systems such as pipelines and conveyer systems		

Source: Authors.

These four main components are interconnected with a wide range of operational and administrative processes, run by port staff and external third-party providers. In the event of a cyber incident occurring to one or more of these components, the consequences can be categorized in four distinct fields: data, human element, physical and environmental.

The vulnerabilities, consequences, and affected fields within the port environment are illustrated in Figure 3. Specifically, each of the sub-components is coloured based on the level of vulnerability (green: low, amber: medium, red: high). Consequently, a triangle colour based on the severity of the consequences is placed on its subcomponents (green: low, amber: medium, red: high). Finally, a circle consisting of the initial of each affected field (D: Data, E: Environment, H: Human, P: Physical) is placed on each sub-component.

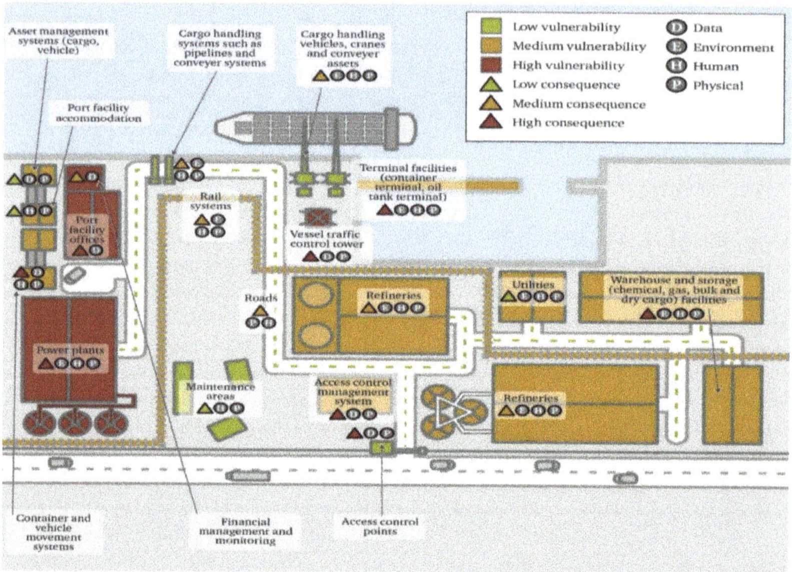

Figure 3. Port Components (Vulnerabilities, Consequences and Affected Fields). Note: This graphical illustration was generated with the support of Chatham House. Source: Authors.

Looking at the four port components initially, the buildings, linear infrastructure, plant and machinery and information and communication systems, as the diagram illustrates, **plant and machinery** used for cargo handling and port management, is the most vulnerable of the four aforementioned components, since it relies heavily on Operational Technology (OT) and, predominantly, Supervisory Control and Data Acquisition (SCADA) systems. These systems are the bulk systems of every port and regardless of the level of vulnerability, the consequences of a cyber attack on most of them are classified as severe, having an impact on more or less all of the aforementioned fields (data, environment, human, physical). Having said that, the most vulnerable sub-systems are the power plants, which are usually continuously connected to the regional main power controlling stations, forming, in this way, a part of the entire Critical National Infrastructure (CNI).

The second most vulnerable component is **the information and communication systems**, which are, a priori, related to data manipulation. Even though all the subcomponents are based on software, most of them are only connected to the internal port IT or OT network. An exception is the financial management and monitoring services used in order to conduct all financial activities of the port-based companies and port authority, which are constantly connected to the internet. Even though limited access to these services and systems may not halt port operations, it will certainly lead to financial instability, which in its turn could lead to other related disruptions, such as reputational damage and third-party compensation. Such an escalation could result in major disruptions of the port's business cycle.

Thirdly, the Vessel Traffic Control Tower (VTCT) is the most vulnerable sub-component of the **Building** category. The VTCT's operations rely on ship-to-shore, ship-to-ship communications, all of which use non-encrypted channels of the electromagnetic spectrum, and vessel management software. Port offices are also vulnerable to cyber-attacks as the entire breadth of port operations is monitored and, in some cases, managed by this facility. The consequences of a cyber attack to these two subcomponents are ranked as severe since such an incident would result in major disruption to port operations, from a seaside and shore side perspective. Port gates in one of Maersk's terminals at a major port in Europe, where all office-based systems were not usable due to the NotPetya malware, were shut as a result, causing lengthy delays and queues of trucks waiting to pick up containers.

Finally, the **linear infrastructure**, consisting of road and rail networks, access control points, utilities, and cargo handling systems, is, as easily understood, the least vulnerable of the four, due to its physical nature and the limited cyber elements incorporated in this. Having said that, even though the physical component of rail systems and utilities are not vulnerable, the monitoring, control and alarm systems encapsulated in them can, potentially, be affected by a cyber incident that could affect their normal operations, from a safety perspective. Such an incident would, consequently, affect the overall port operations.

4.2. Ship's Cyber Ecosystem

The ship, as a cyber environment, is what sets the challenges of the maritime sector apart from other industries. While the other components of the MTS are frequently similar to components present in other industries, the ship is the sector's most valuable asset and one which is, most of the time, operating independently at sea. While a port may operate similarly to other CNI assets, a ship, currently, when at sea does not rely on internet connectivity in order to conduct its main operations, i.e., navigation, engine control and cargo monitoring. There are although several subcomponents, critical to the ship's operations, that have moved from analogue to digital mode of operation. For that reason and due to the importance of the ship as a key asset of the MTS, the same SOSA is adopted.

Unlike the port, where four key components were identified, in the ship environment the various systems used to maintain reliable and consistent ship operability can be categorised in two main components: deck and engine. These two main components consist of 20 sub-components as illustrated in Table 2.

Table 2. Ship ecosystem components.

Deck		Engine	
Signal light column	Anchor and mooring winch control system	Engine Control Room System	Switchboards
Internal communications	TV-Entertainment-Internet Access Systems	Bow Thruster Control System	Water Ingress Detection System
GMDSS Console	Fleet Management system	Alarm Monitoring and Control System	Power Management
Navigation Equipment	Bridge control console	Cabling	
VDR/S-DR	Electrical Crane Equipment		
Reefer Container Monitoring System	Navigation Lights		
Loading and Stability Computer			

Source: Authors.

Based on research findings and as illustrated in Figure 4, there are several subcomponents which present low vulnerability to cyber attacks, a few of the deck's subcomponents are extremely vulnerable. The deck comprises of both IT and OT systems, which are in most cases interconnected. According to the research findings, the most vulnerable subcomponent of the entire ship are the **crew entertainment facilities**, including internet access. Even though it is gradually becoming common practice for most shipping companies to introduce network segregation to their IT infrastructure, separating the business from the crew network, it could be argued that this is not adequate to protect the ship's operations.

Although the two networks may be segregated, they both still use the sole IT infrastructure available on the ship. It only takes a careless seafarer to plug a malware-affected flash drive into the engine control system, to affect the engine management software, which in several cases still runs on a more vulnerable Windows XP environment [25].

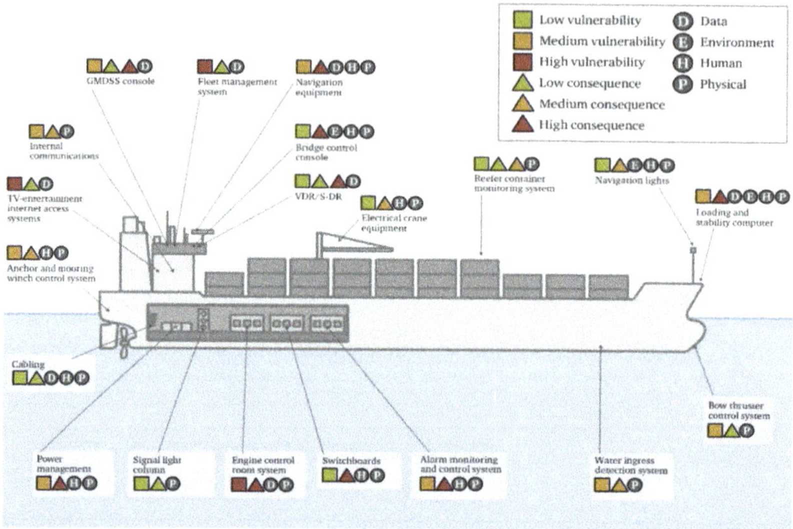

Figure 4. Ship Components (Vulnerabilities, Consequences and Affected Fields). Note: When two triangles exist, they are introduced to indicate the difference in consequences based on the nature of the cyber incident. Source: Authors.

Additionally, the main subcomponent of the ship's deck is the **Bridge Control Console**, where most of the other subcomponents are connected to facilitate the ship's steering, navigation, cargo handling and most of its routine operational activities. Since modern ships operate with no more than three persons on duty at the bridge, it is important for the duty officer to have centralised control and monitoring of the ship's main operations via such a console. This integrated console includes systems that require internet connectivity, such as the Fleet Management System, and others that require frequent updates, such as the Electronic Chart Display and Information System (ECDIS), constituting it as one of the most important and, at the same time, most vulnerable components in a ship's bridge. If one of these components is exposed to malware, the ship's safety can be jeopardised.

The second component, the engine, is primary composed of OT and SCADA systems that provide the ship with electrical power, propulsion, and safety monitoring. Even though most of these subcomponents operate independently, they could still be described as cyber-physical systems, since they are remotely controlled by the **Engine Control Room System** (ECRS) over computer-based software. Following the same centralised principle of operation, the ECRS, is the heart of the entire engine component, making it vital for the ship's undisturbed continuity of operations. In that context, although on most ships, the ECRS is not connected to the internet, lately, at an increasing rate, vendors are requesting access to their shipboard installed systems, to monitor efficiency, performance, and consistency, which in turn, offers a potential back door to any malicious actors in accessing the ship's cyber environment.

These findings illustrate the need for a holistic risk management approach, since the cyber threat landscape is growing in unprecedented rate and, as practice has indicated, no system can ever be 'cyber proof'. This cyber risk management practice should focus on three main areas; (a) advising; (b) threat intelligence support; and (c) training. Cybersecurity is not limited to technology but involves people and business processes. Hackers will take advantage of the power of repetition that every individual is accustomed to and make their way into the enterprise ecosystem. Thus, comprehensive cyber risk management should be practised constantly within the maritime sector.

5. The Socio-Economic Consequences of a Cyber Incident at the MTS: A Case Study of the Persian Gulf

Maritime transportation is carrying more than 80% of the international trade with absolute numbers increasing year on year [26]. Maritime transport plays a key role in the socioeconomic development and wellbeing of most states, including the states around the Persian Gulf. As approximately 30% of the global oil trade and 30.2 million TEU, to and from these states is carried through MTS [27]. Cooperation between the Arab Peninsula's states in protecting and maintaining a sustainable MTS will contribute to people's wellbeing and prosperity.

A unique characteristic of the region is that its trade activities depend heavily on the availability of the Strait of Hormuz. When looking at the geographical distribution of ports in the Persian Gulf, as identified in Figure 5, the importance of the Strait of Hormuz is immediately noticed, since the bulk of the region's seaborne trade is transported through its waters. The Strait has been at the heart of regional tensions for decades. Political instability in the past raised fears around the matter. In 2018, the possibility of closing the Strait from Iran re-emerged in public discourse [28]. It is worth noting that, in the events that occurred in May, June and July 2019, there was a cyber element in them, as ships transiting the Strait experienced GPS interference, bridge-to-bridge communications spoofing, and/or other communications jamming [29].

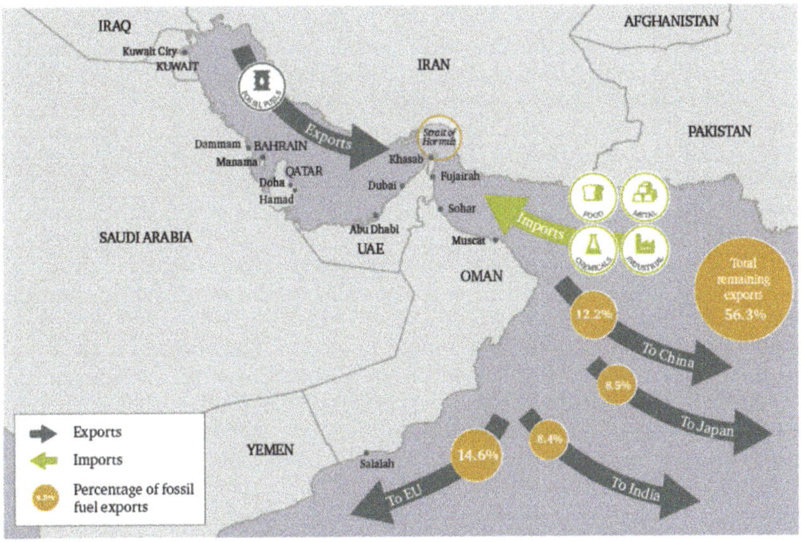

Figure 5. Major Gulf Cooperation Council Ports and the Strait of Hormuz Trade Flows. Source: Authors.

Ports act as an enabler for the region's economic growth, as they are the main gateway for trading industrial and agricultural products, fossil fuels and related by-products, along with facilitating transportation and delivery of services. The Persian Gulf is strategically considered as the starting point of one of the world's most important shipping lanes [30,31] for the transportation of oil and containerised goods. Investment by local and multinational companies in port development in the region [32], with the notable example of the Hamad Port in Qatar [33], also contribute to creating jobs and business opportunities. Shipping is the primary means of transportation for the main export product of the region; fossil fuel. Very Large Crude Carriers (VLCC) or Liquefied Natural Gas (LNG) and Liquefied

Petroleum Gas (LPG) carriers are used to carry oil, gas, and related by-products globally, as the area has the biggest oil reserves, constituting of 47.6% of global share [34].

An attack or any disruption to a port's daily operations or a ship's scheduled trip could result in side effects including loss of revenue, environmental disaster, or even loss of life. From an economic perspective, the late delivery of a ship's cargo consequently can lead to lost revenue across the supply chain, including penalties that must be paid by the shipbroker or the contracting company to its final consumers. The Maersk cyber incident analysed in Section 2 and the closure of Suez Canal by Ever Given on the 23 March 2021 [35] demonstrate some of these consequences.

Apart from the economic impact of such an incident, in the social sphere, a disruption to the MTS affects a wide range of people's livelihoods. Varying from necessary food supplies to spare parts needed for urgent repairs, medical supplies to support medical emergencies, fuel, minerals lubricants and other oil by-products needed to run heavy industry, the consequences of a belated delivery can be easily understood. Even though fossil fuels are the primary export product for the GCC, metals, minerals and primarily agricultural products are mainly imported to support business development and social wellbeing. A disruption to the MTS delivering these products would have serious consequences.

6. Lessons to Be Learned from International Responses, Key Takeaways and Future Steps

This last part of the paper builds on the findings presented in the previous sections and tries to pinpoint lessons learned from use cases on how similar efforts to address cybersecurity concerns have been implemented. It also looks at the key research takeaways, while briefly presenting the most recent, worth noting, initiatives on the matter.

6.1. Legislative Frameworks: The EU Network and Information Security Directive (NISD)

Currently, there is a broad range of frameworks, national legislations, guidelines, best practices, and recommendations regarding cybersecurity, globally. Much of the material is written from an information systems security perspective and needs to be carefully interpreted when applying it to systems and processes used in MTSs around the world. One of the most applicable examples is the NISD adopted by the EU in 2016. This first piece of EU-wide cybersecurity legislation aims to enhance cybersecurity among EU members. One of the Directive's three parts focuses on state supervision of critical sectors, one of which is transport.

Prior to its departure from the EU, the UK was one of the first states to implement NISD. For the maritime transport, namely, ports, the government has assigned Associated British Ports (ABP) as one of the Operators of Essential Services. What was made apparent throughout the research activities was that compliance to NISD and similar mandatory legislation is a stepped approach. In order to meet the objectives of the Directive, ABP has adopted a risk-based approach. In that context, aiming to achieve compliance and cyber resilience, each port operator needs to ensure adequate cyber risk management is practised constantly. This process is monitored by the Cyber Assessment Framework (CAF) [36]. Similar approaches have been adopted by other EU states in the implementation of the Directive [37].

What is worth noting is that, even though the NISD is a good start to improve the cyber risk landscape, it leads to an uneven implementation from each EU state. That is because the various stakeholders are introducing their own requirements, such as state-related security measures, NISD trade bloc size set of requirements, and the IMO set of requirements, creating a very complex policy and security landscape. Overcoming this complexity is a major challenge for regulators and maritime operators.

6.2. The Role of Top Management

What is important to understand is that, as the Maersk cyber case study highlighted, in the event of a cyber incident, probably, the only asset that a company will have at the initial stages of incident response and recovery is people [38]. During the research team's stakeholder engagements there was a consensus on the role of top management on promoting cybersecurity awareness and best practices within each organisation. This can be achieved in several ways. Initially, senior managers should use appropriate language tailored to the responsibilities and background of their staff, to simplify terms and concepts and make it easy for everyone within the organisation to understand what is at stake. Leading by example is certainly the way forward in this case. Managers should follow all existing procedures and best practices in their everyday activities to set acceptable behaviour standards regarding cybersecurity. It should be mentioned that within the maritime sector specifically, the IMO has been promoting the role of top management in achieving promoting cultural changes, in other instances in the past [39].

On the other hand, for such an approach to be effective, senior management should understand what the risk to their organisation is. According to Shaikh [40] "a lot of people {senior managers} generally believe it (cyber threats} is hyped up". A noteworthy initiative is the research project on 'cyber readiness of boards' funded by NCSC and Lloyd's Register Foundation [41]. That project explored the factors shaping UK board (including maritime) decisions around cyber risk and developed interventions to provide guidance and support, bringing together experts from academia and industry.

6.3. The Role of Insurance

The role of insurance in enhancing resilience has been predominant in the maritime sector for decades. Insurance products and strategies are encouraging maritime companies to focus on risk reduction through provision of risk information and premium discounts for risk mitigation. Cyber insurance though, is limited, as there is insufficient data available that will assist insurers in establishing the cost of cyber attacks to set up insurance policies [42]. One way of filling this 'knowledge gap' to better inform insurance premiums, is to carry out a cyber risk assessment to identify the organisation's cyber exposure. Additionally, maritime stakeholders are reluctant in purchasing cyber insurance, since, as highlighted by several interviewees for this project, they do not believe the cyber domain is an area where their business can be interrupted.

Recent examples, presented in Appendix A, indicate that companies should start considering their own cyber insurance strategy. This is facilitated by the fact that Lloyd's of London has introduced cyber risk codes for product development [43]. These are CY- Data and Privacy Breach and CZ—Property Damage. Hence, insurance providers should focus on knowledge and information sharing, so maritime companies can prioritise and inform their decisions as to whether cyber insurance is appropriate for their organisation.

6.4. The Way Forward

Interconnectivity between ports, ships, external providers, and all relevant stakeholders engaged in the maritime industry is increasing. As such, we can arguably assess that the vulnerabilities to cyber breaches of ports' and ships' subcomponents, presented in Figures 2 and 3 will only increase. In shipping specifically, the provision of internet access has reached unprecedented heights, as new technologies become available [44]. Moreover, innovative digital technologies, such as autonomous operations of vessel docking [45], blockchain, automation and robotics [46], are adopted globally to facilitate port operations and maximise productivity. While these developments offer great benefits to the industry, they also introduce new risks, as malicious actors can exploit this increased connectivity to cause harm.

Industry-led initiatives should have cyber resilience policies and mechanisms in place that support business continuity. It should be the role of international bodies, classification societies and national authorities to ensure compliance and full implementation of these measures. Within that scope, initiatives such as the launch of the US National Maritime Cybersecurity Plan to the National Strategy for Maritime Security, in December 2020 [47], should be embraced by the industry.

7. Conclusions

Ports and ships are the hubs of the supply chain and global trade between maritime and land-based transport routes. Considering that most of the global trade is carried by sea, a major disruption of a strategic port or of a maritime chokepoint can have devastating effects. Through the adoption of digitalisation most of the components within ports and ships, as illustrated in Figures 3 and 4, have an IT or OT system in their core operations. Hitherto air gapped, these systems are increasingly utilising the port's or ship's IT infrastructure to operate and monitor activities. The dissemination and analysis of data collected from these components shape command decisions that integrate a physical outcome as well. The intertwinement of these cyber-physical systems adds further complexity in any proposed or implemented cybersecurity policy.

This work has shed light in both the port and ship ecosystems, the two most critical components of the MTS. Adopting a SOSA research, findings highlighted the exposure of both ecosystems to potential malicious cyber related activities. The interesting realisation that, currently, only a small number of subcomponents within both ecosystems are highly vulnerable to cyber threats should not be taken as a reassuring factor. The increasing adoption of digital technologies will, consequently, increase this number in the years to come. Solutions like Port Community Systems and the concept of Smart Ports are notable case studies that should be analysed in future research.

Additionally, a second major realisation from this analysis is that most of the subcomponents in both ecosystems, if affected by a cyber-related disruption, may cause a physical, knock-on, effect. Disrupting the ship's main engine, the navigation aids or the safety thresholds of cranes may lead to accidents with devastating consequences, affecting, pending on their scale, the very socio-economic wellbeing of states and peoples.

Addressing cyber risks in the maritime sector is not solely a technology problem. Maritime stakeholders should act proactively and develop business continuity mechanisms to react efficiently and timely in the event of a cyber attack against any of the components of the MTS. States and national competent authorities should push ship, port and terminal operators to act accordingly, while keeping an eye on relevant international regulations, and geopolitical tensions that could impact regional security.

An attack on a ship is not a distant scenario, but what should be mentioned is that such an attack may not inevitably translate into a halt of operations for the vessel, as there are redundancy measures in place, but it will certainly affect normal operations. Current research and corporate initiatives to address security challenges such as AIS trustworthiness, GPS anti-spoofing techniques, sensor network monitoring for increased onboard safety, container monitoring and tracking, should be coupled with state-driven initiatives aiming to improve the industry's cybersecurity posture.

Future research, based on the findings of this paper, can follow several different angles. We would suggest an analysis of existing findings presented in the paper using FMEA (Failure Mode and Effect Analysis), in order to rank all candidates. In addition, as we recognised that this research was based on a qualitative analysis further research should be carried out bringing to surface more numerical elements through a quantitative methodology. Finally, inclusion of uncertainties in the transport system could provide an interesting angle in the future research.

Author Contributions: Conceptualization, C.K. and S.K.; methodology, T.W., C.K. and S.K.; formal analysis, C.K., S.K. and G.K.; investigation, C.K. and S.K.; resources, C.K.; data curation, C.K.; writing—original draft preparation, C.K., S.K. and T.W. and G.K.; writing—review and editing, C.K., S.K., T.W. and G.K.; visualization, C.K.; project administration, C.K. and S.K.; funding acquisition, T.W. and C.K. All authors have read and agreed to the published version of the manuscript.

Funding: This research was partly funded by the Stavros Niarchos Foundation, as part of the main author's fellowship at Chatham House, and, partly, from the Alan Turing Institute, Department of Defence and Security.

Institutional Review Board Statement: Ethical review and approval were waived for this study, due to methodology limitations.

Informed Consent Statement: Informed consent was obtained from all subjects involved in the study.

Data Availability Statement: Data available on request due to privacy restrictions. The data presented in this study are available on request from the corresponding author. The data are not publicly available due to GDPR restrictions.

Acknowledgments: The authors would like to thank the Stavros Niarchos Foundation for their funding and support of the main author's fellowship at Chatham House. We would also like to thank the Queen Elisabeth II Academy for Leadership in International Affairs and the International Security Department staff for their guidance, contributions, and trust in this work. The authors would also like to express their gratitude to all the people who contributed one way or another to the publication of this paper. Particular thanks go to the people who were interviewed for this project, along with all the participants who attended the roundtable discussions. The valuable feedback received from anonymous peer reviewers was incorporated, as far as possible, into the paper.

Conflicts of Interest: The authors declare no conflict of interest.

Appendix A

The following table summarises the most notable maritime related cyber incidents as identified during the research process. This is informed by the author's previous published work and the Risk Focus Cybersecurity prepared by Thomas Miller and the UK P&I Club [3].

Table A1. Notable Maritime Related Cyber Incidents.

Date	Victim	Incident Description
2010–2011	Greek Shipping Company	For 2 years a Greek shipping company suffered several successful piracy attacks in the Gulf of Aden, since local pirates hired hackers to gain access to the company's HQ and identify the most vulnerable ships along with route timetables. Hackers were able to gain access to the company's IT systems via wi-fi equipment that was installed at the company's offices.
August 2011	Iranian Shipping Line (IRISL)	The servers were hacked resulting in damage to data relating to rates, loading, delivery and location. Consequently, the location of many cargo containers remained unidentified and an undisclosed amount of financial losses were incurred as a result.
2011–2013	Port of Antwerp	The port had been a victim of an APT attack since 2011 commissioned by a drug cartel. The attack targeted terminal systems which were subsequently compromised by hackers and used to release containers without port authorities becoming aware. Illicit drugs and contraband worth approximately USD 365 million, firearms and approximately USD 1.5 million were seized when authorities finally became aware
2012	Australian Customs & Border Protection Service agency	Cargo systems controlled by customs and border protection were hacked in order to determine which shipping containers were suspected by authorities
2012–2014	Danish Port Authority	An email virus spread through the port network that was likely initiated through an infected pdf document. The virus spread and successfully reached other Danish government institutions
April 2016	South Korea	280 ships were forced to return to port due to problems on their navigation systems. The issue was largely blamed on North Korea however this remains unconfirmed

Table A1. *Cont.*

Date	Victim	Incident Description
June 2017	AP Moller Maersk	NotPetya also known as ExPetr ransomware led to outages on A.P. Moller Maersk computer systems impacting both oil and gas production and port operations. Following the incident, Maersk claimed to have changed its IT systems to prevent similar incidents from occurring in the future. The incident resulted in an estimated USD 300 million of losses
June 2017	Ships in Novorossiysk	At least 20 ships in the Black Sea were reporting false data was being transmitted, indicating the ships were 32 km inland of their actual position. It is now believed to have been as a result of a GNSS spoofing attack
November 2017	Clarksons	Perpetrators gained unauthorized access to computer systems, accessing confidential information and threatening to release information unless ransom payment is made. Company share prices decreased by 2.71%
July 2018	Cosco US	Cosco's Shipping Lines suffered from a cyber breach that affected email and network telephone initially in the US, but not ships. This expanded to their Americas facilities hence the company decided to shut down the connections with other regions for further investigation. The company was able to recover within a week. It is believed to have been a malware (ransomware) incident.
September 2018	Ports of Barcelona & San Diego	Within a week both ports suffered from a cyber-related business disruption event. Even though both organizations did not disclose a lot of information, they both claimed that this was not a major disruption and it affected mainly IT systems at shore. It is assumed that it was the same malicious content that affected both ports.
January 2018–September 2019	GPS and AIS interference in Eastern Mediterranean	Several ships and offshore platforms reported GPS interference in the region for a large period of time. NATO Maritime Command was forced to issue a notice requesting additional information by affected parties.
June–August 2019	GPS and communications interference in the Strait of Hormuz	During several maritime related incidents in the Strait of Hormuz, generated over a period of high tensions between Iran and the USA, several ships reported interference in their GPS and communications channels.
May 2020	Shahid Rajaee port, Iran	The Iranian Port of Shahid Rajaee suffered from a cyber attack, allegedly linked to Israel, that affected shipping traffic along with computers that regulate the flow of vessels, trucks and good, creating massive backups on waterways and roads leading to the facility.
Q1 and Q2 2020	Cyber attacks on shipping companies	During the first two quarters of 2020, amidst the COVID-19 pandemic, an increased number of cyber attacks to major shipping companies has been reported. Companies like Mediterranean Shipping Company (MSC), Toll Group (twice) and Carnival Cruises suffered from cyber breaches, where sensitive data of employees, clients, passengers and suppliers were exposed. These breaches appear to have common characteristics as data centers of these companies were mainly affected.

Appendix B. Interview Questionnaire

Q1 How many years of experience do you have in the maritime/cyber-security sector?

- 0–5
- 6–10
- 11–15
- 15–20

Q2 In which field of the maritime sector do you work?

- Shipowners and port operators
- Chief Information Security Officers (CISO)
- In-formation security and technology providers
- Marine insurers
- Brokers

- Academics
- Other/please specify

Q3 Which are the subcomponents of modern ship:

- Signal light column
- Anchor and mooring winch control system
- Engine Control Room System
- Switchboards
- Internal communications
- TV-Entertainment-Internet Access Systems
- Bow Thruster Control System
- Water Ingress Detection System
- GMDSS Console
- Fleet Management system
- Alarm Monitoring and Control System
- Power Management
- Navigation Equipment
- Bridge control console
- VDR/S-DR
- Reefer Container Monitoring System
- Navigation Lights
- Loading and Stability Computer
- Other (please specify)

Q4 Which are the subcomponents of a modern port:

- Port Facility offices
- Cargo handling vehicles, cranes and conveyer assets
- Financial Management and Monitoring
- Port Facility accommodation
- Access control points
- Refineries
- Container and vehicle movement systems
- Warehouse and storage (chemical, gas, bulk and dry cargo) facilities
- Rail systems
- Power plants
- Asset management systems (cargo, vehicle)
- Vessel Traffic Control Tower
- Terminal Facilities (Container Terminal, Oil tank terminal)
- Access control management system
- Maintenance areas
- Cargo handling systems such as pipelines and conveyer systems
- Other (please specify)

Q5 Classify the potential consequences of a cyber incident for the following subcomponents of a modern ship.

Sub Component	High	Medium	Low
Signal light column			
Anchor and mooring winch control system			
Engine Control Room System			
Switchboards			
Internal communications			
TV-Entertainment-Internet Access Systems			

Sub Component	High	Medium	Low
Bow Thruster Control System			
Water Ingress Detection System			
GMDSS Console			
Fleet Management system			
Alarm Monitoring and Control System			
Power Management			
Navigation Equipment			
Bridge control console			
VDR/S-DR			
Reefer Container Monitoring System			
Navigation Lights			
Loading and Stability Computer			
Other (please specify)			

Q6 Classify the potential consequences of a cyber incident for the following subcomponents for a modern port.

Sub Component	High	Medium	Low
Port Facility offices			
Cargo handling vehicles, cranes and conveyer assets			
Financial Management and Monitoring			
Port Facility accommodation			
Access control points			
Refineries			
Container and vehicle movement systems			
Warehouse and storage (chemical, gas, bulk and dry cargo) facilities			
Rail systems			
Power plants			
Asset management systems (cargo, vehicle)			
Vessel Traffic Control Tower			
Terminal Facilities (Container Terminal, Oil tank terminal)			
Access control management system			
Maintenance areas			
Cargo handling systems such as pipelines and conveyer systems			
Other (please specify)			

Q7 Rank the vulnerability against a cyber incident for the following subcomponents of a modern ship

Sub Component	High	Medium	Low
Signal light column			
Anchor and mooring winch control system			
Engine Control Room System			
Switchboards			
Internal communications			
TV-Entertainment-Internet Access Systems			
Bow Thruster Control System			
Water Ingress Detection System			
GMDSS Console			
Fleet Management system			
Alarm Monitoring and Control System			
Power Management			
Navigation Equipment			
Bridge control console			
VDR/S-DR			
Reefer Container Monitoring System			
Navigation Lights			
Loading and Stability Computer			
Other (please specify)			

Q8 Rank the vulnerability against a cyber incident for the following subcomponents of a modern port

Sub Component	High	Medium	Low
Port Facility offices			
Cargo handling vehicles, cranes and conveyer assets			
Financial Management and Monitoring			
Port Facility accommodation			
Access control points			
Refineries			
Container and vehicle movement systems			
Warehouse and storage (chemical, gas, bulk and dry cargo) facilities			
Rail systems			
Power plants			
Asset management systems (cargo, vehicle)			
Vessel Traffic Control Tower			
Terminal Facilities (Container Terminal, Oil tank terminal)			
Access control management system			
Maintenance areas			

Sub Component	High	Medium	Low
Cargo handling systems such as pipelines and conveyer systems			
Other (please specify)			

Q9 Identify the relevant affected fields in case of a cyber incident to each of the following subcomponents of a modern ship

Sub Component	Data	Environment	Human	Physical
Signal light column				
Anchor and mooring winch control system				
Engine Control Room System				
Switchboards				
Internal communications				
TV-Entertainment-Internet Access Systems				
Bow Thruster Control System				
Water Ingress Detection System				
GMDSS Console				
Fleet Management system				
Alarm Monitoring and Control System				
Power Management				
Navigation Equipment				
Bridge control console				
VDR/S-DR				
Reefer Container Monitoring System				
Navigation Lights				
Loading and Stability Computer				
Other (please specify)				

Q10 Identify the relevant affected fields in case of a cyber incident to each of the following subcomponents of a modern port

Sub Component	Data	Environment	Human	Physical
Port Facility offices				
Cargo handling vehicles, cranes and conveyer assets				
Financial Management and Monitoring				
Port Facility accommodation				
Access control points				
Refineries				
Container and vehicle movement systems				
Warehouse and storage (chemical, gas, bulk and dry cargo) facilities				

Sub Component	Data	Environment	Human	Physical
Rail systems				
Power plants				
Asset management systems (cargo, vehicle)				
Vessel Traffic Control Tower				
Terminal Facilities (Container Terminal, Oil tank terminal)				
Access control management system				
Maintenance areas				
Cargo handling systems such as pipelines and conveyer systems				
Other (please specify)				

Q11 Please add any comments, that you might have here. Thank you for your participation to our survey!

References

1. Hellenic Shipping News Worldwide. Available online: https://www.hellenicshippingnews.com/maritime-cyber-attacks-increase-by-900-in-three-years/ (accessed on 7 December 2021).
2. Norma Cyber. Available online: https://www.normacyber.no/news/norma-annual-threat-assessment-2022 (accessed on 30 April 2022).
3. Kapalidis, P. Cybersecurity at Sea. In *Global Challenges in Maritime Security*, 1st ed.; Otto, L., Ed.; Springer: Cham, Switzerland, 2020; Volume 1, pp. 127–143.
4. Mismarine. Available online: https://mismarine.com/benefits-and-challenges-digitising-the-shipping-industry/ (accessed on 1 January 2022).
5. ENISA. Available online: https://www.enisa.europa.eu/publications/enisa-threat-landscape-2021 (accessed on 3 February 2022).
6. Maersk. Available online: https://investor.maersk.com/news-releases/news-release-details/cyber-attack-update (accessed on 10 November 2021).
7. Reuters. Available online: https://www.reuters.com/article/us-cyber-attack-maersk-idUSKBN19I1NO (accessed on 15 November 2021).
8. Digital Guardian. Available online: https://digitalguardian.com/blog/cost-malware-infection-maersk-300-million (accessed on 12 April 2022).
9. Kapalidis, P. 4 Cases of Cybersecurity Failures in Shipping History. *Collab. Shipp. Ind. Innov. Technol.* **2017**, 10–11.
10. Los Angeles Times. Available online: https://www.latimes.com/business/la-fi-maersk-cyberattack-20170817-story.html (accessed on 2 October 2021).
11. CNBC. Available online: https://www.cnbc.com/2018/01/24/cnbc-interview-with-moller-maersk-ceo-soren-skou-from-the-world-economic-forum-2018.html (accessed on 2 February 2022).
12. IMO. Available online: https://www.imo.org/en/OurWork/Security/Pages/Guidance-home.aspx (accessed on 3 January 2022).
13. Jamshidi, M. *Systems of Systems Engineering: Principles and Applications*; CRC Press: Boca Raton, FL, USA, 2017.
14. Boyes, H.; Isbell, R.; Luck, A. *Cyber Security of Ports and Port Systems*; The Institute of Engineering and Technology: Stevenage, UK, 2016.
15. Boyes, H.; Isbell, R.; Luck, A. *Cyber Security of Ports and Port Systems v2*; The Institute of Engineering and Technology: Stevenage, UK, 2020.
16. Ablon, L. Cybersecurity Considerations for the Maritime Environment. In *Issues in Maritime Cybersecurity*, 1st ed.; DiRenzo, J., Drumhiller, N., Roberts, F., Eds.; Westphalia Press: Washington, DC, USA, 2017; Volume 1, pp. 17–24.
17. Allianz. Available online: https://www.agcs.allianz.com/content/dam/onemarketing/agcs/agcs/reports/AGCS-Safety-Shipping-Review-2017.pdf (accessed on 3 October 2021).
18. Global Maritime Forum. Available online: https://www.globalmaritimeforum.org/content/2019/10/Global-Maritime-Issues-Monitor-2019.pdf (accessed on 6 November 2021).
19. Progoulakis, I.; Nikitakos, N.; Rohmeyer, P.; Bunin, B.; Dalaklis, D.; Karamperidis, S. Perspectives on Cybersecurity for Offshore Oil and Gas Assets. *J. Mar. Sci. Eng.* **2021**, *9*, 112. [CrossRef]
20. MITRE. Available online: https://attack.mitre.org/ (accessed on 14 June 2022).
21. The Maritime Executive. Available online: https://www.maritime-executive.com/blog/digitalization-and-ship-connectivity-in-2018 (accessed on 14 November 2021).

22. Allianz. Available online: https://www.agcs.allianz.com/news-and-insights/expert-risk-articles/human-error-shipping-safety.html (accessed on 22 December 2021).
23. ENISA. Available online: https://www.enisa.europa.eu/publications/port-cybersecurity-good-practices-for-cybersecurity-in-the-maritime-sector (accessed on 14 February 2022).
24. Polemi, N. Ports' Critical Infrastructures. In *Port Cybersecurity. Securing Critical Information Infrastructures and Supply Chains*, 1st ed.; Stover, T., Ed.; Elsevier: Amsterdam, The Netherlands, 2018; Volume 1, pp. 3–7.
25. Kapalidis, C. Maritime Threat: Is it Real? *Seatr. Marit. Rev.* **2018**, 115.
26. UNCTAD. Available online: https://unctad.org/system/files/official-document/rmt2021_en_0.pdf (accessed on 16 February 2022).
27. Notteboom, T.; Pallis, A.; Rodrigue, J.P. *Port Economics, Management and Policy*, 1st ed.; Routledge: New York, NY, USA, 2022; pp. 1–104.
28. The Guardian. Available online: https://www.theguardian.com/world/2018/jul/05/iran-retaliate-us-oil-threats-eu-visit-hassan-rouhani-trump (accessed on 7 October 2021).
29. U.S.Department of Transportation Maritime Administration. Available online: https://www.maritime.dot.gov/msci/2019-012-persian-gulf-strait-hormuz-gulf-oman-arabian-sea-red-sea-threats-commercial-vessels (accessed on 15 April 2020).
30. REUTERS. Available online: https://www.reuters.com/article/us-iran-oil-factbox/strait-of-hormuz-the-worlds-most-important-oil-artery-idUSKBN1JV24O (accessed on 9 January 2022).
31. U.S. Energy Information Administration. Available online: https://www.eia.gov/international/content/analysis/special_topics/World_Oil_Transit_Chokepoints/wotc.pdf (accessed on 6 October 2021).
32. Emirates NBD. Available online: http://www.emiratesnbdresearch.com/research/article/?a=gcc-maritime-strategy-enters-a-new-era-969 (accessed on 5 February 2022).
33. REUTERS. Available online: https://www.reuters.com/article/gulf-qatar-port/gulf-crisis-a-blessing-in-disguise-for-qatar-seaport-idUSL8N1JC2LJ (accessed on 8 January 2021).
34. BP. Available online: https://www.bp.com/en/global/corporate/news-and-insights/press-releases.html (accessed on 6 January 2021).
35. Forti, N.; d'Afflisio, E.; Braca, P.; Millefiori, L.M.; Willett, P.; Carniel, S. Maritime Anomaly Detection in a Real-World Scenario: Ever Given Grounding in the Suez Canal. *IEEE Trans. Intell. Transp. Syst.* **2021**, *23*, 13904–13910. [CrossRef]
36. It Governance. Available online: https://www.itgovernance.co.uk/nis-directive (accessed on 10 December 2021).
37. European Commission. Available online: https://digital-strategy.ec.europa.eu/en/policies/nis-transposition (accessed on 16 February 2022).
38. ZDNet. Available online: https://www.zdnet.com/article/ransomware-the-key-lesson-maersk-learned-from-battling-the-notpetya-attack/ (accessed on 1 December 2021).
39. Transport Styrelsen. Available online: https://www.transportstyrelsen.se/en/shipping/ (accessed on 12 January 2022).
40. Qatar tribune. Available online: https://www.qatar-tribune.com/news-details/id/155546 (accessed on 3 January 2022).
41. RISCS. Available online: https://www.riscs.org.uk/project/cr4b-cyber-readiness-for-boards/ (accessed on 3 March 2022).
42. Informa Connect. Available online: https://informaconnect.com/how-will-the-marine-insurance-industry-respond-to-new-sources-of-risk/ (accessed on 2 January 2022).
43. Lloyd's. Available online: https://www.lloyds.com/market-resources/underwriting/risk-codes (accessed on 11 January 2022).
44. Futurenautics. Available online: http://www.navarino.co.uk/wp-content/uploads/2018/04/Crew_Connectivity_2018_Survey_Report.pdf (accessed on 16 October 2021).
45. Athanasoglou, C.; Papoutsidakis, M.; Papachristos, D.; Nikitakos, N. Automated Docking Procedure in Modern Shipping. *Int. J. Comput. Appl.* **2019**, *178*, 55–58. [CrossRef]
46. Zolich, A.; Palma, D.; Kansanen, K.; Fjørtoft, K.; Sousa, J.; Johansson, K.; Jiang, Y.; Dong, H.; Johansen, T. Survey on Communication and Networks for Autonomous Marine Systems. *J. Intell. Robot. Syst.* **2018**, *95*, 789–813. [CrossRef]
47. Maritime Cybersecurity. Available online: https://www.maritime-cybersecurity.com/National_Maritime_Cybersecurity_Plan.html (accessed on 21 October 2021).

Journal of
Marine Science and Engineering

MDPI

Article

Cyber Physical Systems Security for Maritime Assets

Iosif Progoulakis [1,*], Paul Rohmeyer [2] and Nikitas Nikitakos [1]

1 Department of Shipping Trade and Transport, University of the Aegean, Korais St. 2A, GR 82132 Chios, Greece; nnik@aegean.gr
2 School of Business, Stevens Institute of Technology, 1 Castle Point on the Hudson, Hoboken, NJ 07030, USA; prohmeye@stevens.edu
* Correspondence: iprogoulakis@aegean.gr

Abstract: The integration of IT, OT, and human factor elements in maritime assets is critical for their efficient and safe operation and performance. This integration defines cyber physical systems and involves a number of IT and OT components, systems, and functions that involve multiple and diverse communication paths that are technologically and operationally evolving along with credible cyber security threats. These cyber security threats and risks as well as a number of known security breach scenarios are described in this paper to highlight the evolution of cyber physical systems in the maritime domain and their emerging cyber vulnerabilities. Current industry and governmental standards and directives related to cyber security in the maritime domain attempt to enforce the regulatory compliance and reinforce asset cyber security integrity for optimum and safe performance with limited focus, however, in the existing OT infrastructure and systems. The use of outside-of-the-maritime industry security risk assessment tools and processes, such the API STD 780 Security Risk Assessment (SRA) and the Bow Tie Analysis methodologies, can assist the asset owner to assess its IT and OT infrastructure for cyber and physical security vulnerabilities and allocate proper mitigation measures assuming their similarities to ICS infrastructure. The application of cyber security controls deriving from the adaptation of the NIST CSF and the MITRE ATT&CK Threat Model can further increase the cyber security integrity of maritime assets, assuming they are periodically evaluated for their effectiveness and applicability. Finally, the improvement in communication among stakeholders, the increase in operational and technical cyber and physical security resiliency, and the increase in operational cyber security awareness would be further increased for maritime assets by the convergence of the distinct physical and cyber security functions as well as onshore- and offshore-based cyber infrastructure of maritime companies and asset owners.

Keywords: cyber security; cyber physical systems; IT; OT; maritime assets; cyber and physical security convergence; API STD 780; Security Risk Assessment (SRA); Bow Tie Analysis

1. Introduction

This paper explores the cyber physical concepts of cyber security for maritime assets, both onshore and offshore. Cyber physical systems in the maritime sector constitute a complex field that involves the integration of IT (information technology) and OT (operational technology) systems and the interface with human element parameters. This integration, as depicted in Figure 1, defines the concept of cyber physical systems and represents the majority of systems onboard maritime assets. Figure 2 provides a simplified illustration of the communication paths between shore-based and vessel-based stakeholders and IT/OT platforms to illustrate the interconnection of IT- and OT-based communication interactions. Maritime assets operated by humans contain an IT and OT interface that links together processes, systems, components, and the technical and operational performance. A naval vessel is considered a platform of systems of systems, containing IT and OT devices. A vessel's crew represents the operator of these components and processes and is responsible for the operational and performance integrity of the vessel as a whole. In parallel, shipping companies have an IT interface that supports naval vessels at a technical and operational level. Again, the human element is present as the operator of IT platforms which are tools to achieve performance and financially oriented tasks to support the maritime operations. Ports interface with naval vessels on a shore-to-ship and ship-to-shore operational level, either receiving from or loading the maritime commercial goods into naval vessels. In this case, for a port to carry out these loading and offloading tasks, a combination of IT and OT platforms is utilized. From cargo management platforms and cranes to utilities' support systems, IT and OT platforms are used for the support of maritime assets at a technical and operational level. The human element is present as the operator, configurator, and moderator in all cyber physical systems. The maintenance of OT devices and systems is carried out on a physical interface or remotely. The monitoring and configuration of key performance indicators of a vessel's systems is conducted mostly remotely. In all cases, the human element plays an important role as the operational lead.

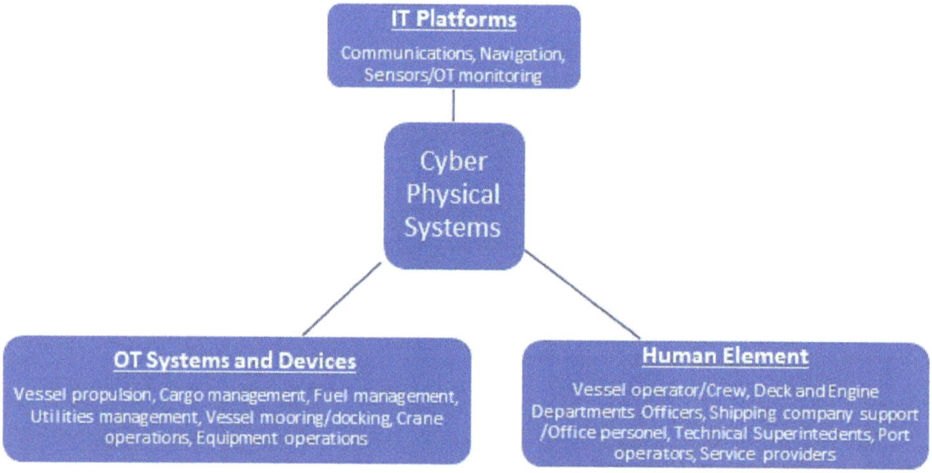

Figure 1. IT, OT, and human element interface in cyber physical systems.

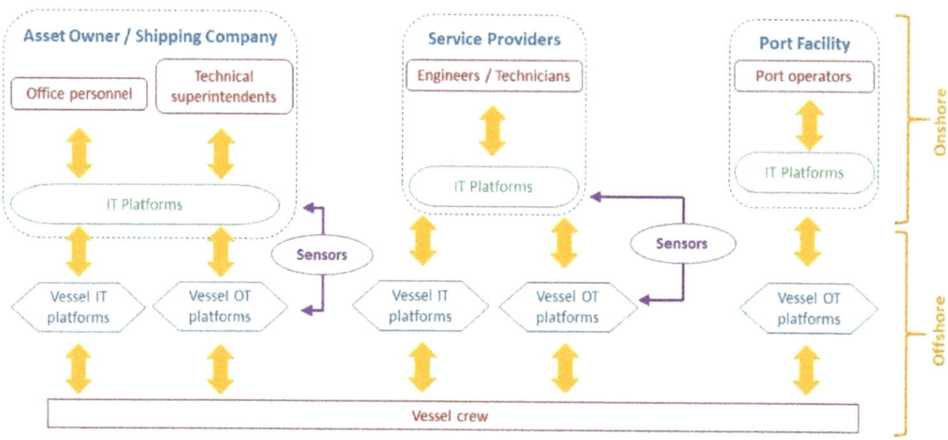

Figure 2. Communication paths of shore-based and vessel-based stakeholders and IT/OT platforms.

Research Methodology and Article Structure

This paper assesses the cyber physical concepts of cyber security for maritime assets, by presenting a comprehensive review of existing industry and governmental policies, directives, and standards that cover the subject of cyber security for maritime assets including cyber physical systems and operations. The main aspects of cyber physical security for the maritime sector in general are discussed in Section 2. The different security threats faced and known vulnerabilities, a brief overview of notable cyber incidents, and the emerging risk factors of maritime assets and their cyber physical systems and functions are discussed in Section 3. An overview of existing industry and governmental initiatives are presented in Section 4 and cover maritime industry organizations, industry standardization organizations, maritime classification societies, government agencies, marine insurance, and P&I clubs. This paper also discusses, in Section 5, two risk assessment methods deriving from different industry sectors and which could be applied in the maritime industry and specifically for the cyber security risk analysis for cyber physical systems. Section 5 also gives some examples of these methods, outlining important conclusions in cyber physical security assessment. The risk treatment process involving the design of cyber security controls for cyber physical systems and functions in maritime assets is discussed in Section 6. Important notes deriving from the review of the presented literature review in Section 4 are discussed in Section 7. Finally, key conclusions are presented in Section 8 along with a discussion on the scientific contribution of this paper and future research.

2. Cyber Physical Security Aspects for the Maritime Sector

The nature and complexities of IT and OT in maritime environments can be analyzed by considering the technical characteristics of vessels, ports, terminals, offshore oil and gas rigs, and other dimensions of maritime asset architecture. Each presents unique attributes and cyber physical risk challenges.

The maritime industry today relies on global communications and would be substantially crippled if communications became unavailable. The sector relies on complex, interconnected, global cyber architectures that contain both IT and OT systems in support of complex process functions. Technology is deployed to achieve greater efficiencies and optimization. Maritime system architectures are undergoing a substantial evolution that includes the rapid, global introduction of a variety of emerging technologies. This includes integration of capabilities associated with Industry 4.0, including the Internet of Things (IoT), distributed cloud computing, data analytics, robotics, embedded software, and other advances that combine to form the emergent systems landscape [1].

Ports facilitate automated operations and processes on a continuous basis. This includes the loading and unloading of cargo, financial transactions, contract management, exchanging supply chain information, environmental monitoring, and other functions. They also need to provide physical security and interact with law enforcement, military, and other regulatory bodies as needed. The move to intelligent, or "smart port", architectures is underway and features the use of data analytics to drive improved decision making, streamline processes, optimize traffic management, and provide enhanced monitoring of essentially all port operations. Vessels, rigs, and other maritime assets address similar functional requirements. In the case of vessels, navigational requirements produce additional system requirements and architectures [2].

The complexity of maritime systems can be further illustrated by considering them "systems of systems" (SoS), with varying degrees of centralization, control, and ownership of the sub-set, constituent systems that comprise the SoS [3]. The SoS context suggests variability in the coordination and alignment of standards, process design, and other aspects of system governance.

Cyber security challenges are presented in all aspects of the maritime cyber physical architecture. Achieving the core cyber security objectives of confidentiality, integrity, and availability throughout the system and its components presents substantial difficulties that continuously evolve. The challenges are notable in ensuring cyber security with respect to

IT and OT systems assets that store, transit, or process operational, financial, personnel, navigation, and other data.

3. Cyber Security Threats and Vulnerabilities

Cyber security risk in all industries reflects threats to the confidentiality, integrity, and availability of systems and data and is a manifestation of the ever-present vulnerabilities in systems and architectural components. We can consider historical breach activity in maritime as well as other industries to illustrate the nature of specific threats and vulnerabilities. Examination of other industries can be useful for identifying threats and vulnerability characteristics that may not have yet been visible in past maritime cyber breach events.

Maritime operations, by nature, have been international and sometimes global for centuries. However, globalization in the present era creates technology risks due to the varying levels of technical sophistication and standards of preparedness across jurisdictions. There can be wide variety in the vintage of onboard and shoreside systems that are increasingly integrated. On a single voyage, a commercial vessel may need to accommodate data transfer in a wide variety of mechanisms including USB flash drives, direct connection of endpoint computers brought on-board by local port authorities, wired connections, and of course wireless of various vintages and security. Technical risk is present in all individual technologies, and new types of risk result from increased interdependency and interconnection.

In general, the identified cyber threats for maritime assets and their cyber physical interface can be classified as internal, external, or colluded [4]. This classification is similar to the physical threats faced. An insider threat can be a vessel crew member or port operator who unintentionally or intentionally allows the penetration of cyber security barriers (in this case IT platforms and tools) by operating in the cyber–physical domain without proper cyber security practices. From the use of a virus-infected USB device to the opening of an unsolicited email infected with malware, external threats can be a number of common cyber criminals, hackers, hacktivists or even state adversaries or terrorists using sophisticated methods to manipulate, degrade, or take control of IT and OT systems. The colluded threats are a combination of internal threat actors acting with the guidance of external sources.

3.1. Maritime Breach Incidents

Examination of recent maritime cyber breach events highlights important characteristics to inform risk assessment, control design, and loss expectations. Breach consequences include various forms of disruption to the mission of providing confidentiality, integrity, and availability to maritime systems. The impact of cyber breaches on maritime can be considered similar to cyber incident experiences across all industries; however, the maritime context does present unique characteristics in breach impact.

Maritime ransomware incidents have been notable. Perhaps the worst maritime cyber event, the NotPetya ransomware event, resulted in losses measured in over hundreds of millions of dollars for Moller–Maersk. The disruption caused by NotPetya caused global disruptions for weeks [5]. The event began with the infection of a single user workstation and took only seven minutes to widely propagate, according to the Maersk Chief Information Security Officer [6]. The Port of San Diego was significantly impacted by Ransomware as well, in a 2019 attack. The breach affected primarily IT systems as well as the San Diego Harbor police systems [7]. Clarksons, the British shipping services company, was victimized in a breach where the perpetrators sought ransom in exchange for the promise of not releasing sensitive information including personal data. Clarksons was not operationally disrupted and refused to pay the ransom [8]. The extent of the resulting data leakage, however, is unclear.

In regard to maritime assets, an attack against a US-flagged, ultra-large container ship using the Emonet malware was noted to have debilitated all operational systems of the vessel [9]. Vessel navigational system malfunctions due to the fact of GPS spoofing have also been reported [10]. Remote cyber-attacks utilizing vulnerabilities in wireless keyboards and printers have also been known [11] to have affected vessel IT and OT systems. Finally, cyber-attacks from state adversaries have also been reported in multiple vessels, resulting in the loss of propulsion and steering [12].

Attacks on industrial control systems (ICS), by nature, clearly illustrate the potential consequences of cyber-attacks on cyber physical systems. In 2019 the US Coast Guard released information about a Ryuk ransomware attack on a facility that was apparently the result of a phishing email. The malware had a sizeable impact on IT but also propagated into the ICS used for facility monitoring and cargo movement [13]. The impact of cyber security attacks on ICS and SCADA (Supervisory Control and Data Acquisition) systems has long been a point of concern despite the relatively lower number of reported breach incidents compared to attacks on IT networks. The recent Colonial Pipeline attack, however, provided a clear demonstration of the potential consequences including widespread disruption throughout integrated supply chains. The nature of attacks on ICS are significantly different than attacks on IT assets. As described by ICS cyber security researcher Joe Weiss, "The scary fact that almost any person in an organization that clicks on a "poison" attachment could cause a problem of the magnitude of what happened with Colonial Pipelines should give everyone pause for thought. The more I dug into the problem, the more complicated it became." [14].

3.2. Emerging Risk Factors

Cyber physical maritime system architecture is rapidly evolving and should be expected to produce or highlight novel risk scenarios. The study of past breaches is invaluable in forecasting risk dimensions in all systems. However, the ongoing infusion of new technology should be anticipated to produce previously unknown risk considerations. Therefore, examination of risk should ideally combine analysis of relevant historical events as well as new largely theoretical insights into potential vulnerabilities, attacker motives, consequences, and other factors.

Industrial Internet of Things (IIoT) reflects the increased integration of IoT into cyber physical system architectures, where we can expect the individual risk dimensions of IoT devices will manifest, in some cases, as substantial system risks. Risk should be anticipated due to the lack of standardization across device designs and protocols, and the expectation that new IoT components will be linked with legacy systems of varying sophistication, vintage, and security. Desires to control early costs in IIoT may contribute to concerning levels of risk acceptance [15]. Additionally, new exposures should be expected within the network communications architectures that will support new IIoT.

Greater process integration is typically expected to result from increased system interconnection. However, a wide base systems development and engineering research revealed the shortcomings of automating legacy processes without considering new capabilities and limitations introduced by the new technology [16]. Integration of artificial intelligence and robotics can similarly present "people and process risk" by removing historically important quality control benefits of manual oversight [17].

Process integration is expected in both horizontal and vertical integration strategies and is recognized as a goal of Industry 4.0 in the maritime industry [1]. Horizontal integrations seek to combine the strengths of entities operating at the same level of a value chain, such as multiple producers of the same component, while strategies of vertical implementation anticipate upstream and/or downstream coordination and, therefore, system connectivity. Human factors can create risk as well. Many cyber physical environments can be considered socio-technical systems, where technology and people interact and may produce substantial risk challenges. This includes personnel vulnerability aspects related to human operators throughout the system and process architecture.

The physical dimension presents unique vulnerability characteristics, as threat actors can take actions that are initially physical intrusions with an ultimate goal of a cyber breach. Physical breaches to systems and information, such as the theft of equipment or perhaps intruders undertaking cyber-attack actions from ship or terminal-based endpoints that are inside the trusted physical and cyber perimeters, and typically signed-in with accounts that possess escalated privileges. Therefore, a sequential physical-to-cyber-attack path represents an important potential threat vector in maritime environments, particularly considering the challenges in physical protection of vessels and rigs, where physical attackers may be motivated to ultimately breach systems on board as well as gain access to remote company and/or government systems.

4. Cyber Security Initiatives for the Maritime Sector

The cyber–physical aspects of security and cyber security in the maritime domain, in general, are covered by numerous publications, directives, guidelines, and standards from the industry and government. These are described in the subsequent subsections with the list not being conclusive.

4.1. Maritime Industry Organizations

Regarding international industry organizations, the International Maritime Organization (IMO) released in 2017 Resolution MSC.428(98) [18] and IMO Guidance MSC-FAL.1/Circ.3 [19]. These address the implementation of maritime risk management in vessels' safety management systems (SMSs) in accordance with the ISM (International Safety Management) Code objectives and requirements. MSC.428(98) and MSC-FAL.1/Circ.3. complement the IMO ISPS (International Ship and Port Facility Security) code for vessels dealing with cyber and physical security issues.

BIMCO (Baltic and International Maritime Council), The International Association for Classification Societies (IACS), The International Association of Independent Tanker Owners (INTERTANKO) along with other industry partners have issued the Guidelines on Cyber Security Onboard Ships [20] (2021) aimed at the implementation of cyber risk management strategies in accordance with relevant industry regulations and best practices on board a ship with a focus on operational and technical processes, equipment, personnel training, and the response to cyber incidents and the subsequent operational recovery.

4.2. Industry Standardization Organizations

The US National Institute of Standards and Technology (NIST) has developed the Cyber Security Framework [21] which is used worldwide in the maritime and general industrial sectors. The NIST series of standards captures the operational and technical cyber security requirements of general industrial and maritime assets. The NIST Cyber Security Framework consists of five functions: (1) cyber security risk identification for systems, assets, data and operations; (2) the implementation of cyber security protection safeguards for assets; (3) cyber security incident detection; (4) cyber security incident response; (5) cyber security incident recovery. The NIST Cyber Security Framework is complemented by NIST Special Publications 800-30 [22], 800-37 [23], and 800-82 [24], which relate to Industrial Control Systems (ICS) and their cyber security risk assessment and management. Specifically, for cyber physical systems, NIST also published Special Publications 1500-201 [25], 1500-202 [26], and 1500-203 [27], which, in three volumes, comprise the NIST Framework for Cyber–Physical Systems. The NIST Framework for Cyber–Physical Systems is derived from industry and academia and applies to various IT and OT systems in industry sectors including transportation and maritime. It also explores the interaction of systems and components within a cyber infrastructure, and it defines the SoS state of IT and OT assets and provides a comprehensive tool for the analysis and description of cyber physical systems.

The International Organization for Standardization (ISO) and the International Electrotechnical Commission (IEC) have published a number of standards applicable to cyber security, which also cover IT and OT systems and ICS in operation in the maritime domain. ISO/IEC 27001 [28] is applicable for the maritime sector and it provides a structured framework for the identification, assessment, and mitigation of cyber security risks. IEC-62443 is a series of standards on industrial communication networks and security of IT systems and networks that cover both technical- and process-related industrial cyber security. From the IEC-62443 series, IEC-62443-4-2 [29] provides the technical requirements for components of Industrial Automation and Control Systems (IACS) and describes mitigation measures for cyber security vulnerabilities. IEC 62443-3-3 [30] describes the requirements of technical control systems and the security levels of their control system capability. ISO/IEC 21827 [31] presents practices used in the industry and describes the Systems Security Engineering—Capability Maturity Model® (SSE-CMM®), outlining the process for organizational security engineering for companies and assets. ISO/IEC 18045 [32] provides guidelines for the evaluation of systems and their IT security. ISO/IEC 15408-1 [33] provides the structure and basic concepts for the cyber security evaluation of IT products, defining the concept of a target of evaluation (TOE) and the context of evaluation of IT systems and components. ISO/IEC 27032 [34] covers the subject of security for information, networks, internet, and critical information infrastructure protection, providing guidance for cyber security improvement.

ASTM International, formerly known as the American Society for Testing and Materials, issued two standards that relate to cyber security in maritime assets. ASTM F3286-17 [35] relates to the organizational need to mitigate the likelihood of cyber security attacks and to reduce the extent of potential cyber incidents through the protection of sensitive data onboard vessels and offshore operations. ASTM F3286-17 also utilizes the NIST Cyber Security Framework for maritime assets and critical infrastructure. ASTM F3449-20 [36] provides guidance, information, and options to the maritime industry in order to incorporate technical and operational cyber elements into vessel safety management systems (SMS). The suggested guidelines are in compliance with the International Safety Management (ISM) Code and other national and international requirements as well as the IMO Resolution MSC.428(98).

4.3. Maritime Classification Societies

In the sector of maritime classification societies, the International Association for Classification Societies (IACS) has issued Recommendation No. 166 on Cyber Resilience, which provides technical requirements for the buildup of necessary cyber resilient infrastructure for vessels. IACS Recommendation No. 166 [37] supports IMO Resolution MSC.428(98) and IMO Guidance MSC-FAL.1/Circ.3.

The American Bureau of Shipping (ABS) published five cyber security-related guidance documents [38–42] that apply to vessel operators, owners as well as construction and integration companies. The ABS guidance documents provide best practices for the implementation of cyber security measures at the operational and technical levels. They also provide guidance on system data integrity and the application of mitigation measures for IT and OT systems for maritime assets.

DNV GL published recommended practices DNVGL-RP-G 496 (2016) [43] and DNVGL-CP-0231 (2018) [44]. These implement the guidelines of the International Electrotechnical Commission's IEC 62443 standard for the cyber security assessment of IT and OT systems and infrastructure as well as industrial automation and control systems onboard maritime assets. The aim of DNVGL-RP-G 496 (2016) and DNVGL-CP-0231 (2018) is the buildup of IT and OT system infrastructure resiliency against various cyber security threats.

Lloyd's Register (LR) issued three guidance notes (2016) [45–47]. These cover the deployment of IT and OT systems in maritime assets and autonomous ships. They also cover the type of approval of cyber-enabled components of vessel IT and OT systems as well as the LR Cyber Security Framework (CSF) for the Marine and Offshore sector, adopting IMO Resolution MSC.428(98) and IMO Guidance MSC-FAL.1/Circ.3.

The Japanese ship classification society Nippon Kaiji Kyokai or Class NK published two guideline documents that apply to cyber security elements for maritime vessels. The Class NK guideline documents "Guidelines for Designing Cyber Security Onboard Ships" (ClassNK 2020) [48] and "Cyber Security Management Systems for Ships" (Class NK 2019) [49] deal with the implementation of operational and technical controls and measures against cyber threats for IT and OT systems on board vessels. They also cover cyber security management of IT and OT systems for companies and maritime assets and their implementation, maintenance, and improvement aiming for safe navigation.

The Croatian Register of Shipping (CRS) issued ISM Code Statutory Newsletter Number 03.08.2020 [50], which outlines the cyber risk management policies and procedures for maritime assets in accordance with the International Ship and Port Facility Security Code (ISPS Code), International Safety Management Code (ISM Code), IMO Resolution MSC.428(98), and IMO Guidance MSC-FAL.1/Circ.3.

The Indian Register of Shipping (IRCLASS) published the Maritime Cyber Safety Guidelines (IRS-G-SAF-02—2018) [51] that provide the requirements for evaluation and management of cyber risk in ships. In addition, the Guidelines on Certification of Software for Computer Based Control Systems (IRS-G-DES-01—2019) [52] were published. These guidelines outline the quality assurance certification requirements for software for computer-based control systems and their shipboard applications.

The Russian Maritime Register of Shipping published the Guidelines on Cyber Safety (ND No. 2-030101-040-E—2021) [53] which is based on IACS Recommendation No. 166 (Recommendation on Cyber Resilience) and implements the provisions of IMO Resolution MSC.428(98) and IMO Guidance MSC-FAL.1/Circ.3. These guidelines provide recommendations on the design, manufacture, maintenance, and testing of the shipboard computer-based systems as well as applicable recommendations for the vessel's safety management systems (SMS).

The International Registries and Maritime Administrator of The Republic of the Marshall Islands issued Marine Guideline No. 2-11-16 (2018) [54], which outlines the necessary resources for the establishment of policies and procedures for the mitigation of maritime cyber risks in accordance with IMO Resolution MSC.428(98) and IMO Guidance MSC-FAL.1/Circ.3. The aim of Marine Guideline No. 2-11-16 is for asset owners to develop the safeguards against cyber risks for Republic of the Marshall Islands (RMI)-flagged vessels. In addition, the Maritime Administrator of The Republic of the Marshall Islands issued Ship Security Advisory No. 13-20 (2020) [55], which recommends the use of the Guidelines on Cyber Security Onboard Ships from BIMCO as well as USCG Work Instruction CVC-WI-027 on Vessel Cyber Risk Management.

Bureau Veritas issued two Rule Note documents related to cyber security aspects in the maritime sector. Rule Note NR 642 (2018) [56] deals with cyber security related to manufacturing and technical requirements of components and systems to be utilized on board ships. Rule Note NR 659 (2020) [57] applies to the design, construction, commissioning, and maintenance of computer-based systems (CBS) and IT and OT systems to be utilized on board vessels and which are required to be certified as part of the classification of the maritime asset.

4.4. Government Agencies

In the governmental sector in the US, the US Congress (2020) issued Bill S. 4023 "Enhancing Maritime Cyber Security Act of 2020" [58]. Bill S. 4023 assigns to the US Cyber Security and Infrastructure Security Agency (CISA) and the Maritime Administration (MARAD), the implementation of cyber security strategies and measures. The US Coast Guard (2020) issued Navigation and Vessel Inspection Circular (NVIC) 01-20 [59], titled "Guidelines for Addressing Cyber Risks at Maritime Transportation Security Act (MTSA) Regulated Facilities". NVIC 01-20 provides guidance to MTSA-regulated facilities for the assessment, documentation, and addressing of computer system and network cyber vulnerabilities in their assets. NVIC 01-20 covers maritime assets and facilities in the outer continental shelf and offshore operations and encourages the implementation of the National Institute of Standards and Technology (NIST) Framework for Improving Critical Infrastructure Cyber Security and NIST Special Publication 800-82. The USCG has also issued Vessel Cyber Risk Management Work Instruction CVC-WI-027 (rev.2, 2021) [60], which provides guidance for the assessment of vessels' cyber risks in order to avoid posing risk to the Marine Transportation System (MTS) in the case of a cyber breach scenario.

In the United Kingdom (UK), the Institution of Engineering and Technology (IET), the Department for Transport (DfT), and the Defense Science and Technology Laboratory (Dstl) issued the Code of Practice for Cyber Security for Ships (2017) [61]. This Code of Practice provides a management framework for the reduction of the risk of cyber incidents that could affect a ship's safety or security, its crew, passengers, or cargo. The UK DfT, the Defense Science and Technology Laboratory (Dstl), the National Cyber Security Centre (NCSC), and the Institution of Engineering and Technology (IET) also published a Good Practice Guide in Cyber Security for Ports and Port Systems (2020) [62]. This Good Practice Guide is applicable for port and port systems and facilities and is aimed at the integration of cyber security into their overall security planning.

In the European Union, cyber security for the maritime industry is addressed by the European Union Maritime Security Strategy (EUMSS) Action Plan (2018) [63] that aims to reinforce and enhance the EU's capabilities to mitigate security challenges and improve the protection and resilience of maritime systems and infrastructure. Regulation (EU) 2016/679 [64], called the General Data Protection Regulation (GDPR), protects the processing of personal data for all industry sectors, including the maritime sector. The cyber security of IT networks and the delegation of operational cyber security to the European Union Agency for Network and Information Security (ENISA), are addressed by directive 2016/1148/EU [65] and the EU Cyber security Act (2019/881/EU) [66]. The European Union's cyber security strategy JOIN/2013/01 [67] was also developed to strategically implement mitigation technologies and policies and to raise cyber resilience and security levels. ENISA also published reports related to cyber risk management for ports (2020) [68] and port cyber security (2019) [69].

Finally, in Singapore, the Maritime and Port Authority of Singapore published Shipping Circular No. 15 [70] on Maritime Cyber Risk Management. This Circular applies the requirements of IMO Resolution MSC.428(98) and IMO Guidance MSC-FAL.1/Circ.3.

4.5. Marine Insurance Companies and P&I Clubs

The financial loss and asset damage coverage due to the fact of a cyber-attack is a complex insurance issue that is addressed by marine insurers and P&I clubs in various ways. While marine insurance companies provide vessel structure and equipment coverage to asset owners and cargo coverage to cargo owners, P&I clubs cover third-party liabilities [71]. A cyber-attack breach, however, could potentially lead to a third-party liability that may or may not be covered by the insurance contract. Currently, maritime asset insurance includes the Institute Cyber Attack Exclusion Clause (CL 380) 10/11/2003. An Exclusion Clause (CL 380) is included in all marine insurance policies and covers liabilities for the failure of computer equipment and systems. It does not cover, however, losses, damages, or liabilities that are caused by computer systems or equipment when used as a means of aggression or attack.

The Institute Cyber Attack Exclusion Clause (CL 380) 10/11/2003 is incorporated into contract insurance clauses by the following marine insurers and P&I Clubs [71]:

- Exclusion Clause (CL 380) Exclusion Clause (CL 380) The International Group of Protection and Indemnity Clubs (IGPANDI) and every member of this Group (13 members);
- Gard P&I (Bermuda) Ltd.;
- The London Steam-Ship Owners' Mutual Insurance Association Limited';
- The Ship-owners Mutual Protection and Indemnity Association (Luxembourg);
- Assuranceforeningen Skuld;
- The West of England Ship Owners Mutual Insurance Association (Luxembourg);
- United Kingdom Mutual Steamship Assurance Association (Bermuda) Ltd;
- The Steamship Mutual Underwriting Association (Bermuda) Limited.

Similar to the above, The Japan Ship Owners' Mutual Protection and Indemnity Association does not cover insurance liabilities due to the fact of cyber risks, but it does not explicitly apply the Institute Cyber Attack Exclusion Clause (CL 380) 10/11/2003 [71]. The Standard Club Ltd. and Sveriges Ångfartygs Assurans Förening/The Swedish Club [72] exclude liabilities deriving from computer equipment, systems, or software.

The North of England P&I Association recognizes the cyber security threats faced by the maritime sector as well as the existing operational and technical vulnerabilities and has issued a number of publications, briefings and circulars [71–75], for their customers. They also recommend customer compliance to IMO Resolution MSC.428(98) and IMO Guidance MSC-FAL.1/Circ.3 as well as the BIMCO Guidelines on Cyber Security Onboard Ships (2021) (Loss Prevention Briefing, 2016). The use of a cloud-based software platform for their customers is also available and recommended for them in order to carry out assessments of their organizational, operational and technical cyber capabilities and vulnerabilities (Circular 2020/02). In the case of a cyber breach due to the fact of a computer virus (and the subsequent non-existence of suitable IT system antivirus protection), however, there is no coverage for losses, liabilities, or costs suffered (Circular 2021/06).

Finally, the Institute Cyber Attack Exclusion Clause (CL 380) 10 November 2003 is not applied by The American Steamship Owners Mutual Protection and Indemnity Association, Inc.; The Britannia Steam Ship Insurance Association Limited; Sveriges Ångfartygs Assurans Förening/The Swedish Club [71].

5. Risk Analysis Methods for Cyber Physical Security

5.1. API STD 780 SRA (Security Risk Assessment)

The Security Risk Assessment (SRA) methodology was developed for primary use in the oil and gas, chemical, and industrial sectors. SRA is described by API (American Petroleum Institute) standard (STD) 780 (2013) [76] and was developed to be applied in a wide spectrum of applications in the security risk assessment of facilities and assets. API STD 780 SRA can be used to assess security risks from various physical security threats, such as terrorism, piracy, theft, sabotage, for fixed and mobile assets/facilities. The application of SRA can cover both localized areas and functions of assets as well as greater assemblies of facilities/assets at a macroscopic scale. This makes SRA applicable to cyber physical security applications, as it can capture both the physical aspect of security breaches and vulnerabilities as well as the interface of assets with IT/OT systems and infrastructure.

The API STD 780 SRA methodology assesses and allows for the management of security risks through a risk-based, performance-oriented management process aimed at the protection and security of assets, people, and the environment. The SRA is a five-step process as shown in Figure 3.

Application of the API STD 780 SRA Method in a Maritime Asset

In order to prove the case that the API STD 780 SRA can be applied in cyber physical security applications for a maritime asset, an FPSO (floating production storage and offloading) vessel was analyzed in different security incidents. For this analysis, the main maritime features and functions of the vessel were analyzed, excluding the complex processing systems used in the upstream oil and gas sector. For the implementation of the API STD 780 SRA, the available vessel's technical designs and PIDs (process Instrumentation diagrams) were studied.

The five steps of the security assessment described in API STD 780 were followed in order to analyze the asset components, its threats, vulnerabilities, and credible security risks and to propose risk mitigation measures.

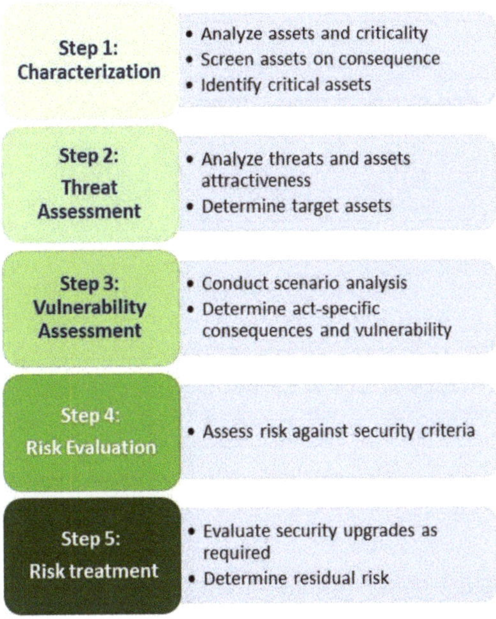

Figure 3. API SRA process (source: [76], with information elaborated by authors).

In Step 1 of the SRA process, the following occur:

i. FPSO vessel asset components are identified;
ii. Interdependencies of the assets are determined;
iii. Existing security safeguards (internal and external) are identified;
iv. Consequences to human life, environment, and business continuity are determined;
v. Criticality ranking based on severe consequences is assigned.

Critical assets that have been identified include vessel engine room, electrical switch board rooms, emergency generators, electrical frequency converters, HVAC equipment, radio/communication equipment, bow and aft thrusters, main propulsion systems, and motor control centers (MCCs). These have been ranked based on their criticality in vessel operational performance, security, and safety. Existing cyber-related measures that have been identified include the physical cyber-separation in some vessel systems and the existing schedule for system software upgrades.

In Step 2 of the SRA process, the following are carried out:

i. Adversaries are identified and evaluated;
ii. Threat ranking for each adversary is assigned;
iii. Security countermeasures are applied;
iv. Identified assets' attractiveness ranking is analyzed and assigned.

The adversaries identified include criminals, interstate adversaries, disgruntled employees, and terrorists. The potential security breach scenario was identified as the planting of malware into control systems and a direct remote cyber-attack to vessel OT and IT systems. Asset attractiveness was assigned based on asset, component, and process interconnections.

In Step 3 of the SRA process, the following are accomplished:

i. Security breach scenarios are defined, and their consequences are evaluated;
ii. Security breach scenario sequence and consequences are evaluated;
iii. Effectiveness of existing security barriers are evaluated;
iv. Vulnerabilities are identified, considering recovery capability and estimating vulnerability degree;

v. Severity ranking of scenario-specific consequences is allocated.

The two main security breach scenarios identified are: (a) the intentional remote planting of malware, corruptive software virus, etc.; (b) the unintentional introduction of malware through a remote USB device. Both scenarios result in the disruption of operations and cause system failure. The main vulnerabilities for IT/OT and control systems identified are (a) the remote connectivity of controls with company's maintenance hub; (b) the existence of obsolete or vulnerable hardware or software that can lead to cyber contamination of the network and vessel systems.

In Step 4 of the SRA process, the following tasks are completed:

i. Conditional likelihood of scenarios is evaluated;
ii. Initial risk ranking is assigned;
iii. Risk is prioritized.

Considering the existing history of cyber breach incidents of maritime assets, the scenarios of intentional remote cyber-attacks and cyber contamination from an external adversary, such as a criminal and a vessel crew member, received a higher risk ranking. The unintentional cyber contamination of IT/OT systems from a vessel crew member was ranked equally in risk level.

Finally, in Step 5 of the SRA process, the following are achieved:

i. The necessity for countermeasures is evaluated along with the recommendations for specific assets;
ii. Likelihood of attack and severity of scenario consequences are recalculated;
iii. Residual risk is determined;
iv. Recommended measures are prioritized.

The main countermeasures determined against the identified cyber security breach scenarios included (a) upgrading the electronic/IT protection systems; (b) installation of firewalls for all segmented systems and subsystems; (c) assignment of an on-board IT expert, as the likelihood of an intentional or unintentional cyber breach scenario from an external (i.e., criminal) or internal (i.e., vessel crew member) threat still ranked high. The implementation of a rigorous IT and OT software and hardware upgrade, the execution of a cybersecurity training program for vessel crew members and the allocation of a specialized IT technician on board the vessel were considered high priority. The residual risk for cyber breach scenarios remained, however, due to the following reasons:

- The anticipated difficulty in the implementation of the recommended measures due to the presence of financial and technical constraints;
- The continuously evolving sophistication and ingenuity of cyber-attacks.

From the above, it can be determined that while the API STD 780 SRA method applied was not evaluated in-depth for all technical cyber physical systems, since it was applied at a macroscopic level, it is obvious that it is a capable tool that assists in the security analysis of complex industrial as well as maritime assets such as those found on an FPSO vessel.

5.2. *Bow Tie Analysis (BTA)*

Bow Tie Analysis (BTA) is a qualitative Process Safety Management (PSM) method primarily used in the oil and gas, chemical, and processing sectors for the proactive and reactive review of safety incidents. Bow Tie Analysis is used for the definition of risks, hazards, and consequences of safety incidents in systems, equipment, processes, and operations. Bow Tie Analysis is also applicable in the maritime sector, as it can be used to analyze the interconnection of marine equipment, systems, and processes in vessels and other maritime assets in the case of safety incidents. For physical and cyber security applications, Bow Tie Analysis can be used to identify the applicable security barriers and mitigation measures for IT/OT assets and processes at the micro- and macro-scales.

- Micro: components, equipment, sub-assemblies, and instruments;
- Macro: assemblies, assets, larger equipment, and operations.

The bow tie diagram, as shown in Figure 4, has the following elements [77]:

- Hazard: a condition that can potentially cause damage;
- Top Event: The event that can occur due to the loss of control of the hazard;
- Threats (depicted on the left side of the bow tie diagram): events or scenarios that could cause the loss of control of a hazard and lead to a top event;
- Consequences (depicted on the right side of the bow tie diagram): unwanted harmful effects that can affect the assets, system, process, people, or the environment;
- Prevention Barriers (on the left side of the diagram): prevention barriers that stop threats from resulting in the top event;
- Mitigation Barriers (to the right of the top event): mitigation barriers that mitigate the top event (i.e., reduce the scale of and possibly stop undesired consequences);
- Degradation Factors: factors that can be applied to both prevention and mitigation barriers and can lead to damage or failure of the barrier to which they are attached;
- Degradation Controls: Control barriers that act to mitigate the degradation factors, assuring the function of the preventive or mitigation barriers. Barriers and degradation controls are illustrated to show these as fundamental elements of the safety management system. Altogether, the diagram provides a holistic picture of the risk management system.

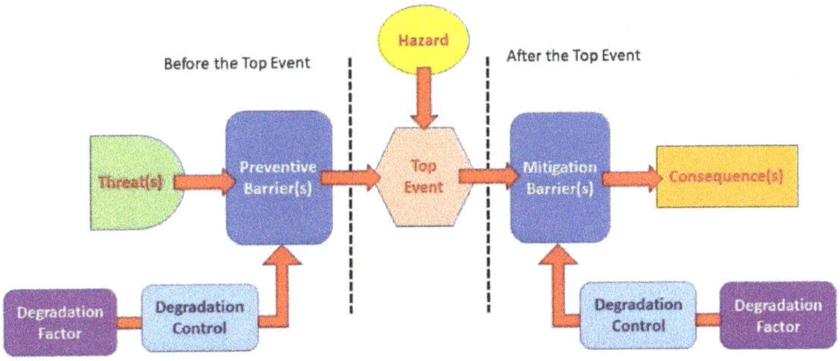

Figure 4. Bow Tie Analysis diagram.

Bow Tie Analysis is an effective methodology to assess and manage security hazards, risks, consequences, and their mitigation. In general, it has the following advantages [78]:

- Ease of use;
- Communicates to all levels of an organization, being highly visual and without excessive or complicated documentation;
- Forces the user to think in terms of risk assessment and barrier-based risk management from the outset;
- Builds system robustness in a logical, structured, and incremental manner;
- Enforces the importance of effective barrier and mitigation management;
- Easily identify areas of weakness in any system;
- Quickly prioritize safety and operational and business critical tasks and objectives;
- Readily integrates with existing management systems;
- Provides assurance via a visual link between bow tie and business processes;
- Promotes awareness of hazards, risks, and management methods;
- Supports brainstorming and hazard identification;
- Effective demonstration of hazard management;
- High-level hazard identification and risk assessment;
- Complements incident investigation and safety and risk regulations.

The application of the Bow Tie Analysis method in cyber security has been demonstrated in general industrial sectors by third-party engineering and cyber security consultants [79,80] and industry organizations [81]. While the usage of the Bow Tie Analysis method is not directly applied in the maritime sector, due to the complexity of the systems involved, it is determined as also being applicable for maritime assets. Considering the cyber physical systems found in maritime assets, Shuang-Hua et al. [82] developed a harmonizing methodology to analyze safety and security risks simultaneously for industrial applications and cyber physical systems. Their methodology incorporated the Bow Tie Analysis method and security assessment via examining preventive barriers. Similarly, Meland et al. [83] determined that the Bow Tie Analysis method is suitable for the analysis of security and safety of cyber and physical systems, identifying a wide range of proactive (preventive) and reactive mitigating barriers. Abdo et al. [84] developed another methodology that integrates the attack tree and Bow Tie Analysis methodologies for a combined safety and security industrial risk analysis including cyber security applications. Using this method, the representation of risk scenarios and quantification of likelihood provide the right control measures. Finally, Bernsmed et al. [85] use the Bow Tie Analysis method as a tool to analyze and visualize physical and cyber security risks and presented their approach using an example in the maritime sector involving navigational communication systems. In addition to the above, Bow Tie Analysis was proposed by DNVGL-RP-G496 [43] as an assessment tool of preventive and reactive security barriers of vessels and mobile offshore units' cyber systems. Similarly, ISO/IEC 31010 [86] mentions Bow Tie Analysis as a risk management tool for cyber systems and cyber security applications.

Application of the Bow Tie Analysis Method in a Maritime Cyber Physical Asset

In order to illustrate the applicability of Bow Tie Analysis as a methodology for risk assessment and management for cyber and physical security, an example involving the malware contamination of a vessel's engine human–machine interface (HMI) system is be described below. The scenario involves a typical IT–OT interface through network communication influenced by data transferred via external sources (network or external device) which leads to the malware contamination of the engine's HMI and the incorrect display of the engine's performance data. The IT–OT communication flow path through the vessel's ICS is shown in Figure 5.

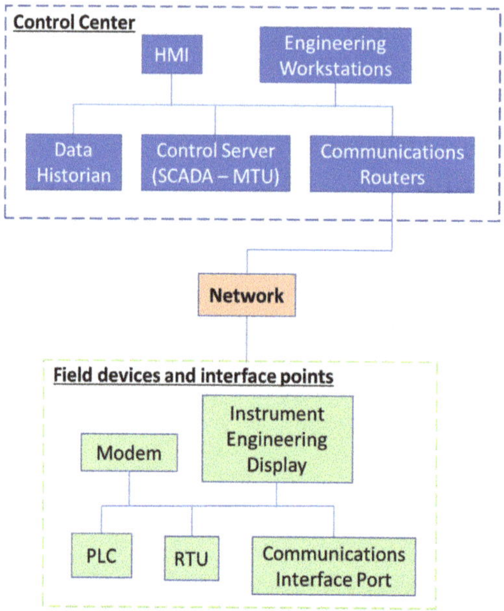

Figure 5. Vessel's ICS communication data flow.

Basically, the vessel's engine control center communicates to field devices through network data flow. The control center includes the HMI, engineering workstation(s), server data historian, control server, and communication routers and interface ports, among other systems. The field devices include interface ports, PLCs (Programmable Logic Controllers), RTUs (Remote Terminal Units), and Instrument Engineer Displays (IEDs). The Bow Tie Analysis example for this scenario is shown in Figure 6.

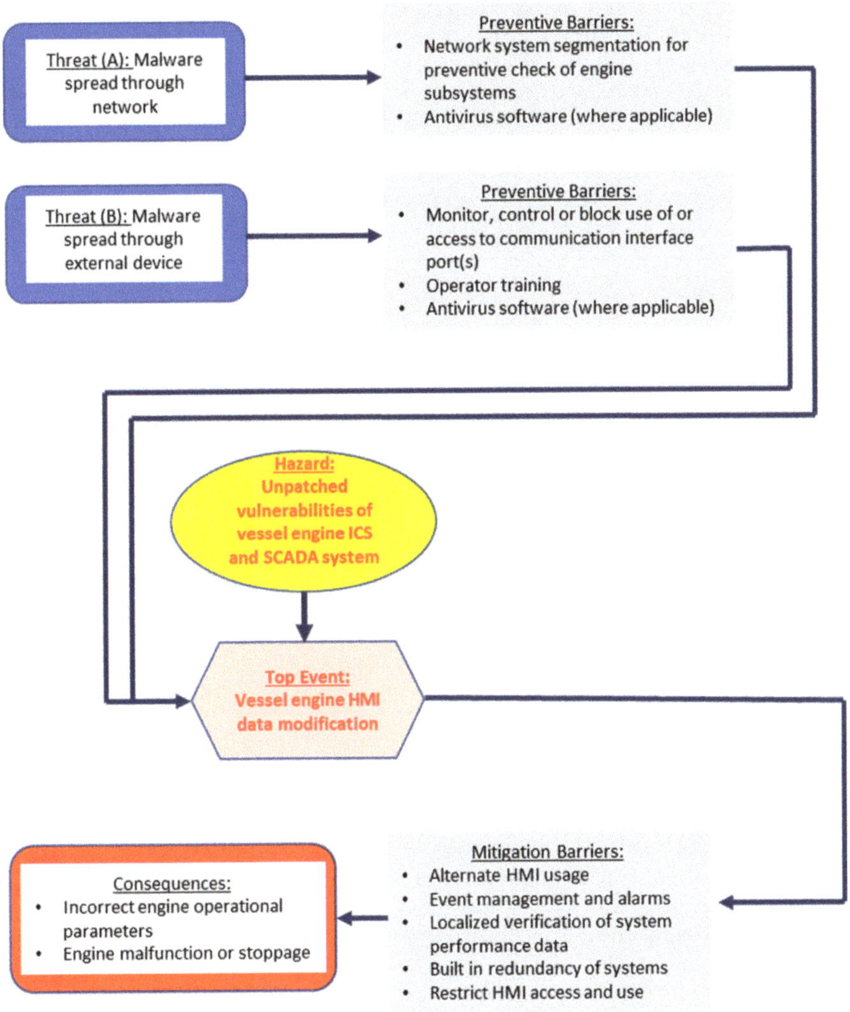

Figure 6. Bow Tie Analysis for vessel engine HMI data modification due to the fact of malware contamination.

Figure 6 shows the malware contamination threat being analyzed involving two different sub-scenarios. In the first threat scenario, the malware spread is caused by network data flow. In this case, one of the preventive barriers for such a threat is network system segmentation, limiting or controlling data flow and allowing preventive checks of the engine's subsystems. The second threat scenario involves engine HMI malware contamination due to the fact of its connection to an external device. Such a device can be a technician's or engineer's laptop, or an external storage drive used to update ICS software during maintenance operations. One of the preventive barriers in this case would be the monitoring, control, or blockage of the use of or access to the communication interface ports. In this way, unintentional or intentional transfer of malware to systems is eliminated. Another preventive barrier would be operator training, which would raise cyber-awareness and cyber hygiene practices and improve the monitoring of critical operational parameters and intervention in the case of excessive deviations being noticed. The use of an antivirus software is also a useful tool and a preventive barrier for both cases, eliminating the installation of malware. Unfortunately, it is not always applicable to vessels' systems due to the varied type and age of IT and OT systems and software on board vessels.

In the post-event side of the bow tie diagram, the mitigation barriers that could be used after the HMI data modification would include the design and integration of redundant devices to enable the use of substitute systems such as an alternate HMI. Event management and alarm notifications could also be included in the system assuming they are integrated within and do not overlap or duplicate such existing systems. The localized verification of the operational parameters of engine systems would also be a reliable barrier to monitor the overall system performance and act against any noted operational malfunctions or abnormalities. For such a method, however, the operator's in-depth knowledge of system performance and operations would be necessary to ensure the proper interpretation of system data. The control and restriction of HMI access could also be implemented as a physical security method to ensure the unauthorized access to systems data is prevented. The ultimate consequence of the failure of all preventive and mitigating barriers would be the display of incorrect operational and performance data through the vessel's engine HMI, resulting in the vessel's engine malfunction, damage, and possible complete stoppage.

Finally, it should be emphasized that the example given highlights the human factor (HMI operator) to machine (OT systems and devices, e.g., HMI) interface and the deriving degradation factors. Human factors (operator use and knowledge of systems) can degrade any preventive or mitigating barriers applied, leading to destructive consequences. Human factors are at the center of the risk management process applied in IT/OT systems. IT/OT system operators and end users need to understand the importance of their roles in maintaining preventive and mitigating barriers. The Bow Tie Analysis method can be a valuable tool to prepare and educate IT/OT system operators and users to ensure cyber physical security breach scenarios are minimized, eliminated, or contained by the use of proper barriers.

6. Risk Treatment—Design of Cyber Security Controls

Substantial guidance exists to support the design of effective technical risk treatments and cyber security controls. This includes well-developed frameworks and other references from government and standards-setting bodies as well as a large base of industry best practices. While individual control requirements should be expected to present some degree of uniqueness in relation to unique risk characteristics, there can be substantial value achieved by the application of a standards-based approach to establish the majority of the controls architecture.

There has been substantial coalescence around the NIST cyber security standards across industries. The US Coast Guard (USCG) has collaborated with the US National Cyber Security Center of Excellence (NCCOE) to specifically address cyber concerns related to maritime, providing guidance related to many of the industry's dimensions described previously in this paper. This includes application of the NIST Cyber Security Framework (NIST CSF) to specific maritime challenges including maritime bulk liquids transfer, offshore operations, and passenger vessel operations [87].

NIST CSF incorporates technical and process dimensions to provide a basis for a comprehensive controls design that established major functional groupings of cyber security activities with subsequent decomposition into categories and subcategories. The first function, Identify, is intended to guide an enterprise to inventory all relevant assets including technologies, data, and personnel. Attention is then turned to the Protect function that includes the design of safeguards relevant to the identified asset base. Next is analysis of the Detect functions including automated monitoring and associated incident alerting. Detect activities then feed the Respond function, outlining the essential system and personnel reactions to observed events. Lastly, the functional requirements to Recover the organization and guide resumption of normal function are described. The result is a robust framework that interrelates ongoing controls design and operation along with necessary response steps when incidents are suspected [88].

NIST CSF describes how the controls design should ideally be guided by the organization's risk management processes and perspectives, which should be derived from the enterprise's risk management posture. Organizational factors, such as risk appetite, legal frameworks, and organizational business or process needs, combine to form the risk posture. NIST CSF illustrates the "notional information and decision flows within an organization", describing the interaction of operational, organizational management, and senior executives. These information and decision flows provide a basis to govern the degree of interaction between IT and OT systems [88].

A significant characteristic that adds complexities to control designs, however, is cyber physical systems, that by definition, reflect the convergence of the cyber and the physical. Therefore, unlike other contexts, the designer of cyber physical controls must consider the unique aspects of threat, vulnerability, and impact. For example, application of the NIST CSF to information systems often manifests as exercises in logging and alerting, whereas cyber physical controls will also potentially require incorporation of physical reactions such as ceasing forward motion, stopping the flow of materials, or ensuring physical aspects "fail closed" to prevent potentially disastrous physical outcomes. The NIST CSF and other industry-neutral cyber security standards do not reflect such risk dimensions.

Once implemented, the controls design requires ongoing oversight using continuous monitoring technologies as well as a robust program of audits and assessment. Technical monitoring control designs are also described in the NIST CSF and provide the basis for the deployment of Security Incident and Event Management (SIEM) architectures that can help the organization manage the continuous inflow of event data, which increasingly present as high-frequency data streams. Advances in artificial intelligence and machine learning technologies offer some promise for helping organizations navigate the substantial event data generated.

The application of available standards, such as the NIST CSF, establishes a framework for the design of adequate controls that will require periodic evaluation to ensure controls are operationally effective. This includes the completion of manual and automated controls testing and using techniques, such as penetration testing, to simulate adversary conduct. The periodic execution of controls effectiveness audits and tests is a staple in other industries and should be considered essential in the maritime context.

7. Discussion

The review of available industry and governmental directives and standards has proved the fact that cyber physical security in maritime assets and, in effect, OT cyber security, while although it is recognized as critical in the overall cyber security of assets, it is not covered adequately to provide solutions for the maritime industry. The majority of available standards and directives focus primarily on the IT side of systems and operations and fail to define mitigation measures and procedures that would guide asset owners and operators in confronting the credible threats faced. Those that do cover cyber physical security are either prescriptive or performance based and do not highlight the need for corporate policies and procedures that would apply to maritime assets both onshore and offshore.

Considering the physical security element involved for cyber systems, asset owners and operators need to realize the need for the actual physical protection and security required for such systems, components, and processes including portable and personal devices used for business or operations on board vessels. Vessel and facility physical security while enforced by the ISPS (International Ship and Port Facility Security) Code, remains partly applied to the physical security of critical cyber assets despite the cyber risk management imposed by the implementation of IMO Resolution MSC.428(98) and IMO Guidance MSC-FAL.1/Circ.3. The enforcement of IEC-62443 could provide mitigation measures and procedures that if followed by asset owners and operators, could provide superior protection of critical cyber systems and functions of assets. This could be through the establishment of complementary physical and cyber security policies that would address both physical and cyber security in the protection of maritime assets in the microscopic scale (systems, components, and processes on board vessels). The application of physical security controls and barriers for specific components or systems of a vessel or maritime facility could control and mitigate unauthorized access to protected assets. This could be further controlled and enforced by the provision of specific access and entry controls for each security barrier or control established.

The auditing of the cyber security and operations of critical components and processes of maritime assets should also be considered by asset owners and operators. This auditing should look into both the technical design and performance of as well as the service provision to maritime assets and their systems and components. This auditing process, if applied by asset owners and operators, could provide the following:

(a) Identification of remote access to the asset, its systems, and functions;
(b) The definition of the existing level of hardening of cyber systems and processes;
(c) The verification of performance and integrity of service providers to the maritime asset owner (shipping company) and operator (vessel crew);
(d) The suitability and adequacy of existing corporate contract documents for the provision of services considering cyber physical security requirements;
(e) The identification of critical systems and functions onboard the vessel and within the shipping company (asset owner);
(f) The asset posture for operational and technical resiliency;
(g) The existing system notification sequence and mechanisms in place that would alert the asset owner and operator in case of a cyber breach incident.

The above could be achieved by the use of security risk assessment methods from the industrial sectors such as the security risk assessment method from API STD 780 and the Bow Tie Analysis method. Considering the advancement of cyber physical security in the general industrial and oil and gas sectors for SCADA systems, the implementation of assessment tools outside the maritime industry could provide advanced capabilities for complex OT systems onboard vessels.

Similarly, the MITRE ATT&CK Threat Model [89] could be used to preemptively classify cyber-attacks and assess organizational risks [90]. It could assist in the understanding of the attack behavior, tactics, and techniques of cyber adversaries and allow for the proper organization of such data [91] for use by the corporate (shipping company or asset owner) chief information security officer (CISO) and cyber mitigation team. Due to the versatility of the MITRE ATT&CK Threat Model for the assessment of enterprise IT networks and cloud, mobile devices, and industrial control systems (ICS) [90], its use for the characterization and verification of post-compromise adversary behavior while operating within maritime asset IT and OT systems and infrastructure and networks could prove to be a useful vulnerability assessment tool.

The maritime industry, equally to other industries, also appears to tackle cyber physical security considering two distinct operational and technical functions, the physical security function, as dictated by the ISPS code, and the cyber security function. This division of cyber and physical security functions necessitates the distinct roles of a CISO and the "asset" physical security officer which, according to the ISPS code, entails the combination of a ship security officer (SSO), vessel security officer (VSO), port facility security officer (PFSO), and company security officer (CSO). This segmentation of duties includes the segmentation of security (cyber and physical) operational policies and procedures that basically lead to "siloed" security functions and, in turn, to hindered collaboration and visibility of interconnected cyber–physical assets and processes and the reduced capacity to tackle complex security threats (physical, cyber, or hybrid). The solution for this issue would be the convergence of security functions at a corporate and operational level as also suggested by CISA [92]. This convergence of cyber and physical security functions would lead to a fully integrated security perception at an operational and technical level, enhanced security of assets (components, systems, and operations), and reinforced organizational and operational security and awareness.

Similar to the segmentation of cyber and physical functions as described previously, the maritime industry also appears to address cyber security considering two distinct operational and technical entities: the shore-based infrastructure and the maritime-based asset infrastructure. The shore-based infrastructure, as described in Section 1, includes the infrastructure supporting the maritime asset and operations and focuses primarily on IT operations and systems. The maritime-based asset infrastructure consists of the vessel's cyber systems, components, and functions that enable, monitor, and control its maritime operations. This corporate approach to cyber segmentation leads to the deconstructed security approach between IT processes and systems ashore and the OT systems and processes onboard maritime assets offshore. This leads further to the diminished corporate understanding of IT and OT functions and vulnerabilities onboard maritime assets and the cyber and physical security threats faced. In this case, maritime asset owners need to invest in the general cyber awareness of their assets and operations by following a holistic approach that includes both OT and IT systems and functions for both their corporate organizations and their maritime assets.

Finally, considering the human element interface in cyber physical systems, as highlighted in Section 1, the reinforcement of cyber security skills for personnel active in the maritime domain should be considered of paramount importance and should be implemented accordingly. Significant investment in the training of maritime vessel operators (crew) and shipping company personnel should be dedicated. Training should be implemented for all personnel across the corporate hierarchy, from field personnel (vessel crew) to company executives. Cyber security-related training should be customized for specific maritime operational stakeholders such as all officers of the engine and deck departments as well as technical superintendents. These stakeholders should be educated to the level of the IT and OT systems utilized on board vessels so that they can contribute to the design, implementation, monitoring, and control of cyber physical security barriers and mitigation measures. As the cyber security threats and IT/OT technologies are evolving, the provided training needs to be revised and re-applied at a recurring rate.

8. Conclusions

The following key points and conclusions were derived from the analysis undertaken in this paper:

(a) The majority of available standards, policies, and directives focus primarily on the IT side of systems and operations and fail to tackle the interoperability and threats to OT systems and components in the maritime sector;
(b) The importance of the physical protection of IT and OT assets needs to be recognized by asset owners, operators, and integrators. The enforcement of physical protection and security of portable devices and critical IT/OT infrastructure onshore and offshore can mitigate imminent cyber threats;
(c) The auditing of cyber security and operations of critical components and processes of maritime assets should be realized by asset owners and operators, considering their technical design, performance, and service provisions. Auditing can expose unknown vulnerabilities in systems, processes, and infrastructure while also verifying the effectiveness and status of already in place corporate and technical measures and procedures;

(d) The use of "out-of-the-industry" security risk assessment methods should be examined. Existing security assessment methods, such as the API STD 780 SRA and the Bow Tie Analysis methods, have shown their capability to tackle complex industrial systems in the processing and oil and gas industries, similar to the ones found onboard maritime assets;
(e) The convergence of shore-based and vessel-based cyber and physical security should be pursued by maritime asset owners and operators. This could enhance cyber and physical security policies' implementation and improve threat mitigation;
(f) Cyber security knowledge for maritime vessel operators (crew) and shipping company personnel should be further improved by the implementation of proper training. This can be achieved only through the allocation of adequate funding, corporate commitment, and the realization of a holistic stakeholder involvement and management process for cyber-related functions at a corporate and field level.

The submitted paper is considered novel in its concept and contributes to the scientific and academic fields because it focuses on IT/OT system interfaces and cyber physical security specifically for the maritime sector. The industry and governmental standards, policies, and directives reviewed make this paper a single source of information where academics and industry associates can educate themselves and understand the complexity of cyber physical and IT/OT systems in the maritime sector as well as the available directives that they ought to conform to. The presentation of two security risk assessment methods, the API STD 780 SRA and the Bow Tie Analysis methods, is also novel in its concept, because it presents "outside-the-box" thinking for sourcing risk assessment methods from the oil and gas and process safety sectors that can be applied in maritime cyber physical security applications. These methods have proved their usability for complex systems and assets and can handle the IT/OT system interface and complexity for maritime assets.

Finally, further research in the field of cyber physical security could expand in the use of other assessment methods from the accident investigation and security assessment sectors. The use of methods, such as STAMP (Systems-Theoretic Accident Model And Processes), ACCIMAP, Tripod Beta, CARVER (Criticality, Accessibility, Recuperability, Vulnerability, Effect, and Recognizability), RAMCAP (Risk Analysis and Management for Critical Asset Protection), or MBVA/MBRA (Model-Based Vulnerability Assessment/Model-Based Risk Assessment) could be compared to BTA and SRA in possible cyber physical security breach scenarios. Further analysis of more complex IT and OT systems used in maritime assets could be also carried out to identify the weaknesses in system architecture and possibly improve system design and operation.

Author Contributions: I.P. conceived the original title and layout of the paper and wrote and edited the draft and final manuscript; P.R. conceived the idea and revised title and contributed to the layout of the paper, co-wrote manuscript sections, and co-edited the draft and final manuscript. N.N. provided supervision, direction on the layout and structure of the paper, vessel operational information and reviewed the draft and final manuscript. All authors have read and agreed to the published version of the manuscript.

Funding: This research received no external funding.

Institutional Review Board Statement: Not applicable.

Informed Consent Statement: Not applicable.

Conflicts of Interest: The authors declare no conflict of interest.

References

1. Zarzuelo, I.; Soeane, M.; Bermudez, B. Industry 4.0 in the port and maritime industry: A literature review. *J. Ind. Inf. Integr.* **2020**, *20*, 100173.
2. Svilicic, B.; Rudan, I.; Jugovic, A.; Zec, D. A Study on Cyber Security Threats in a Shipboard Integrated Navigational System. *J. Mar. Sci. Eng.* **2019**, *7*, 364. [CrossRef]

3. Dahman, J.S.; Baldwin, K.J. Understanding the current state of US defense systems of systems and the implications for systems engineering. In Proceedings of the 2nd Annual IEEE Systems Conference, Montreal, QC, Canada, 7–10 April 2008.
4. American Petroleum Institute (API). *Security Vulnerability Assessment Methodology for the Petroleum and Petrochemical Industries*; American Petroleum Institute: Washington, DC, USA, 2004.
5. Mathews, L. NotPetya Ransomware Attack Cost Shipping Giant Maersk Over $200 Million. Available online: https://www.forbes.com/sites/leemathews/2017/08/16/notpetya-ransomware-attack-cost-shipping-giant-maersk-over-200-million/?sh=4a211fae4f9a.2017 (accessed on 10 June 2021).
6. Parizo, E. Maersk CISO Says NotPeyta Devastated Several Unnamed US Firms. Available online: https://www.darkreading.com/threat-intelligence/maersk-ciso-says-notpeyta-devastated-several-unnamed-us-firms/a/d-id/1336558.2019 (accessed on 10 June 2021).
7. Nguyen, A. Port of San Diego Hit with Ransomware; Hackers Demanded Payment in Bitcoin. 2019. Available online: https://www.nbcsandiego.com/local/port-of-san-diego-hit-with-ransomware-hackers-demanded-payment-in-bitcoin/50375/ (accessed on 10 June 2021).
8. Schuler, M. Clarkson Plc Reveals Details of 2017 Cyber Security Incident. 2018. Available online: https://gcaptain.com/clarkson-plc-reveals-details-of-2017-cyber-security-incident/ (accessed on 10 June 2021).
9. Rundle, J. Coast Guard Details February Cyberattack on Ship. Available online: https://www.wsj.com/articles/coast-guard-details-february-cyberattack-on-ship-11564133401 (accessed on 6 August 2021).
10. Goward, D. New GPS 'Circle Spoofing' Moves Ship Locations Thousands of Miles. Available online: https://www.gpsworld.com/new-gps-circle-spoofing-moves-ship-locations-thousands-of-miles/ (accessed on 15 June 2021).
11. Bush, D. Ethical Hacker Says Ships Are Wide Open to Cyber Attack. Available online: https://lloydslist.maritimeintelligence.informa.com/LL1136933/Ethical-hacker-says-ships-are-wide-open-to-cyber-attack (accessed on 20 August 2021).
12. Bebbington, T. Cyberattack or Coincidence? Available online: https://www.seatrade-maritime.com/opinions-analysis/cyberattack-or-coincidence (accessed on 6 August 2021).
13. USCG. Cyberattack Impacts MTSA Facility Operations. 2019. Available online: https://www.dco.uscg.mil/Portals/9/DCO%20Documents/5p/MSIB/2019/MSIB_10_19.pdf (accessed on 10 June 2021).
14. Weiss, J. The Colonial Pipeline Cyberattack—Did IT/OT Convergence Contribute to the Attack. 2021. Available online: https://www.controlglobal.com/blogs/unfettered/the-colonial-pipeline-cyberattack-did-itot-convergence-contribute-to-the-attack (accessed on 15 May 2021).
15. Gold, J. What is the Industrial Internet of Things? Essentials of IIoT. 2018. Available online: https://www.networkworld.com/article/3243928/what-is-the-industrial-internet-of-things-essentials-of-iiot.html (accessed on 6 June 2021).
16. Henshell, A. Taylorism and the History of Processes: 6 Key Thinkers You Should Know. 2018. Available online: https://www.process.st/taylorism/ (accessed on 15 June 2021).
17. Bhatti, B. 7 Types of AI Risk and How to Mitigate their Impact. 2020. Available online: https://towardsdatascience.com/7-types-of-ai-risk-and-how-to-mitigate-their-impact-36c086bfd732 (accessed on 5 June 2021).
18. International Maritime Organization (IMO) Resolution MSC. *Maritime Cyber Risk Management in Safety Management Systems*; International Maritime Organization: London, UK, 2017; Volume 428.
19. International Maritime Organization (IMO) Resolution MSC-FAL.1/Circ.3. *Guidelines on Maritime Cyber Risk Management*; International Maritime Organization: London, UK, 2017.
20. BIMCO, CLIA, ICS, INTERCARGO, INTERTANKO, OCIMF AND IUMI, "The Guidelines on Cyber Security Onboard Ships", Version 3.0. 2021. Available online: https://www.ics-shipping.org/wp-content/uploads/2021/02/2021-Cyber-Security-Guidelines.pdf (accessed on 15 August 2021).
21. National Institute of Standards and Technology (NIST) Cyber Security Framework. Available online: https://www.nist.gov/cyberframework (accessed on 25 May 2021).
22. National Institute of Standards and Technology (NIST) Special Publication 800-30. Guide for Conducting Risk Assessments. 2012. Available online: https://nvlpubs.nist.gov/nistpubs/Legacy/SP/nistspecialpublication800-30r1.pdf (accessed on 13 May 2021).
23. National Institute of Standards and Technology (NIST) Special Publication 800-37. Risk Management Framework for Information Systems and Organizations—A System Life Cycle Approach for Security and Privacy. 2018. Available online: https://nvlpubs.nist.gov/nistpubs/SpecialPublications/NIST.SP.800-37r2.pdf (accessed on 23 May 2021).
24. National Institute of Standards and Technology (NIST) Special Publication 800-82. Guide to Industrial Control Systems (ICS) Security. 2015. Available online: https://csrc.nist.gov/publications/detail/sp/800-82/rev-2/final (accessed on 20 May 2021).
25. National Institute of Standards and Technology (NIST) Special Publication 1500-201. Framework for Cyber-Physical Systems: Volume 1, Overview. 2017. Available online: https://nvlpubs.nist.gov/nistpubs/SpecialPublications/NIST.SP.1500-201.pdf (accessed on 25 May 2021).
26. National Institute of Standards and Technology (NIST) Special Publication 1500-202. Framework for Cyber-Physical Systems: Volume 2, Working Group Reports. 2017. Available online: https://nvlpubs.nist.gov/nistpubs/SpecialPublications/NIST.SP.1500-202.pdf (accessed on 29 May 2021).
27. National Institute of Standards and Technology (NIST) Special Publication 1500-203. Framework for Cyber-Physical Systems: Volume 3, Timing Annex. 2017. Available online: https://nvlpubs.nist.gov/nistpubs/SpecialPublications/NIST.SP.1500-203.pdf (accessed on 20 May 2021).

28. International Organization for Standardization/ International Electrotechnical Commission standard ISO/IEC 27001. *Information Technology—Security Techniques—Information Security Management Systems—Requirements*; International Organization for Standardization: Geneva, Switzerland, 2013.
29. International Electrotechnical Commission standard IEC-62443-4-2. *Security for Industrial Automation and Control Systems: Technical Security Requirements for IACS Components*; International Organization for Standardization: Geneva, Switzerland, 2019.
30. International Electrotechnical Commission standard IEC 62443-3-3. *Security for Industrial Automation and Control Systems Part 3-3: System Security Requirements and Security Levels*; International Organization for Standardization: Geneva, Switzerland, 2013.
31. International Organization for Standardization/ International Electrotechnical Commission standard ISO/IEC 21827. *Information Technology—Security Techniques—Systems Security Engineering—Capability Maturity Model®(SSE-CMM®)*; International Organization for Standardization: Geneva, Switzerland, 2008.
32. International Organization for Standardization/ International Electrotechnical Commission standard ISO/IEC 18045. *Information Technology—Security Techniques—Methodology for IT Security Evaluation*; International Organization for Standardization: Geneva, Switzerland, 2008.
33. International Organization for Standardization/ International Electrotechnical Commission standard ISO/IEC 15408-1. *Information Technology—Security Techniques—Evaluation Criteria for IT Security*; International Organization for Standardization: Geneva, Switzerland, 2009.
34. International Organization for Standardization/ International Electrotechnical Commission standard ISO/IEC 27032. *Information Technology—Security Techniques—Guidelines for Cybersecurity*; International Organization for Standardization: Geneva, Switzerland, 2012.
35. American Society for Testing and Materials standard ASTM F3286-17. *Standard Guide for Cybersecurity and Cyberattack Mitigation*; ASTM International: West Conshohocken, PA, USA, 2017.
36. American Society for Testing and Materials standard ASTM F3449-20. *Standard Guide for Inclusion of Cyber Risks into Maritime Safety Management Systems in Accordance with IMO Resolution MSC.428(98)—Cyber Risks and Challenges*; ASTM International: West Conshohocken, PA, USA, 2020.
37. International Association for Classification Societies (IACS) Recommendation No. 166: "Recommendation on Cyber Resilience". 2020. Available online: https://www.iacs.org.uk/publications/recommendations/161-180/rec-166-new-corr1/ (accessed on 15 August 2021).
38. American Bureau of Shipping (ABS). *Guidance Notes on the Application of Cybersecurity Principles to Marine and Offshore Operations—ABS CyberSafety*; American Bureau of Shipping (ABS): Houston, TX, USA, 2016; Volume 1.
39. American Bureau of Shipping (ABS). *Guide for Cybersecurity Implementation for the Marine and Offshore Industries—ABS CyberSafety*; American Bureau of Shipping (ABS): Houston, TX, USA, 2018; Volume 1.
40. American Bureau of Shipping (ABS). *Guidance Notes on Data Integrity for Marine and Offshore Operations—ABS CyberSafety*; American Bureau of Shipping (ABS): Houston, TX, USA, 2016; Volume 3.
41. American Bureau of Shipping (ABS). *Guide for Software Systems Verification—ABS CyberSafety*; American Bureau of Shipping (ABS): Houston, TX, USA, 2016; Volume 4.
42. American Bureau of Shipping (ABS). *Guidance Notes on Software Provider Conformity Program—ABS CyberSafety*; American Bureau of Shipping (ABS): Houston, TX, USA, 2016; Volume 5.
43. DNV GL. (Det Norske Veritas-Germanischer Lloyd) Recommended Practice DNVGL-RP-G496. In *Cyber Security Resilience Management for Ships and Mobile Offshore Units in Operation*; DNVGL-RP-0496; DNV GL: Oslo, Norway, 2016; Volume 5.
44. DNV GL. (Det Norske Veritas-Germanischer Lloyd) Class Programme Type Approval DNVGL-CP-0231. In *Cyber Security Capabilities of Control System Components*; DNVGL-RP-0496; DNV GL: Oslo, Norway, 2018.
45. Lloyd's Register Guidance Note. *Cyber-Enabled Ships—Deploying Information and Communications Technology in Shipping—Lloyd's Register's Approach to Assurance*; Lloyd's Register: London, UK, 2016.
46. Lloyd's Register Guidance Note. *Cyber-Enabled Ships—ShipRight Procedure—Autonomous Ships*; Lloyd's Register: London, UK, 2016.
47. Lloyd's Register Guidance Note. *Cyber-Enabled Ships—Type Approval of Cyber Enabled Systems Components*; Lloyd's Register: London, UK, 2016.
48. Class, N.K. *Guidelines for Designing Cyber Security Onboard Ships*, 2nd ed.; Class N.K.: Tokyo, Japan, 2020.
49. Class, N.K. *Cyber Security Management Systems for Ships*, 1st ed.; Class N.K.: Tokyo, Japan, 2019.
50. Croatian Register of Shipping (CRS) ISM Code Statutory Newsletter Number 03.08.202. *Maritime Cyber Security Risk Management*; Croatian Register of Shipping: Split, Croatia, 2020.
51. Indian Register of Shipping (IRCLASS) Guidelines IRS-G-SAF-02 -2018. *Maritime Cyber Safety*; Indian Register of Shipping: Mumbai, India, 2018.
52. Indian Register of Shipping (IRCLASS) Guidelines IRS-G-DES-01—2019. *Certification of Software for Computer Based Control Systems*; Indian Register of Shipping: Mumbai, India, 2019.
53. Russian Maritime Register of Shipping ND No. 2-030101-040-E—2021. *Guidelines on Cyber Safety*; Russian Maritime Register of Shipping: Saint Petersburg, Russia, 2021.

54. The International Registries and Maritime Administrator of The Republic of the Marshall Islands Marine Guideline No. 2-11-16. *Maritime Cyber Risk Management*; The International Registries and Maritime Administrator of The Republic of the Marshall Islands: Reston, VA, USA, 2018.
55. The International Registries and Maritime Administrator of The Republic of the Marshall Islands Ship Security Advisory No. 13-20. *Cyber Risk Management—Revised Industry Guidelines and United States Port State Control Measures*; The International Registries and Maritime Administrator of The Republic of the Marshall Islands: Reston, VA, USA, 2020.
56. Bureau Veritas, Rule Note NR 642 DT R00 E. *Cybersecurity Requirements for Products to be Installed On-Board Naval Ships*; Bureau Veritas: Neuilly-sur-Seine, France, 2018.
57. Bureau Veritas, Rule Note NR 659 DT R01. *Rules on Cyber Security for the Classification of Marine Units*; Bureau Veritas: Neuilly-sur-Seine, France, 2020.
58. U.S. Congress Bill. S. 4023 Enhancing Maritime Cybersecurity Act of 2020. 22 June 2020. Available online: https://www.govtrack.us/congress/bills/116/s4023/text (accessed on 19 May 2021).
59. U.S. Coast Guard; U.S. Department of Homeland Security. Navigation and Vessel Inspection Circular (NVIC) 01-20 Guidelines for Addressing Cyber Risks at Maritime Transportation Security Act (MTSA) Regulated Facilities. 26 February 2020. Available online: https://www.dco.uscg.mil/Portals/9/DCO%20Documents/5p/5ps/NVIC/2020/NVIC_01-20_CyberRisk_dtd_2020-02-26.pdf?ver=2020-03-19-071814-023 (accessed on 10 May 2021).
60. U.S. Coast Guard; U.S. Department of Homeland Security. Office of Commercial Vessel Compliance (CG-CVC) Mission Management System (MMS) Work Instruction (WI) CVC-WI-027 "Vessel Cyber Risk Management Work Instruction. 18 February 2021. Available online: https://www.dco.uscg.mil/Portals/9/CVC-WI-27%282%29.pdf (accessed on 23 May 2021).
61. UK Institution of Engineering and Technology (IET) Guidance Document. *Code of Practice: Cyber Security for Ships*; UK Institution of Engineering and Technology: London, UK, 2017.
62. UK Institution of Engineering and Technology (IET) Good Practice Guide. *Cyber Security for Ports and Port Systems*; UK Institution of Engineering and Technology: London, UK, 2020.
63. Council of the European Union, 10494/18. *Council Conclusions on the Revision of the European Union Maritime Security Strategy (EUMSS) Action Plan*; Council of the European Union: Brussels, Belgium, 2018.
64. European Union, Directive 2016/679/EU on the Protection of Natural Persons with Regard to the Processing of Personal Data and on the Free Movement of Such Data, and Repealing Directive 95/46/EC (General Data Protection Regulation). Available online: https://eur-lex.europa.eu/legal-content/EN/TXT/?qid=1599862836456&uri=CELEX:32016R0679 (accessed on 10 May 2021).
65. Directive (EU) 2016/1148 of the European Parliament and of the Council of 6 July 2016 Concerning Measures for a High Common Level of Security of Network and Information Systems across the Union. Available online: https://eur-lex.europa.eu/eli/dir/2016/1148/oj (accessed on 23 May 2021).
66. European Union, Directive 2019/881/EU on ENISA (the European Union Agency for Cybersecurity) and on Information and Communications Technology Cybersecurity Certification and Repealing Regulation (EU) No 526/2013 (Cybersecurity Act). Available online: https://eur-lex.europa.eu/eli/reg/2019/881/oj (accessed on 14 April 2021).
67. Joint Communication to The European Parliament, The Council, The European Economic And Social Committee And The Committee Of The Regions Join/2013/01 Cybersecurity Strategy of the European Union: An Open, Safe and Secure Cyberspace. Available online: https://eur-lex.europa.eu/legal-content/EN/TXT/?qid=1553779410177&uri=CELEX:52013JC0001 (accessed on 25 April 2021).
68. ENISA Report. Cyber Risk Management for Ports: Guidelines for Cybersecurity in the Maritime Sector. 2020. Available online: https://www.enisa.europa.eu/publications/guidelines-cyber-risk-management-for-ports (accessed on 27 May 2021).
69. ENISA Report. Port Cybersecurity: Good Practices for Cybersecurity in the Maritime Sector. 2019. Available online: https://www.enisa.europa.eu/publications/port-cybersecurity-good-practices-for-cybersecurity-in-the-maritime-sector/at_download/fullReport (accessed on 28 May 2021).
70. Maritime and Port Authority of Singapore. *Shipping Circular No. 15 of 2020: Maritime Cyber Risk Management*; Maritime and Port Authority of Singapore: Singapore, 2020.
71. Alifragki, M.E. Cyber-Attacks: The new type of piracy in the Maritime World. Master's Thesis, Department of Maritime Studies, University of Piraeus, Pireas, Greece, 2019. Available online: https://dione.lib.unipi.gr/xmlui/handle/unipi/12503 (accessed on 15 May 2021).
72. The North of England P&I Association. Cyber Risks & P&I Cover. Available online: https://www.nepia.com/cyber-risks-pi-cover/ (accessed on 2 May 2021).
73. Circular 2021/06: Class War Risks—Renewals 2021/2022, The North of England P&I Association. 16 February 2021. Available online: https://www.nepia.com/circulars/class-war-risks-renewals-2021-2022/ (accessed on 11 May 2021).
74. Circular 2020/02: Cyber Security: Kick Start—New Member Benefit for Cyber Security Compliance, The North of England P&I Association. 10 January 2020. Available online: https://www.nepia.com/circulars/cyber-security-kick-start-new-member-benefit-for-cyber-security-compliance/ (accessed on 25 May 2021).
75. *Loss Prevention Briefing: Cyber Risks in Shipping*; North of England P&I Association: Newcastle upon Tyne, UK, 2016.
76. American Petroleum Institute (API), Standard (STD) 780. *Security Risk Assessment Methodology for the Petroleum and Petrochemical Industries*; American Petroleum Institute: Washington, DC, USA, 2013.

77. Center for Chemical Process Safety (CCPS), The Energy Institute. *Bow Ties in Risk Management: A Concept Book for Process Safety*; John Wiley & Sons Inc.: Hoboken, NJ, USA, 2018.
78. American Bureau of Shipping (ABS). American Bureau of Shipping (ABS) Technical Report. In *Bowtie Applications for the Marine and Offshore Industries*; American Bureau of Shipping (ABS): Houston, TX, USA, 2013.
79. DRAGOS Inc and OSIsoft Inc White Paper: Using Bow Tie Risk Modeling for Industrial Cybersecurity. DRAGOS Inc., 2021. Available online: https://www.dragos.com/resource/using-bow-tie-risk-modeling-for-industrial-cybersecurity/ (accessed on 15 November 2021).
80. aeBlogs: The Benefits of Visualizing CyberPHAs Using Bowtie Diagrams. aeSolutions Inc. Available online: https://www.aesolutions.com/post/The-benefits-of-visualizing-cyberphas-using-bowtie-diagrams (accessed on 10 September 2021).
81. SANS Institute Information Security Reading Room White Paper: Evaluating Cyber Risk in Engineering Environments: A Proposed Framework and Methodology. Rebekah Mohr, 2016. Available online: https://www.sans.org/white-papers/37017/ (accessed on 10 September 2021).
82. Ji, Z.; Shuang-Hua, Y.; Yi-jia, C.; Yuchen, W.; Chenchen, Z.; Liang, Y.; Yinqiao, Z. Harmonizing safety and security risk analysis and prevention in cyber-physical systems. *Process. Saf. Environ. Prot.* **2021**, *148*, 1279–1291. [CrossRef]
83. Meland, P.H.; Bernsmed, K.; Frøystad, C.; Li, J.; Sindre, G. An experimental evaluation of bow-tie analysis for security. *Inf. Comput. Secur.* **2019**, *26*, 536–561. [CrossRef]
84. Abdo, H.; Kaouk, M.; Flaus, J.; Masse, F. A safety/security risk analysis approach of Industrial Control Systems: A cyber bowtie—combining new version of attack tree with bowtie analysis. *Comput. Secur.* **2018**, *72*, 175–195. [CrossRef]
85. Bernsmed, K.; Frøystad, C.; Meland, P.H.; Nesheim, D.A.; Rødseth, Ø.J. Visualizing Cyber Security Risks with Bow-Tie Diagrams. In *Graphical Models for Security, GraMSec 2017*; Lecture Notes in Computer, Science, Liu, P., Mauw, S., Stolen, K., Eds.; Springer: Cham, Switzerland, 2018; Volume 10744. [CrossRef]
86. International Organization for Standardization/ International Electrotechnical Commission standard ISO/IEC 31010. *Risk Management—Risk Assessment Techniques*; International Organization for Standardization: Geneva, Switzerland, 2019.
87. Available online: https://www.nccoe.nist.gov/projects/use-cases/maritime-ong (accessed on 15 May 2021).
88. Available online: https://nvlpubs.nist.gov/nistpubs/CSWP/NIST.CSWP.04162018.pdf (accessed on 15 May 2021).
89. The MITRE Corporation. "MITRE ATT&CK®", The MITRE Corporation. 2016. Available online: https://attack.mitre.org/ (accessed on 4 June 2021).
90. Georgiadou, A.; Mouzakitis, S.; Askounis, D. Assessing MITRE ATT&CK Risk Using a Cyber-Security Culture Framework. *Sensors* **2021**, *21*, 3267. [CrossRef] [PubMed]
91. MITRE Report MP180360R1. MITRE ATT&CK®: Design and Philosophy. 2020. Available online: https://attack.mitre.org/docs/ATTACK_Design_and_Philosophy_March_2020.pdf (accessed on 4 June 2021).
92. US DHS CISA Cybersecurity and Physical Security Convergence Guide. Available online: https://www.cisa.gov/publication/cybersecurity-and-physical-security-convergence (accessed on 4 June 2021).

Journal of
Marine Science and Engineering

MDPI

Article

Maritime Cyber Security: A Global Challenge Tackled through Distinct Regional Approaches

Stavros Karamperidis [1,*], Chronis Kapalidis [2,3] and Tim Watson [2]

[1] Department of International Shipping, Logistics and Operations, University of Plymouth, Plymouth PL4 8AA, UK
[2] Warwick Manufacturing Group, University of Warwick, Coventry CV4 7AL, UK; chronis.kapalidis.1@warwick.ac.uk (C.K.); tw@warwick.ac.uk (T.W.)
[3] Information Security Forum (ISF), London EC3M 1AJ, UK
* Correspondence: stavros.karamperidis@plymouth.ac.uk

Abstract: Maritime cyber security is an emerging issue that requires immediate attention, according to the International Maritime Organization (IMO). Feedback received from global shipping professionals indicate that a common threat to the industry, such as cyber security, is dealt with differently among industry practitioners around the globe. Data collected from two targeted focus groups (one in Europe and the second in Asia, two leading groups in the maritime transport sector) demonstrated that, based on technology adoption maturity, cyber security is perceived differently between these groups. The COVID-19 pandemic has highlighted these differences. Our findings lead to useful intelligence that will inform key maritime decision makers, both in meeting the IMO requirements and preparing the organization to address cyber risks.

Keywords: maritime cyber security; cultural differences; maritime transport sector

1. Introduction

Innovative technologies have found their way to the maritime transport sector as they minimize the costs and maximize the benefits in everyday operations. At the same time, these new technologies enhance the interconnectedness of core port and shipping operations to the whole supply chain. As such, any interruption to the core of these operations may have a consequent knock-on effect to the wider economy and industries related to the supply chain, as illustrated in the CyRiM Report [1].

Despite the increasing numbers of cyber incidents to corporate networks and data, the maritime transport sector is rather slow in addressing cyber risk [2]. Well-established regulations and guidelines have been implemented for decades on topics such as environmental and crew safety and, more recently, on ballast water management; these risks are tangible. However, cyber risks are different. Their intangible nature means that their consequences are not palpable; therefore, it is difficult for them to be initially identified and addressed. Infected applications, computers in the office, or operational technology (OT) systems on board may continue to operate without any noticeable performance issues. Unlike any other risk, when a cyber breach occurs, it can affect the entire infrastructure of an organization, including its fleet and offices around the world.

This threat landscape will only grow, as ships at sea increase their connectivity, exchanging data so they can increase supply chain visibility and performance. Unfortunately, while connectivity solutions have evolved, achieving greater resilience, a single specific vulnerability in one industry or organization can swiftly cascade to affect other industries and organizations due to the lack of appropriate security controls [1].

Maritime transport companies are part of a complex supply chain that, at present, is digitalized. Digitalization is taking place so that performance in the overall supply chain can be improved [3]. At the same time, shipping companies are providing remote access to on-board systems to third party service providers and vendors for software updates, performance monitoring, and maintenance [4]. These two intertwined activities increase the cyber-attack surface. It is not a farfetched scenario in which a critical cloud service provider or even a satellite communications services provider could be interrupted or ceased altogether because of a cyber breach. Such incidents could cascade on a global scale and impact all economic activities. The most profound example is the NotPetya malware attack in July 2017, which had a huge impact on global economic activities, costing approximately $892.5 million [5]. Research institutes and regulatory authorities are struggling to model, let alone quantify, cyber risk, mainly due to the lack of relevant data. When the threats move at the speed of light, they can be hard to comprehend.

In other leading industries, such as banking, manufacturing, retail, and healthcare, boards of directors and executive leadership are gradually becoming more aware of the cyber threat to their businesses and the need to manage cyber security at the enterprise level. A 2020 survey conducted by the Information Systems Audit and Control Association (ISACA) [6] found that 82% of respondents understood the board of directors to be "concerned" or "very concerned" about cyber security. Similarly, in shipping, as our research identified and demonstrates in Figures 1 and 2, more than 90% of experts consider cyber security to be either very or somewhat important for their working environment. However, this concern does not always align with how board members allocate resources to tackle cyber risk. Hence, security professionals are tasked to address this emerging threat without the necessary tools in place [7].

Even though the industry is cognizant of the importance of cyber security, confusion persists about how serious the cyber threat actually is, the risks that it poses to their enterprises, and the prioritization it demands. Going beyond the technical interpretation of cyber security, this paper aims to tackle the following research aim: *Identify how cultural differences affect the level of understanding on maritime cyber security.*

It should be highlighted that the paper is constructed based on the data extracted during two high-profile industry workshops. Members of the research team were invited to present in these two workshops. As such, an opportunity arose to put together a set of questions to be shared with the participants to extract valuable data from experts with deep knowledge of the maritime transport sector. Due to that restriction, the research environment was not fully controlled by the research team. Research was not conducted in the typical academic way, but it was a great opportunity to extract raw data and present it to an academic audience seeking rare true insight knowledge. Therefore, the research team extracted, analyzed, and present the findings in this paper, so they could provide rare evidence for an issue of paramount importance for the maritime transport sector. This paper makes a valuable contribution to current literature by focusing on outcomes extracted from experts and links them with existing literature.

Following the introduction, the related literature review on the key stakeholders involved in the maritime transport sector and the illustration of maritime cyber-crime importance for the sector is presented in Section 2. Section 3 describes the theory that supports the work conducted from the research team and shows how the data was collected through a survey. Outcomes of the survey are thoroughly discussed and presented in Section 4. Section 5 presents a conclusion and proposal for future research.

2. Literature Review

The adoption of new technologies in the global shipping sector, such as office and shipboard information technology (IT) systems, advanced supervisory control and data acquisition (SCADA) systems, and industrial control systems (ICS) enhance maritime operations that support 90% of world trade [8]. At the same time, these technologies, despite their benefits, introduce new vulnerabilities and threats to day-to-daily maritime transport operations. The rapid growth of technology-intense solutions in most cases does not take security into consideration [9]. In order to deliver transportation services to their customers, maritime transport stakeholders are now required to take cyber security into consideration, as their customers expect them to operate in a secure, digitalized environment. Hence, it is critical for maritime stakeholders to rise to the challenge of the newly introduced cyber threat landscape in the maritime transport sector.

2.1. Maritime Transport Sector

The maritime transport sector consists of three main components: (1) mobile assets, (2) infrastructure, and (3) financial activities. Specifically mobile assets include: ships and auxiliary platforms. Infrastructure often described as maritime critical infrastructure includes: port infrastructure, offshore energy infrastructure, safety and security controls, navigation aids, communication systems (onshore, offshore, satellite), underwater cables, and pipelines. Finally, financial activities include: insurance agencies and vendors, booking/charter agencies, and banking-economic transactions [10].

Raising awareness is a challenge in the maritime transport sector, as senior maritime transport stakeholders are struggling to understand the changing risk environment. Identifying how cyber threats are affecting their organizations and allocating the necessary resources to address this is not straightforward. Staff engaged with cyber security are struggling to provide concrete evidence regarding the return on investment (ROI) for cyber security measures. This is becoming more difficult due to the abstract nature of cyber threats. Unlike piracy or extreme weather conditions, the consequences of a cyber breach are not always visible.

Safety and security at sea has paramount importance for shipping, a sector with a good safety and security record of accomplishment. Cyber security has been identified as a key part for maintaining safety and security for ships and port operations. Managing cyber threats, according to the Maritime Security Strategy [11] and the related Action Plan of The European Union (EU), is integral to achieving maritime transport security. NATO acknowledged the importance of the maritime cyber security back in October 2016, when it organized the first dedicated maritime cyber conference at the NATO Maritime Interdiction Operations Training Centre (NMIOTC) in Crete [12]. A key initiative aiming to contribute to the better understanding of cyber risk is the International Maritime Organization (IMO) guidelines on maritime cyber risk management (MSC-FAL.1/Circ.3) and the consequent IMO resolution MSC 428(98) (July 2017) on maritime cyber risk management in safety management systems [13,14]. The latter document suggests that, as of 1 January 2021, all stakeholders engaged in the industry should demonstrate cyber capability.

As cyber risk management is now a requirement of the IMO, it would be expected that all maritime transport stakeholders should take similar actions in meeting this requirement and, consequently, address cyber risk within their operating environment. This is not the case, though. Maritime transport stakeholders around the world adopt a different posture to the issue. One key reason for this is that the origin of each individual reflects their level of understanding of a specific topic, including cyber security, as demonstrated in Section 3. Specifically, the maritime transport sector is multinational in nature as several stakeholders from various countries are involved. Typically, a ship may be registered in a particular flag state but may have a crew of many nationalities, none of whom may be from that country. The owners may be from different countries and the vessel operator may be from yet another country.

2.2. Flag States

Flag states are national authorities representing the respective country responsible for establishing and overseeing the regulatory regime governing ship operations within that country's registry. Flag states operate their registries according to their domestic laws and regulations while complying with a set of international codes [15].

The predominant regulatory entity issuing such codes is the IMO. The IMO is a specialized agency of the United Nations with the authority to regulate maritime affairs, including international shipping, port safety, and security. Most flag states transpose the IMO-generated international requirements in their national legislation.

However, the most popular flag states are considered "open registries" or "flags of convenience" that do not require a meaningful economic or financial tie to the country of registry [16]. One of the attractions of open registries is their lower fee structure related to registering and complying with the flag state's requirements. They are also less stringent in their staffing requirements and allow for the hiring of cheaper labor. Ships registered under these schemes are, in general, considered "lightly regulated", although some of them are well managed and highly reputable.

In the wake of the IMO's decision to incorporate cyber risk management under the International Safety Management (ISM) Code, a very small number of flag states have started addressing cyber risk management requirements. An apt example is the USA, where the US Coast Guard [17] issued the Navigation and Vessel Inspection Circular No. 01-20 "Guidelines for addressing cyber risk at maritime transportation security act (MTSA) regulated facilities", providing specific guidance for the incorporation of cyber risks in safety management systems. Because cyber risks represent a current, clear, and persistent threat to the maritime transport sector, overall, shipping companies are strongly encouraged to go above and beyond the compliance requirements and lay the foundations for comprehensive cyber security strategy and plans as early as possible.

2.3. *Classification Societies*

Classification societies are private, not-for-profit organizations that provide quality assurance to the maritime transport sector. The International Association of Classification Societies (IACS) is the prominent organization where most classification societies are members. While these societies were historically based in and focused on certain countries and markets, they have gradually increased their global representation. Furthermore, those societies organically grow in countries by establishing offices there and have also followed the merger and acquisition approach in which two classification societies merge and grow. The most prominent example is the merge of Norwegian DNV (Det Norske Veritas) with the German classification society (Germanischer Lloyd) to form a multinational classification society [18]. There is a large number of additional classification societies that function similarly.

In line with the IMO resolution regarding cyber risk management, classification societies have started to introduce detailed guidelines, assessing the cyber security posture of ships and shipping companies. For example, DNV, ClassNK, and ABS have developed relevant cyber security guidelines [19–21]. At the same time, classification societies are responsible for auditing ISM code implementation, which now includes cyber security.

2.4. *Maritime Cyber Crime*

Having looked at the structure of the maritime transport sector and briefly examined its response to the IMO cyber regulation, it is important to understand how the sector's transition to the digital era can, at the same time, introduce cyber related risks. To illustrate, one cyber threat campaign, referred to by researchers as the Daily Show, began as a phishing attack against the shipmaster of a tanker operator [22]. Since the initial infection, which involved key logger malware, the campaign spread around the world, affecting more than 50% of the IMO member nations, and infected not only unknown numbers of vessels and port facilities but also oil/gas, manufacturing, customs agencies, logistics companies, and banks.

As stated previously, maritime transport companies are constantly adopting new systems, platforms, and technologies to achieve greater levels of capability and efficiency, which introduces new risks to their operations. These risks range from email and IT-office environments to OT systems. The latter is the most important in shipping operations as it affects ships' seaworthiness [10]. It can be argued that, in the maritime transport sector specifically, cyber threats are becoming more frequent and sophisticated, as individual hackers, well-funded and organized criminal networks, nation states, and others target ports, shipping companies, vessels, and shore-side facilities. Predominantly, during the

COVID-19 pandemic, according to MTS-ISAC [23], an increase in cyber attacks in shipping took place, with some additional reporting from the Nautical Institute, demonstrating that this increase may have reached up to 900% [24].

Due to its nature, the cyber domain knows no boundaries. Cyber threat actors are less constrained by geography than connectivity. As shipping is increasingly reliant on connected technologies, every vessel, shore-based facility, and office represents a potential target. Vessels are particularly vulnerable to cyber threats, since ship operators (including management and third-party vendors) regularly access the vessel's networks and related operational systems, both physically and remotely. Unpatched operating systems on vessels with poorly configured networks, open serial ports, and legacy applications represent a considerable asymmetrical risk to shipping and land-based operations (ports, offices, and supply chains) [10].

The maritime transport industry has been slow to respond to the growing cyber threat [2,25]. However, this belated response is not granular throughout. The reasons for this anomaly are identified and analyzed in this paper.

3. Methodology

3.1. Theory That Supports Our Work

One of the main findings of the analysis relates to the perception of the topic of cyber security. This is evident throughout the responses collected for the various questions of the survey, as presented in Section 4. As such, the research team tried to identify a conceptual framework that justifies this approach. The most apt piece of academic work that closely aligns with this observation is the book by Nisbett [26], entitled: *"The geography of thought: How Asians and Westerners think differently ... and why"*. According to Nisbett [26], different cultures perceive specific topics in a different way. Specifically, and related to the composition of participants of this research, East Asian thought is "holistic", while Westerners focus on specific subjects. To elaborate, East Asians interpret specific subjects as part of the whole, examining the relations between objects and events within that discipline. Applying this concept to maritime cyber security, East Asians perceive cyber security as another risk factor, part of the overall aggregated risk affecting maritime transport operations. This argument is reinforced as it is demonstrated from the findings of this research, presented in Section 4. In contrary, the West approach emphasizes notable subjects aiming to tackle any challenges related to this through specific attribution. With Nisbett in mind, applying this concept to maritime cyber security, Western managers perceive cyber security as a standalone risk element to be dealt with by IT professionals. That demonstrates a clear differentiation between the mindsets and approaches that the maritime transport practitioners who participated in the two workshops (Asia and Western representatives) undertake when dealing with cyber security in the maritime transport sector.

Shipping is a truly global industry, as Kumar and Hoffmann [27] (p. 36) state: *"A Greek owned vessel, built in Korea, may be chartered to a Danish operator, who employs Philippine seafarers via a Cypriot crewing agent, is registered in Panama, insured in the UK, and transports German made cargo in the name of a Swiss freight forwarder from a Dutch port to Argentina, through terminals that are concessioned to port operators from Hong Kong and Australia"*. As such, maritime transport stakeholders, irrespective of their physical location or their racial decent, should develop a common, fundamental mindset that could grasp the risk factor called cyber security. That is something difficult to achieve as the sector needs time to develop and adopt that common mindset. The first step towards that direction is for regulatory bodies to introduce relevant documentation, either mandatory or advisory. An apt example is the IMO, with its guidelines and consequent resolution on cyber risk management [13,14], as analyzed above.

3.2. Data Collection

A unique opportunity was offered to test the theory presented above through direct engagement with industry experts from Asia and Europe. Data for this paper was collected by the authors during two industry-focused workshops. The workshops were designed for tackling issues related to cyber security in shipping; they were not designed for collecting academic research data. This was a limiting factor when trying to conduct further statistical analysis of the collected data, as the research team had no control over participants' demographics. However, as several key industry experts participated, it was a great opportunity to collect data for such a contemporary issue as the challenges posed by cyber threats in the maritime transport sector. Both events took place in December 2020, virtually, allowing for stakeholders, from a large number of countries, to participate. Data were collected with the use of an online tool; it was anonymized and securely stored. Each participant was able to submit one or multiple responses, as indicated in each question.

The first workshop was conducted by a large Chamber of Commerce based in East Asia. As aforementioned, data collection was conducted during the workshop and the questionnaire was designed and tested prior to that. However, the overall event was not purposely conducted for data collection; as such, data collected was a "by-product" of the workshop. Therefore, detailed statistics for the demographics of the respondents are missing. The research team only have information related to the overall number and country of origin of participants of the workshop. Additionally, during the workshop, a lively discussion with the participants took place, which indicated their willingness to share information with authors. During the workshop two authors were main speakers in the event.

The second workshop was conducted by a large shipping association based in Greece with representation throughout the EU. Unlike the previous workshop, one of the authors was the main speaker, presenting the same questionnaire during the session. Similar to the previous workshop, a lively discussion with the participants took place. Participants were willing to share additional information with us. Further information is demonstrated in the following section, where the survey is presented in detail.

Data from both workshops were analyzed instantly from the research team. The analysis was conducted on a regional level but also in combination, so that a better understanding of the overall responses of the maritime sector could be obtained. The findings are presented in Section 4. Prior to that, it is demonstrated in the following section how the survey was designed.

3.3. Survey

As was evident from the aforementioned literature review, cyber attacks in the maritime transport sector have increased substantially over the last decade. For that reason, the introductory question posed to participants of this survey aims at highlighting the importance of cyber security in daily operations for the maritime transport sector. Table 1 presents the list of questions used in the survey.

Apart from the initial question, which attempted to understand the importance of cyber security in the maritime transport sector, the questions composing the survey were grouped in two themes: (A) How the industry is coping with the new IMO 2021 regulation requirements; (B) how shipping companies experienced changes in daily operations due to COVID-19 from the spectrum of cyber security.

The survey was available for responses only during the presentations delivered by the research team, which helped the participants get a better understanding of each question. Additionally, clarification was provided, when needed, as a live Q + A session was available, assisting the submission of full questionnaires, as in some cases participants drop out from a survey or they leave some questions blank if they do not fully understand them.

This was a unique, impromptu, opportunity to examine experts, and as such, the survey did not collect participants demographics. Instead, an investigative lead approach was adopted. Participants represented the whole spectrum of the maritime transport sector (e.g., port operators, shipping companies, consultants, ship management companies, technology solution providers, and academics).

Table 1. Questions asked during the surveys.

	Questions
A	**How the industry is coping with the new IMO 2021 regulation requirements**
1	How important is information (cyber) security to your daily job and activities?
2	Which agency is more suitable to assist and guide the maritime transport sector in addressing cyber security?
3	Is the IMO 2021 Cyber regulation the answer to cyber security for shipping?
4	How prepared are shipping companies in meeting the IMO 2021 requirements and consequently addressing cyber security?(Difference in perception in what cyber security entails)
B	**How shipping companies experienced changes in daily operations due to COVID-19 from the spectrum of cyber security**
5	Has the COVID-19 pandemic affected the maritime cyber security landscape?
6	Have you experienced a surge in cyber attacks in your organization during the pandemic?
7	How effective have shipping companies been in adapting to the new norm due to the COVID-19 Pandemic?

Workshop 1: Up to 200 participants partially attended the event, while half of them (100) attended the entire event. Participants represented 30 countries.

Workshop 2: The total number of participants was 42, representing a total number of 15 European countries, whilst the majority of the participants were from Greece (a higher representation from Greece is considered as normal, due to the high representation of Greeks in the shipping sector).

The findings extracted during the two workshops are presented in the following section.

4. Findings and Discussion

This section presents the data collected, analyzes the findings, and discusses key points by linking them to the theoretical approach presented in Section 3.

Based on the responses of the introductory question, it became evident that there is a consensus between maritime professionals regarding the importance of cyber security in the maritime transport sector. Specifically, the research team examined the importance of information or cyber security in the daily maritime transport operations, both in Europe and in Asia, as demonstrated in Figure 1a,b. When European and Asian responses were combined, it was observed that more than 90% of experts considered cyber security to be either very or somewhat important for their working environment, as demonstrated in Figure 2. This evidence shows that cyber security is more relevant presently for the maritime transport sector.

As shipping is already a heavily regulated industry, where the compliance mindset remains predominant, the IMO introduced specific guidelines to address cyber security in shipping. The nature of these guidelines is at a rather high level without offering any specific insights on their implementation. As such, theme A of this survey aims to explore: How the industry is coping with the new IMO 2021 regulation requirements. Through this theme, a novel knowledge approach was introduced, which is needed for future policy guidance and clarifies any doubts in the long-run.

It was deemed necessary, for the first question, to identify the competent authority that could provide guidance and assistance on the implementation of the IMO requirements. As presented in the literature review, the maritime transport sector consists of several entities that play a specific role in its successful development over the last century. Classification societies, P&I clubs, governments, flag states, marine insurance, and regional and international organizations now have to understand their role in addressing cyber security in shipping. Unlike the consensus unveiled in the previous question, the responses collected from the two working groups for this question highlight a significant difference regarding the competent authority for shipping. As demonstrated in Figure 3b, the majority of maritime transport professionals in Asia (53%) state that classification societies are the predominant authority, suitable to assist and guide the maritime transport sector in addressing cyber security, followed by governments with 27% and flag states with 10%. This perception is not shared with their colleagues in Europe, where the IMO, classification societies, and P&I clubs share equal representation (approximately 25%), as demonstrated in Figure 3a. At the same time, flag states seem to get similar percentages in both working groups. An aggregated response of both European and Asian is demonstrated in Figure 4. As both groups consider classification societies to be the authority to assist the maritime transport sector in tackling cyber security, it is therefore the highest scoring sector, with 40%, while the second most suitable organizations for tackling cyber security in maritime transport are governments, with 20%. Therefore, as demonstrated in Figure 4, classification societies have more "authority" than governments due to the multinational nature of shipping.

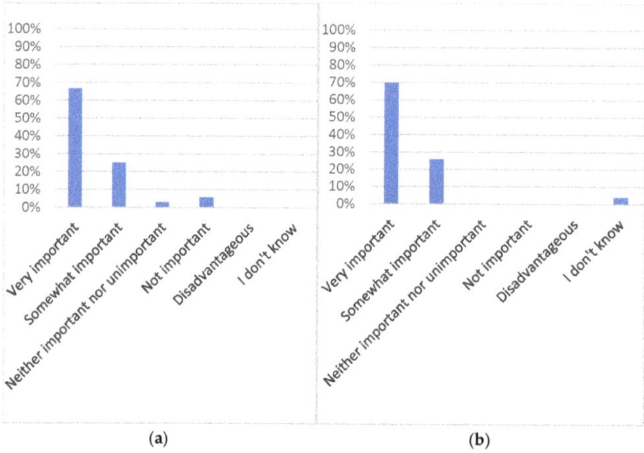

Figure 1. How important is information (cyber) security to your daily job and activities? (**a**) Europe; (**b**) Asia.

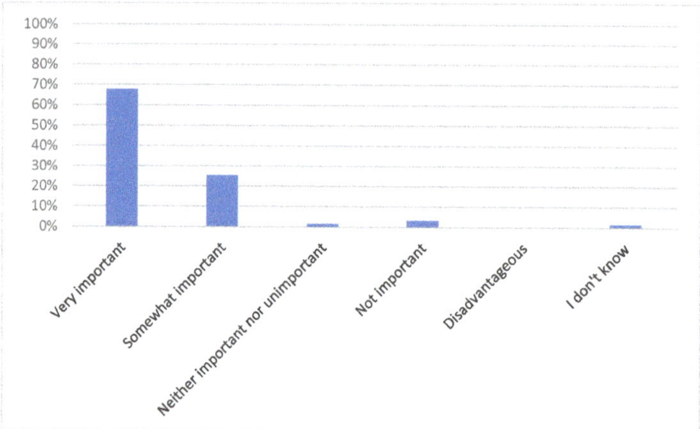

Figure 2. How important is information (cyber) security to your daily job and activities? (Asia and Europe).

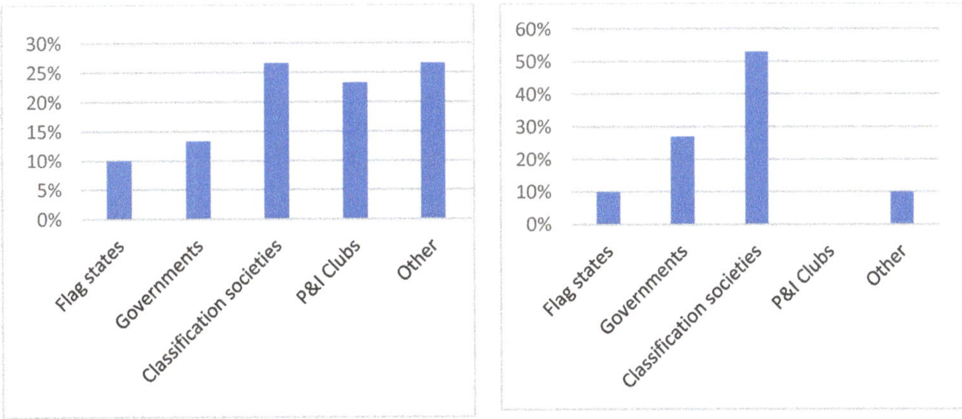

(**a**) (**b**)

Figure 3. Which agency is more suitable to assist and guide the maritime transport sector in addressing cyber security? (**a**) Europe; (**b**) Asia.

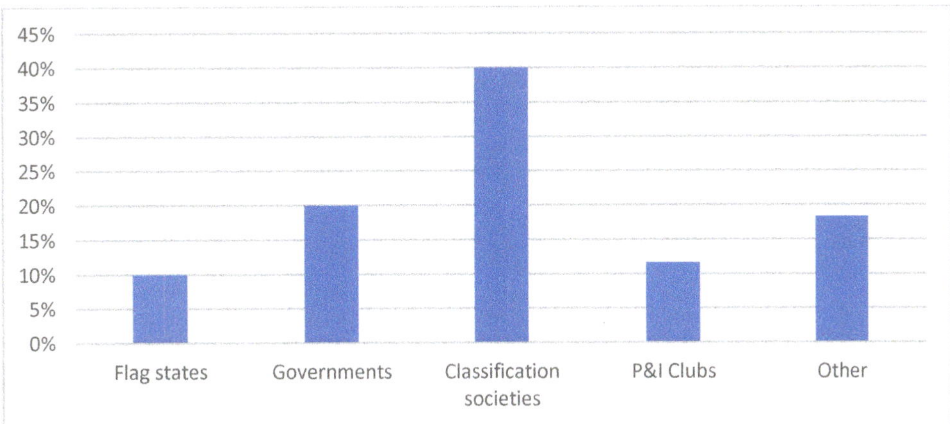

Figure 4. Which agency is more suitable to assist and guide the maritime transport sector in addressing cyber security? (Asia and Europe).

Aiming to understand the diversification between the responses in the two groups, examination of the broader picture should not be neglected. Going beyond cyber security, shipping professionals in Europe rely on the IMO's authority to regulate the industry. Similarly, P&I clubs, created by shipping professionals themselves, have a long-standing tradition in handling claims. On the contrary, classification societies are the de-facto entities that define responses to all, short of regulatory requirements in Asia. As such, cyber security would be no different.

Based on the findings of the previous questions, question 3 from Table 1 goes beyond the regulatory requirements and explores if the IMO resolution can effectively protect shipping from cyber breaches. As presented in the literature review, industry specific press has criticized the IMO cyber resolution as high level and not offering any tangible outcomes for its implementation, let alone specific steps for its inclusion in shipping companies' safety systems. Outcomes presented in Figure 5a,b are aggregated in Figure 6, and demonstrate findings regarding the IMO cyber resolution per region, Europe and Asia. The responses in Figure 6 are split, with half of the participants agreeing that the

IMO 2021 cyber resolution is the answer to cyber security for shipping and the other half either disagreeing or not being able to agree or disagree.

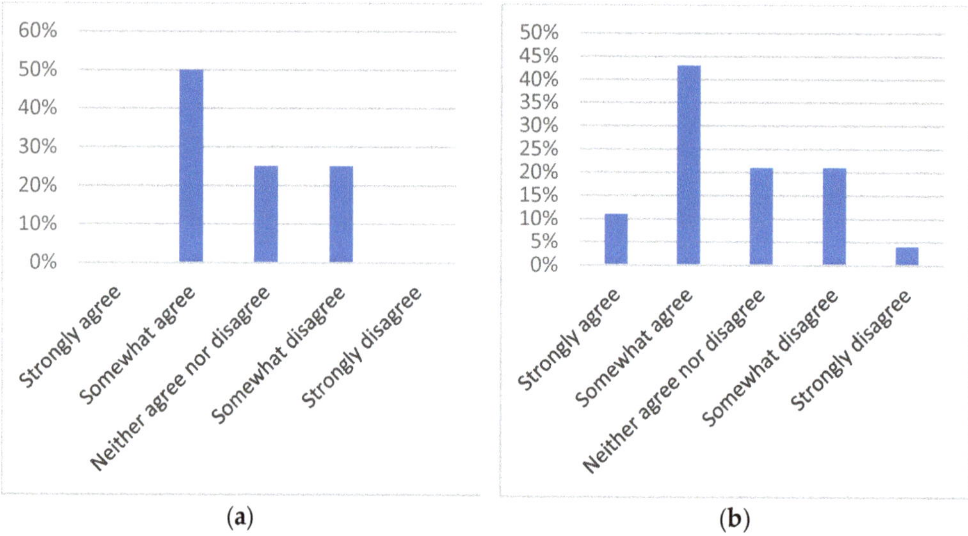

Figure 5. Is the IMO 2021 Cyber regulation the answer to cyber security for shipping? (**a**) Europe; (**b**) Asia.

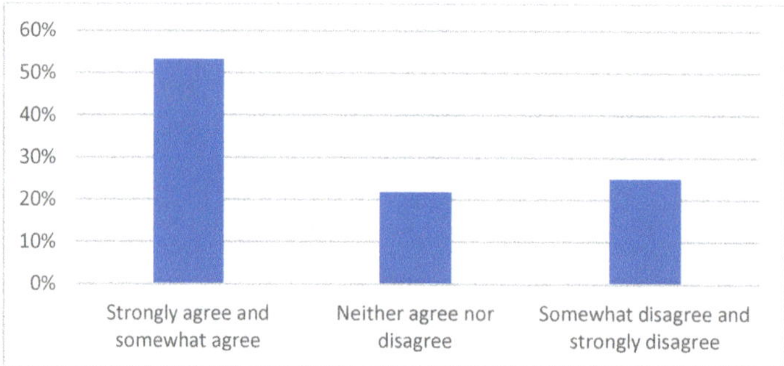

Figure 6. Is the IMO 2021 Cyber regulation the answer to cyber security for shipping? (Asia and Europe).

This split in the responses, which is similar for both the European and Asian participants of the survey, highlights the issue that the IMO 2021 cyber resolution, unlike other IMO guidelines, does not provide a clear answer to the sector's needs. For example, the ballast water management and sulfur cap regulations provide clear instructions on their applications, followed by technical specifications. As demonstrated from the findings of the survey, the IMO cyber resolution does not clearly pass the message required to the sector. The reason for that is because the IMO resolution is more descriptive rather than prescriptive.

Expanding on the previous question, the analysis explores the level of preparation for shipping companies when meeting the IMO 2021 resolution and, consequently, addressing cyber security. According to the practitioners who participated in the survey, and in contrast to the previous question, a difference between the responses in Europe and Asia was observed. While nearly half of the respondents in Europe believed that the sector is somewhat prepared (Figure 7a), a roughly similar percentage in Asia believed the opposite (that the sector is somewhat unprepared, Figure 7b). This difference became evident when the results from Europe and Asia were combined, as demonstrated in Figure 8.

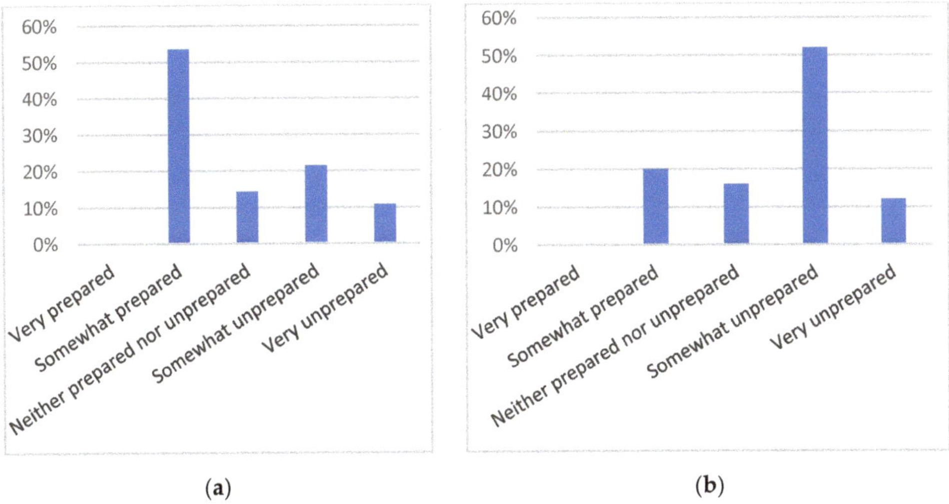

Figure 7. How prepared are shipping companies in meeting the IMO 2021 requirements and consequently addressing cyber security? (**a**) Europe; (**b**) Asia.

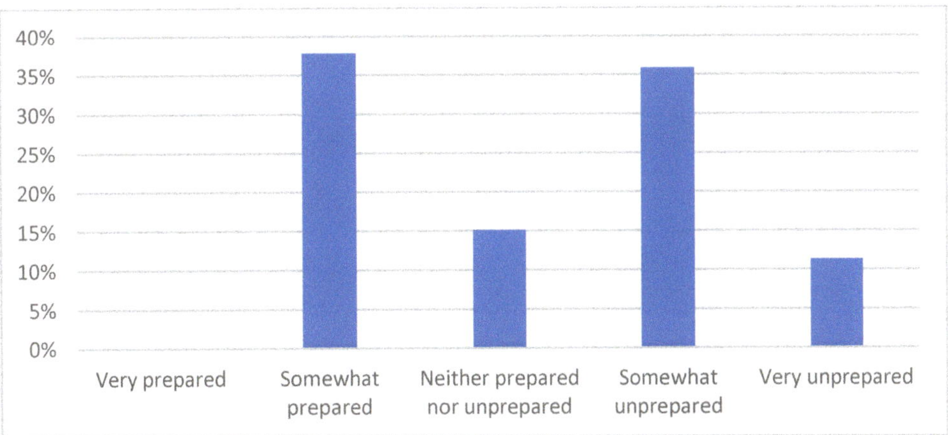

Figure 8. How prepared are shipping companies in meeting the IMO 2021 requirements and consequently addressing cyber security? (Asia and Europe).

The observed difference in perception on the level of preparedness for the IMO 2021 requirements is derived from two main attributes. The first attribute, as presented in the analysis of the findings of the previous questions, is that the IMO resolution does not offer clear guidance to the sector. Therefore, as it is demonstrated in Figure 7a,b that maritime transport stakeholders perceive differently what is required to meet the IMO resolution and address cyber security in the maritime transport sector. The second attribute is that there is a difference in perception in what cyber security entails. That perception derives from the increased technical savviness (as demonstrated in the literature review) compared to European counterparts, who are more "traditional" in operating the sector. Over the years, Asian maritime transport stakeholders are amongst the first to apply technology solutions that improve operations (e.g., minimize costs, etc.) [28,29].

Following the preliminary analysis regarding the industry's response to cyber security, predominantly concerning the IMO 2021 resolution, the second part of the questionnaire examined the industry's perception of the impact of the COVID-19 pandemic in day-to-day maritime transport operations.

The first question of theme B explores if the COVID-19 pandemic affected the maritime cyber security landscape. Even though this is not a binary question, for matters of simplicity, the question was structured as such (yes or no). Responses collected from both focus groups in Europe and Asia, as illustrated in Figure 9a,b and summarized in Figure 10, concur that the cyber security of the maritime transport sector was affected by the COVID-19 pandemic.

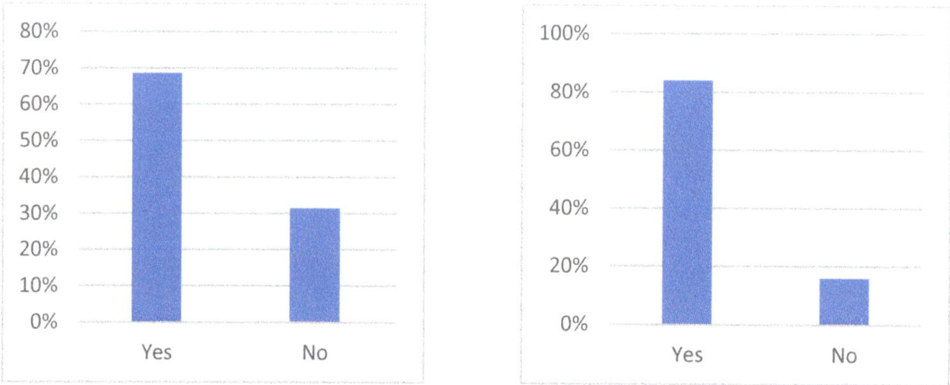

(**a**) (**b**)

Figure 9. Has the COVID-19 pandemic affected the maritime cyber security landscape? (**a**) Europe; (**b**) Asia.

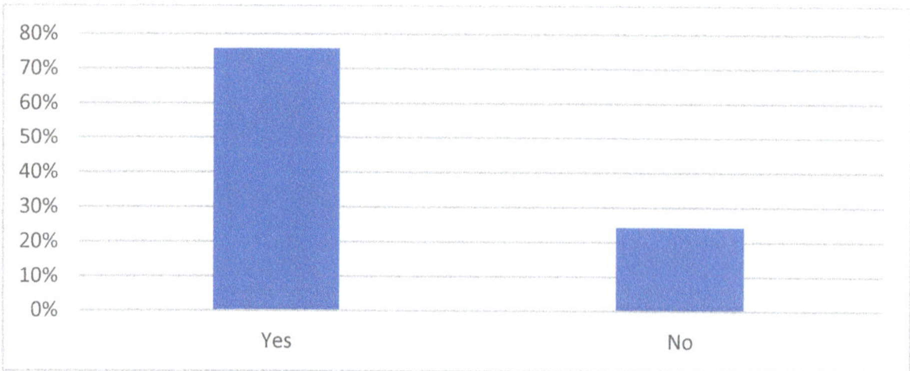

Figure 10. Has the COVID-19 pandemic affected the maritime cyber security landscape? (Asia and Europe).

This comes as no surprise, since, as noted previously, the number of cyberattacks since the appearance of the pandemic has globally, remarkably increased [23,24,29]. However, when trying to identify the effect of the pandemic on the maritime cyber security landscape, it became evident that respondents from Europe who believed that the industry's cyber security landscape was not affected, were twice as many as those from Asia, enhancing the points discussed in Figure 7a,b regarding the misconceptions about cyber security. As aforementioned, this originated from the fact that Asian maritime transport stakeholders were amongst the first to apply technology solutions. Therefore, they were agnostic to the benefits of digital transformation in the maritime transport sector, along with the consequent potential cyber risks that they may face.

The next question of the survey examined whether the participants experienced a surge in cyberattacks in their organizations during the pandemic. This question aimed to narrow down the analysis, drawing from participants' direct engagement within their organizations. While in Europe the responses were split (with 53% mentioning that they did not experience a surge in cyber attacks in their organization and 47% mentioning the opposite), in Asia, two thirds of the respondents mentioned that they did not experience any surge in cyber attacks in their organizations during the pandemic. Figure 11a,b results are aggregated in Figure 12, demonstrate that there was an increase by 40% in cyber attacks in maritime transport organizations during the pandemic, reinforcing the argument that there is a need to increase cyber resilience in the maritime transport sector.

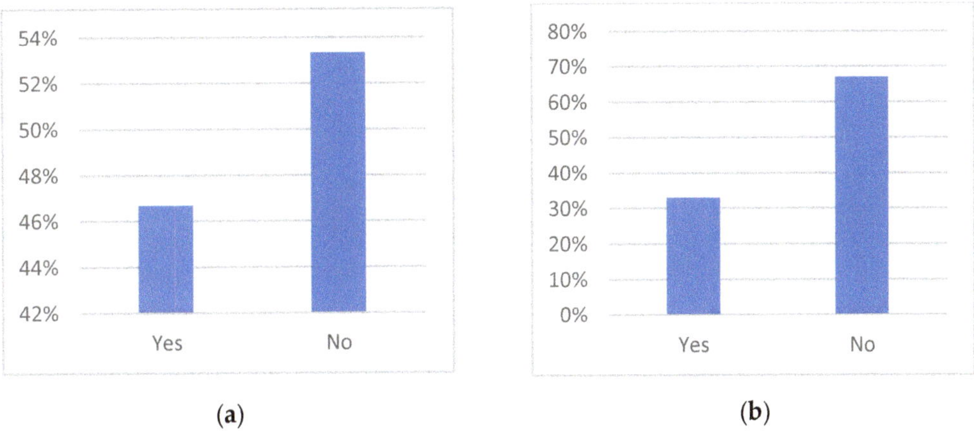

Figure 11. Have you experienced a surge in cyber attacks in your organization during the pandemic? (**a**) Europe; (**b**) Asia.

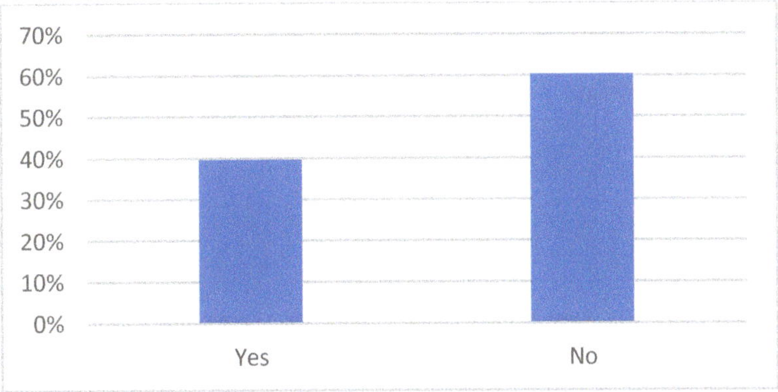

Figure 12. Have you experienced a surge in cyber attacks in your organization during the pandemic? (Asia and Europe).

Looking at the responses presented in Figure 11a,b, an interesting realization that reaffirms what has been previously mentioned was identified. Asian maritime experts were more advanced in terms of security-minded technology solutions applied within their maritime transport organizations, compared to their European counterparts. As such, they were able to mitigate incoming malicious content. Having the appropriate measures in place and not allowing an increased number of attacks to penetrate their organizations ecosystem justified the responses presented in Figure 11b.

Similar to question 4, presented in Table 1 and analyzed above, the last question of the survey tried to identify the level of effectiveness of shipping companies in adapting to the new norm, due to the COVID-19 pandemic. This new norm entailed: (A) an increased number of employees working remotely (from home), (B) the adoption of digital solutions to facilitate this transition, and (C) uptake in number of cyber attacks affecting the industry.

As demonstrated in Figure 13a,b, 86% of European participants mentioned that shipping companies were either very or somewhat effective in adapting to the new norm. In comparison, opinions on the same matter from Asian participants were limited to almost half of the percentage demonstrated above (46% were either very or somewhat effective). Overall, Figure 14 illustrates that a strong majority of industry practitioners (68%) believed that the industry was effective in adapting to the new norm due to the COVID-19 pandemic.

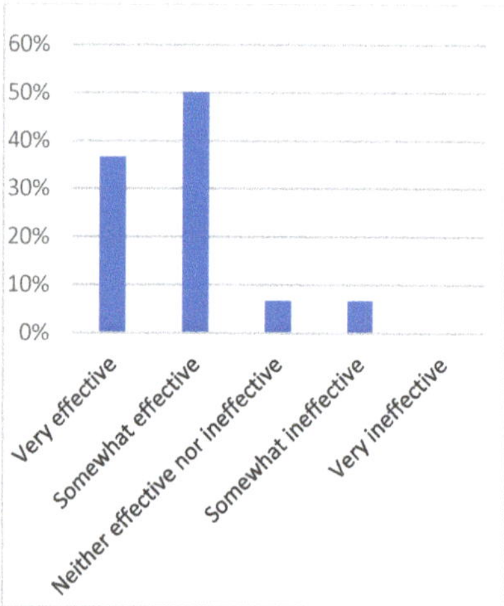

(a)

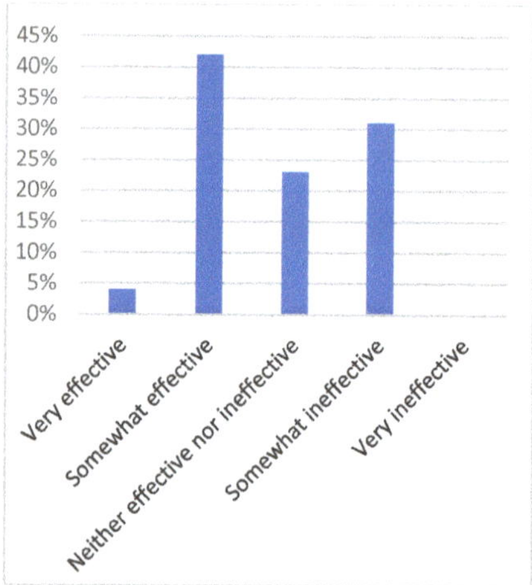

(b)

Figure 13. How effective have shipping companies been in adapting to the new norm due to the COVID-19 Pandemic? (**a**) Europe; (**b**) Asia.

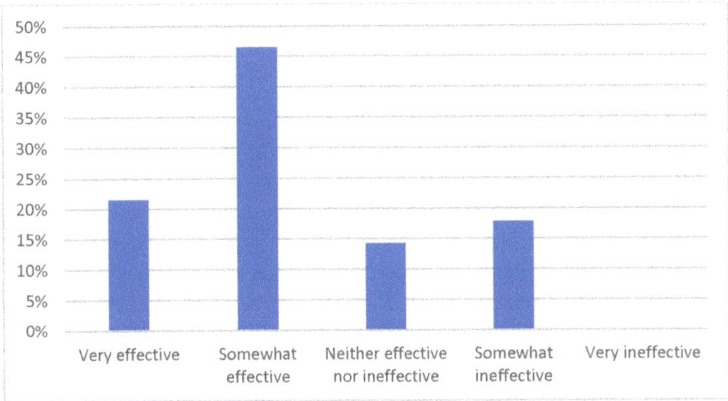

Figure 14. How effective have shipping companies been in adapting to the new norm due to the COVID-19 Pandemic? (Asia and Europe).

The difference in responses illustrated in Figure 13a,b reaffirms the trend identified throughout this survey regarding the level of maturity and understanding of what cyber security entails between the European and Asian participants. With reference to the findings of Figure 14, this comes as no surprise, as, unlike other industries, shipping, due to its nature, is used to having its most valuable assets operating remotely. Ships operating globally, thousand miles away from their shipping company's offices, have established procedures and technology solutions for decades.

In this section, the findings of the survey were demonstrated and discussed. Section 5 summarizes the key points of this research.

5. Conclusions

It became apparent throughout the paper that the increased adoption of digital solutions in the maritime transport sector introduces an insidious threat in cyber space. Ports and ships, being two of the most vital components of the supply chain, are vulnerable to cyber breaches, due to their complex operational environments. Both ships and ports have IT and OT systems composed by various third-party vendors, which, in most cases, require remote access, increasing cyber attacks to surface. Thus, maritime transport stakeholders have to take prompt actions in order to mitigate cyber risk. In order for this to happen, first and foremost, maritime transport stakeholders should understand what cyber security is; how it may affect their business; and the specific countermeasures that are suitable for their organization, and consequently adopt them to tackle these threats.

This research, unlike mainstream academic approaches, was initiated from the data collected during two targeted workshops, with the participation of more than 250 senior maritime transport practitioners. Due to the limitation of the research environment and the level of control available to the research team, it was decided not to over interpret the data with the use of quantitative statistical methods. The research team believes that its analysis will satisfy social scientists in terms of interpreting data collected in this activity by using a qualitative lens.

This impromptu engagement was commonly themed in order to address one key industry concern: whether cultural differences affect the level of understanding of maritime cyber security. As the two workshops were conducted virtually in Asia and Europe, it was realized that the predominant perception for cyber risk differs in these two geographical areas. The main analysis has reiterated that each group had a different understanding. A similar observation was conducted by Nisbett [26], who mentions that different cultures perceive specific topics in a different view. As stated by Nisbett and reaffirmed by our findings, East Asians think "holistically", while Westerners focus on specific subjects. This differentiation can also be attributed to the level of maturity regarding cyber security, as presented in the findings section. As such, Asians understand cyber security challenges better and consequently incorporate them in their aggregated business risk management. In contrast, the less mature Western maritime transport stakeholders perceive cyber security as an impartial risk factor to be dealt in isolation.

Findings presented in this research highlight that many maritime transport stakeholders are not aware of what cyber security entails and do not fully realize the degree of dependence of their businesses on software-enabled systems, platforms, and services. While they might acknowledge the existence of cyber threats in general terms, as apparent in Section 4, they miss important details; understanding how these cyber threats can affect their organizations' daily operations. A holistic approach to cyber risk management begins at the senior management level and extends downwards to the entire organization.

Future Research

Acknowledging that this paper was elaborated based on an opportunity that arose from the authors' engagement with the industry, we believe that further, targeted, academic research is to be conducted, with statistical tools, such as ANOVA, in mind. To achieve that, the research outline should include details of targeted audience profiles, along with a questionnaire, which will be designed to collect information related to participants and their demographics. Such an approach would facilitate the implementation of aforementioned statistical tools and thus present a statistical analysis of the results. Specifically, future research should explore two main topics: (1) revalidate our findings as to how cultural differences affect the level of understanding of a specific topic, in this case maritime cyber security, with further research, such as longitudinal research, which could enhance our findings; (2) explore whether the surge in cyber attacks, partially due to pandemic, affected the sector's response to the IMO cyber requirements. This research should take place in a more academic style, where the theory should be tested against our main finding, that there is a different perception of cyber risk based on cultural background. To achieve this, enhanced collaboration between all key stakeholders (academics, cyber experts, maritime transport sector stakeholders) should take place. As we initiated this piece of research, due to our wide links with maritime transport stakeholders, we would happily participate in any future discussions, which we are sure will take place soon due to the urgent need to tackle such an important issue as maritime cyber risk management.

Author Contributions: Conceptualization, S.K. and C.K.; methodology, S.K., C.K. and T.W.; software, S.K. and C.K.; validation, S.K. and C.K.; formal analysis, S.K. and C.K.; investigation, S.K., C.K. and T.W.; resources, S.K. and C.K.; data curation, C.K. and S.K.; writing—original draft preparation, S.K. and C.K.; writing—review and editing, S.K., C.K. and T.W.; visualization, S.K. and C.K.; supervision, S.K. and C.K.; project administration, S.K. and C.K.; funding acquisition, T.W. All authors have read and agreed to the published version of the manuscript.

Funding: This research received no external funding.

Institutional Review Board Statement: Ethical review and approval were waived for this study, due to methodology limitations.

Informed Consent Statement: Informed consent was obtained from all subjects involved in the study.

Data Availability Statement: Data available on request due to privacy restrictions. The data presented in this study are available on request from the corresponding author. The data are not publicly available due to GDPR restrictions.

Acknowledgments: The authors would like to thank the Greek Chamber of Commerce in Hong Kong and the International Propeller Club of Piraeus for organizing the events that led to this paper. Without their invaluable support, this piece of research would not have been possible.

Conflicts of Interest: The authors declare no conflict of interest.

References

1. CyRiM Report. Shen Attack Cyber Risk in Asia Pacific Ports. Available online: https://apps.cambridgeriskframework.com/erm/portal/publication/shen-attack (accessed on 31 October 2019).
2. Kuhn, K.; Kipkech, J.; Shaikh, S. Maritime Ports and Cybersecurity. In *Maritime Transport and ITS Solutions in Port Logistics*; Fiorini, M., Gupta, N., Eds.; IET: London, UK, 2021.
3. Polemi, N. *Port Cybersecurity: Securing Critical Information Infrastructures and Supply Chains*; Elsevier: Amsterdam, The Netherlands, 2017.
4. Peake, M. Cybersecurity Looks to the Cloud to Protect Data at Sea. Available online: https://www.wartsila.com/insights/article/cybersecurity-looks-to-the-cloud-to-protect-data-at-sea (accessed on 11 June 2021).
5. Greenberg, A. *Sandworm: A New Era of Cyberwar and the Hunt for the Kremlin's Most Dangerous Hackers*; Doubleday: New York, NY, USA, 2020.
6. ISACA State of Cybersecurity 2020. Available online: https://www.isaca.org/go/state-of-cybersecurity-2020 (accessed on 5 June 2021).
7. Fielder, A.; Panaousis, E.; Malacaria, P.; Hankin, C.; Smeraldi, F. Decision support approaches for cybersecurity investment. *Decis. Support Syst.* **2016**, *86*, 13–23. [CrossRef]
8. UNCTAD. *Review of Maritime Transport*; UNCTAD: Geneva, Switzerland, 2018.
9. Miller, C.; Stuart Wells, F. Balancing security and privacy in the digital workplace. *J. Chang. Manag.* **2007**, *7*, 315–328. [CrossRef]
10. Kapalidis, C. Cyber securityat Sea. In *Global Challenges in Maritime Security*; Otto, L., Ed.; Springer: Cham, Switzerland, 2020; pp. 127–143.
11. European Union. *EU Maritime Security Strategy*; European Commission: Brussels, Belgium, 2014.
12. NMIOTC. 1st NMIOTC Cybersecurity Conference in the Maritime Domain 2016 (04-05 OCT 2016), Chania Crete. Available online: https://nmiotc.nato.int/transformation/conferences/cyber-security-conference/ (accessed on 30 June 2021).
13. IMO. *Guidelines on Maritime Cyber Risk Management*; IMO: London, UK, 2017.
14. IMO. *Resolution MSC. Maritime Cyber Risk Management in Safety Management Systems*; IMO: London, UK, 2017.
15. UNCLOS (United Nations Convention on the Law of the Sea). United Nations Convention on the Law of the Sea of 10 December 1982 Overview and Full Text. Available online: https://www.un.org/depts/los/convention_agreements/convention_overview_convention.htm (accessed on 3 June 2021).
16. UNCTAD. *Review of Maritime Transport*; UNCTAD: Geneva, Switzerland, 2020.
17. US Coast Guard. *Navigation and Vessel Inspection Circular No. 01-20: Guidelines for Addressing Cyber Risks at Maritime Transportation Security Act (MTSA) Regulated Facilities*; US Coast Guard: Washington, DC, USA, 2020.
18. DNV. Our History. Available online: https://www.dnv.com/about/in-brief/our-history.html (accessed on 4 June 2021).
19. ABS Group. IMO Cyber Risk Management. Available online: https://www.abs-group.com/What-We-Do/Safety-Risk-and-Compliance/Cybersecurity/Maritime-Cybersecurity/IMO-Cyber-Risk-Management/ (accessed on 4 June 2021).
20. ClassNK Consulting Service, Cybersecurity Management System (CSMS) Construction Support. Available online: https://www.classnkcs.co.jp/en/cyber_security/index.html (accessed on 4 June 2021).
21. DNV. Maritime Cybersecurity. Available online: https://www.dnv.com/maritime/insights/topics/maritime-cyber-security/index.html (accessed on 4 June 2021).
22. Hall, C. The Daily Show Agenda, 27st Annual Berlin First Conference, Wapack Labs. Available online: https://www.first.org/resources/papers/conf2015/first_2015_-_hall-_chris_-_daily_show_agenda_20150618_fw.pdf (accessed on 30 June 2021).
23. MTS-ISAC (Maritime Transportation System—Information Sharing and Analysis Center). 2020 Annual Report. Available online: https://www.mtsisac.org/post/2020-mts-isac-annual-report (accessed on 7 June 2021).
24. MarineLink, Maritime Cyber Attacks Increase 900%. Available online: https://www.marinelink.com/news/maritime-cyberattacks-increase-480311 (accessed on 30 June 2021).
25. Kapalidis, C. Maritime Cybersecurity: No Substitute for Testing, Chatham House. Available online: https://www.chathamhouse.org/expert/comment/maritime-cyber-security-no-substitute-testing (accessed on 11 June 2021).
26. Nisbett, R. *The Geography of Thought: How Asians and Westerners Think Differently... and Why*; Free Press: New York, NY, USA, 2003.
27. Kumar, S.; Hoffmann, J. Globalization: The maritime nexus. In *The Handbook of Maritime Economics and Business*; Grammenos, C., Ed.; Lloyd's: London, UK, 2002; pp. 35–62.
28. Watson, R.; Lind, M.; Haraldson, S. Physical and digital innovation in shipping: Seeding, standardizing, and sequencing. In Proceedings of the 50th Hawaii International Conference on System Sciences, Hilton Waikoloa Village, HI, USA, 4–7 January 2017.
29. Chan, H.K.; Dai, J.; Wang, X.; Lacka, E. Logistics and supply chain innovation in the context of the Belt and Road Initiative (BRI). *Transp. Res. Part E Logist. Transp. Rev.* **2019**, *132*, 51–56. [CrossRef]

Journal of
Marine Science
and Engineering

MDPI

Article

Additive Manufacturing and Maritime Spare Parts: Benefits and Obstacles for the End-Users

Evanthia Kostidi *, Nikitas Nikitakos and Iosif Progoulakis

Department of Shipping Trade and Transport, University of the Aegean, Korai 2a, GR82132 Chios, Greece; nnik@aegean.gr (N.N.); iprogoulakis@aegean.gr (I.P.)
* Correspondence: e.kostidi@aegean.gr

Abstract: 3D printing or additive manufacturing (AM) (in the industrial context) is an innovative, as opposed to subtractive, technology, bringing new opportunities and benefits to the spare part supply chain (SPSC). The aim of this work is to capture the views of the stakeholders at the end of the chain, extruding factors that will benefit the end-user and the factors that are likely to be an obstacle, by employing the questionnaire method. Company objectives regarding spares (cost reductions, improvement of services, space reduction) have been prioritized differently by the stakeholders. The most important barriers according to the participants are the quality assurance of the spare parts made by the new technology followed by the know-how and skills of staff. Other views such as suitable parts are suggested. The practical value of this work, in addition to assessing the readiness of the industry, is that it provides guidance for the successful implementation of AM in the maritime industry.

Keywords: 3D printing; additive manufacturing; spare part supply chain; maritime

1. Introduction

Machinery (in general) is subject to gradual deterioration that makes preventive or corrective maintenance a necessity, thus requiring the availability of spare parts. The maritime industry is characterized by the heavy utilization of equipment and machinery under harsh and corrosive operating conditions and that makes the requirements of maintenance particularly critical due to the detrimental financial and safety consequences of downtime. The availability of parts at the point of consumption makes inventory necessary for suppliers to respond quickly to demand. Keeping an inventory of spare parts causes a financial burden to maritime companies and parts suppliers, mainly in capital cost expenditure as well as logistical storage. Spare parts supply chain (SPSC) plays an important role in keeping the maritime assets operating efficiently and safely. Maintenance networks involve many actors such as the owners of the assets, system integrators, original equipment manufacturers (OEM) and the service providers and their logistics service providers [1].

3D printing or additive manufacturing (AM) (n the industrial context) (The terms "additive manufacturing" and "3D printing" are used interchangeably in the literature although there are distinct differences as defined in ISO/ASTM 52900(en) [2]. Additive manufacturing is used in the industrial context but 3D printing is the more commonly known term as it is used in popular media. This is why the term "3D printing" was used in the communication with the stakeholders.) is an innovative, as opposed to subtractive, technology, bringing new opportunities and benefits to the spare parts supply chain because it enables distributed manufacturing in small batches, thus minimizing the inventory and reducing delays and costs. It is based on the principle of the construction in layers by directly converting the 3D data into physical objects [3]. As AM provides the ability to manufacture finished components in a decentralized manner, many researchers argue in favor of digitizing spare parts and their decentralized construction [4,5]. With the possibility of end-use components being created with AM, several researchers have proposed that spare parts could be digitized and additively manufactured [4–6]. Maritime is not to be omitted from technological developments. Top global marine propulsion engine vendors motivated by the benefits of AM have introduced this new technology not only in prototyping but also in their production process. Initiatives such as the "Pilot Project 3D printing of Marine Spares" [7], "Green Ship of the Future" [8] and "Maritime and Port Authority of Singapore" [9] aim to explore the practical use of AM. Academic research aims to explore new materials and production methods for the maritime environment. The use of innovative manufacturing routes is based on the use of innovative materials such as the application of fiber-reinforced composite materials in the marine environment; the latest advances in this area are presented by Rubino et al. [10]. The article by Parmar et al. depicts the present and future scenarios related to the application of Industry 4.0 concepts to polymer composite (fiber-reinforced) manufacturing and the recent and potential developments in robotic and additive manufacturing [11]. The shipping industry has already entered the era of digitalization and an optimized flow of information by integrating different stakeholders, people at sea as well as technology applications and processes should be viewed as a very crucial strategic asset for any modern shipping company [12]. Salmi et al. [6] investigated the impact of digital spare parts on operating models using five general supply chain models ranging from fully concentrated to fully distributed.

The impacts of AM technologies are not partial or isolated disruptions but rather encompass a range of changes whose overall consequences spread across different phases of the supply chain [13]. To determine how AM could benefit the supply chain, executives should examine all the levels from the product design (product functional requirement and design parameters) to the manufacturing lines as well as the location and layout of the physical facilities [14].

The spare parts supply chain with the introduction of AM has attracted the interest of a number of researchers. Most of the studies cover the following industries: aerospace/aviation [15–19], marine and offshore [16,20], automotive [21,22], mechanical engineering [17] and medical [17,23]. The reason for the interest in this particular supply chain is that spare parts differ from consumer products in terms of demand (i.e., sporadic and difficult to predict) [24–26] and in many cases there are relatively limited historical demand data available [27]. Transporting means operating away from their base and, in many cases, in random paths, presenting additional difficulties in being supplied with the necessary spare parts for their operability. The shipping cost in several cases (i.e., isolated or difficult to reach areas) may be higher than the part itself in addition to the downtime cost.

Despite its importance in worldwide trade, the maritime industry did not evolve in regard to the SPSC in comparison with other industries, mainly because of its smaller footprint. For this reason, for the application of any lessons learned from other industries, the specific characteristics of the maritime domain should be taken into account. For example, the vessel is in an isolated environment, the environment is highly corrosive, extreme and frequent turbulence and vibrations are present, vessels (as well as land offices) are staffed by technicians and information systems have been long applied [28].

Although, as mentioned above, there are several studies on the implementation of AM technology in areas similar to the maritime industry (with moving assets or isolated environments), studies specifically on the maritime industry are scarce. The aim of this work is to expose the factors that may push the digitization process in the spare part supply chain with the introduction of AM and the factors that are likely to be an obstacle to it. The use of the technology in question or similar previous technologies, technological infrastructures and technology awareness are an indication for seedbed ground. This work captures the views and expectations of the stakeholders in the lower end of the SPSC (end-users) in the maritime industry and to the best of our knowledge no similar work has been done before. The rest of the paper is structured as follows: first, the methodology used is presented. The literature review then follows in order to formulate the questionnaire. Next, the collected views of the involved stakeholders at the end of the SPSC (end-users) are analyzed and the results are presented. Finally, the results are discussed and conclusions are drawn.

2. Methodology

As our research aimed to explore the potential of introducing AM in the spare parts supply chain in the maritime industry, the questionnaire method was selected for capturing the perceptions of the involved stakeholders in order to get an understanding of motives and fears. The selected target group interviewees were at the end of the SPSC: people onboard a vessel (bridge and engine crew) and in a land office (ship owners or chief executive officers; staff in technical, procurement, training, research and development departments). Most of them had no or limited AM knowledge, so a short presentation of the basics of AM took place. The shipping companies visited for interviews were not restricted to a particular type of vessel. The term "3D printing" was used instead of the term "Additive Manufacturing" throughout the interviews because it was more understandable by survey participants due to its use in the media.

A literature review was carried out in order to formulate the questionnaire, mainly by searching in Google Scholar for the terms "Additive Manufacturing and the spare part supply chain", "benefits and Additive Manufacturing" and "barriers/challenges and Additive Manufacturing". In order to formulate the questions, preliminary meetings were arranged with representative people of all the mentioned groups. During these meetings, the available technology, its capability and the benefits to the SPSC were presented. The skepticism of meeting participants was captured in order to record the challenges that such a change would face. The preliminary meetings gave us an idea how to structure the questionnaire. In order to get as many as possible views, two well-known marine exhibitions were selected because they attracted all the targeted stakeholders. The views of academics were recorded at relevant scientific conferences.

The interview participants were asked to rank the objectives of the company (regarding spares) such as cost reductions, improved service, storage space reduction or any other company values that would be enhanced with the introduction of the proposed technology. All changes (including technological ones) encounter obstacles and realizing what barriers are to be expected when a part is made by the use of 3D printing will help for better preparation. The criteria for selecting parts to be made by 3D printing may differentiate from the criteria met in the literature due to the nature of the maritime industry.

3. Literature Review in Order to Formulate the Questionnaire

An extensive literature review was undertaken to identify suitable academic and industry research and standards that could assist in the formulation of the survey questionnaire.

3.1. Main Positive Effects of Additive Manufacturing and the Spare Part Supply Chain

The motives to introduce new technology vary considerably by industry. In aerospace, components need to be geometrically optimized to result in lighter structures and thus reduce aircraft weight and fuel consumption. In the automotive industry, an important motive is the rapid creation of prototypes and tools as well as the supply of spare parts for old vehicles. These incentives lead to research in the aerospace and automotive industries to exploit the potential of new technology. As AM matures in shipping, specific incentives will emerge. Meanwhile, as the supply of spare parts is important for vessel maintenance, shipping can learn from studies in other industries. The main positive effects of the application of AM in the SPSC in the area of moving assets stated by authors in the literature are (Table 1) asset availability [19], economic efficiency [17,20], reduced inventory [16,18,29], improved service [15,18,29], lead time reduction [22], better supply chain performance [21] and decreasing carbon emissions [17,18].

Table 1. Main positive effects of additive manufacturing on the spare part supply chain.

Influence on the Supply Chain	Author(s)								
	Holmström et al. (2010) [29]	**Khajavi, et al. (2014) [15]**	**Liu, et al., (2014) [16]**	**Ratnayake, (2016) [22]**	**Li, et al. (2017) [17]**	**Ghadge, et al. (2018) [18]**	**Khajavi, et al. (2018) [19]**	**Westerweel, et al. (2018) [20]**	**Delic, et al. (2019) [21]**
Asset availability							√	√	
Economic efficiency					√		√	√	
Reduced inventory	√		√			√			
Improved service	√	√				√			
Lead time reduction				√					
Supply chain performance									√
Environmental					√	√			

Company objectives lead company decisions and concentrate efforts in obtaining them. The SPSC in the maritime industry involves many stakeholders whose objectives may differ from each other so any exploration of the potential of additive manufacturing improvements should start with noting the objectives of each stakeholder.

Q1. *What are the expected benefits from the introduction of 3D printing as viewed by the stakeholders at the end of the SPSC (end-users)?*

3.2. Barriers toward the Introduction of AM

All changes (including technological ones) encounter obstacles and must be properly mitigated. Vogelsang et al. [30] analyzed the barriers toward the digitalization of enterprises in manufacturing and grouped them into individual, organizational, environmental and technical barriers as well as skills shortage. Barriers such as standardization, quality assurance, intellectual property rights protection, the legacy of non-digitalized products and component information have to be further investigated [31]. Among major technical barriers, a low number of usable materials, insufficient accuracy levels, tolerances and the build chamber size limitation [32] need to be overcome. Another barrier is that the industrial design methodologies applied are for traditional manufacturing and do not take into account the capabilities of AM. CAD (computer-aided design) files often do not exist and technical drawings can be difficult to locate [33].

The need to ensure the functionality and compatibility of products as well as consumer safety are addressed by already published and upcoming standards [34]. Certification and quality assurance are required in order to give confidence to all interested parties that a product, process or service fulfils the specified requirements [35]. Small batch production or the customization of products by AM require new quality assurance (i.e., non-destructive) and certification methods for the production of procedures and the products. This need aims to cover the published guidance notes by classification and certification societies such as Lloyd's Register and TWI (The Welding Institute) [36], DNV (Det Norske Veritas) [37], ABS (American Bureau of Shipping) [38] and the Bureau Veritas Marine and Offshore [39].

Additive manufacturing (AM) is revolutionizing the industrial landscape and these changes require new engineering and management skills to exploit fully and sustainably the benefits offered by these advanced technologies. New technology creates a lack of talent, a need for training in specializations and a workforce adaptable to change [40].

Although there are existing legal [31], cyber security [41], manufacturing time and cost issues, the most frequently met barriers in the literature to produce a part by the use of 3D printing are:

1. Technology know-how;
2. Staff skills;
3. Quality assurance.

Knowing the expected obstacles when a part is made by the use of 3D printing will help for a better preparation.

Q2. *What are the barriers influencing the application of 3D printing, as viewed by the stakeholders at the end of the SPSC?*

The established company policy and culture toward the use of a part made in a different way than the usual procedure (genuine spare) may influence the successful introduction of the new technology. OEM, even in cases that obtain parts through subcontractors, control the spare part flow toward the end-user. In a few cases, spare part sales are a considerable source of income for OEM. Constraints such as machinery guarantees may also influence the adaptation of the AM technology.

Q2.1. *What is the attitude of each stakeholder toward the use of a part made in a different way than the usual procedure (3D printed)?*

3.3. Characteristics of the Suitable Spare Part

The characteristics of the suitable spare part as proposed from initiatives in the area of the maritime industry [7,9] can provide a picture of the parts that may have the potential for 3D printing. Parts that are infrequently demanded but must be available in case they are needed, especially those of a high supply risk, must be in the inventory. Parts that are ordered in small quantities usually have quite an expensive overhead cost. Keeping high value parts in the inventory ties up capital. If parts are of a high design complexity, they are difficult to be made otherwise (i.e., in a lathe) or they are made up of many components that require assembly. For practical reasons, the chosen parts must be of such a size that can fit in the process chamber. These are presented in Table 2.

Table 2. Suitable part characteristics for 3D printing for the maritime industry.

Suitable Part Characteristics
Small size that can fit in the process chamber
Small order quantity
Infrequently demanded
High value
High design complexity
High supply risk

Not all parts can be made by AM for various reasons. That is why criteria are set in selecting parts for 3D printing. Obviously, an automated process of selection is fast and economic but even in cases where appropriate digital data exist, the human view is valuable. Various methods have been proposed to assist in that process, each one having advantages and disadvantages. The approach by Knofius et al. [42] relies on spare part information that is easily retrievable from the company databases. The trade-off methodology by Lindemann et al. [43] involves AM experts and part owners. Overall, prior methods on part candidacy identification may differ in strategies but mostly show a deficiency in comprehensiveness, efficiency and objectiveness. Therefore, a fast decision support system that requires the least user inputs and minimum AM expertise but computationally produces reliable recommendations of AM eligibility is demanding in the task planning and conceptual stage of the product redesign [44].

Although AM brings benefits to spare part production, it is an expensive way to fabricate components. That is why it is subjected to high value items (that lead to a high inventory cost). The proposed characteristics of the suitable spare part in the literature suggest high value parts. Due to maritime nature, a low-cost item to reach a vessel may have a huge transportation cost in addition to the damage or safety from a missing part, thus leading to the high value of the part.

Q3. *Do you think 3D printing could be applied for low-cost items (such as an elastic sealing ring)?*

Knowledge of the technology in question or similar previous technologies, technological infrastructures and technology awareness were an indication for seedbed ground.

3.4. Seedbed Ground Elements

3D printing is becoming a term increasingly known through media reports, which present impressive constructions. They also show the advantages and this makes the decision makers think of applying it to their own needs. Technology awareness, technological infrastructures and the use of the technology in question or similar previous technology was an indication for seedbed ground (Figure 1).

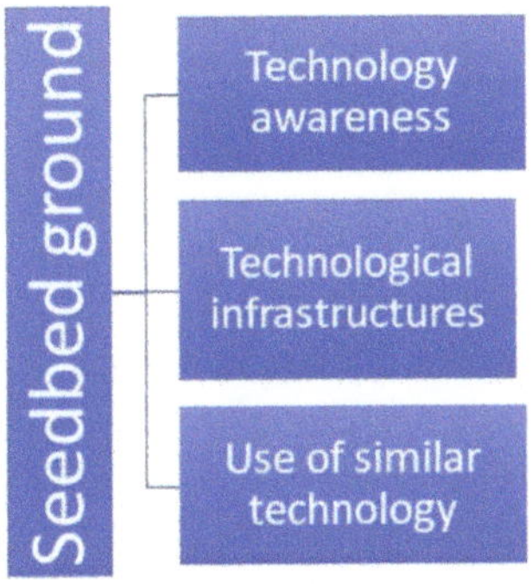

Figure 1. Seedbed ground for the acceptance of new technology.

Q4.1. *Is your company thinking of/examining the use of 3D printing?*

A computerized manufacturing process such as computerized numerical control (CNC) is based on pre-programmed software and code that control the movement of the production equipment. This technology is similar to that of a 3D printer, which has a complexity and requires advanced skills. A company or a person familiar with a CNC operation could easily move to 3D printing.

Q4.2. *Is your company using CNC (computer controlled machine) technology?*

Making or refurbishing spare parts on board by the use of a lathe, for example, or another technique depends on the company policy as it requires trained personnel and may involve the risk of failure, accidents or the prohibition of warranty commitments.

Q4.3. *Is your company making or refurbishing spare parts on board (i.e., use a lathe or some other technique to make a sealing ring)?*

It is well-known that manufacturers secure a portion of their profits (perhaps the largest) from spare parts sales. They require proper maintenance and genuine spare parts from the users of their machines in order to maintain the warranty. For these reasons, but also to protect themselves from imitations, they are protective of the technical information they provide, limiting it to what is absolutely necessary.

Q4.4. *What information is available for manufacturing spares?*

4. Research Results Analysis

The total volume of the survey questionnaires that were collected was 140. The target number was 200 but several, although kind enough to contribute to a free talk, did not fill in the questionnaire. The participants were grouped as crew engine, crew bridge, technical, ship owners/CEOs (Chief Executive Officers) and other R&D (Research & Development), training, academic]. It should be noted that the majority of survey participants were male (83%) as the male population is by far higher than the female population in the maritime industry.

4.1. Expected Benefits

4.1.1. Expected Contribution to Cost Reductions

The participants ranked how 3D printing would contribute to the company objectives of cost reductions, improving services and storage space reductions. The results are presented in Figure 2.

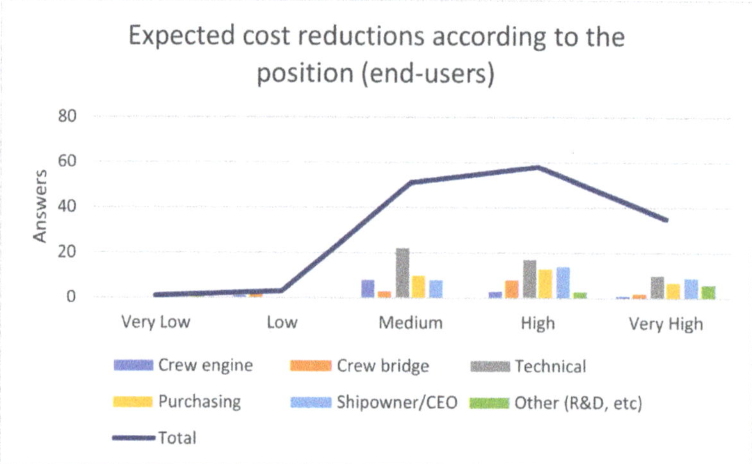

Figure 2. Expected cost reductions according to the position in the company.

Cost reductions seemed to vary as an objective for end-user participants even within their company role. Engine crew as well as technical participants valued cost reductions as of medium importance, with land office technical personnel appearing to be more concerned about cost. This is totally understandable because spare parts should be found when needed despite the cost but the budget must be kept within the set limits. On the other hand, shipowner/CEO participants, bridge crew and purchasing participants were more cautious regarding cost reductions as they preferred to save money.

4.1.2. Expected Improvement in Services According to the Role in the Company

Regarding the improvement of services as a whole for end-user participants, these were regarded to be of medium importance. As far as the role of the company for end-user participants, the improvement of services seemed to vary as an objective. This is depicted in Figure 3.

4.1.3. Expected Storage Space Reduction by the Role in the Company

As far as the role in the company for end-user participants, storage space reductions seemed to vary as an objective once again. Crew engine, crew bridge and technical participants believed that the storage space reduction objective was of medium importance as they were more concerned about the final parts themselves. The same applied for shipowner/CEO participants as well as purchasing personnel. This is depicted in Figure 4.

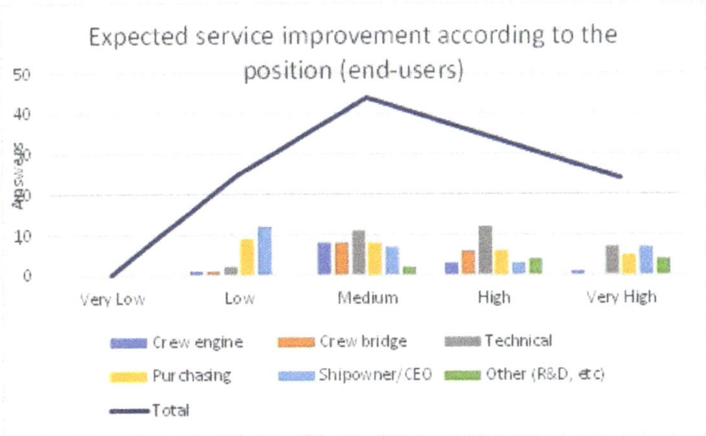

Figure 3. Expected service improvement according to the position.

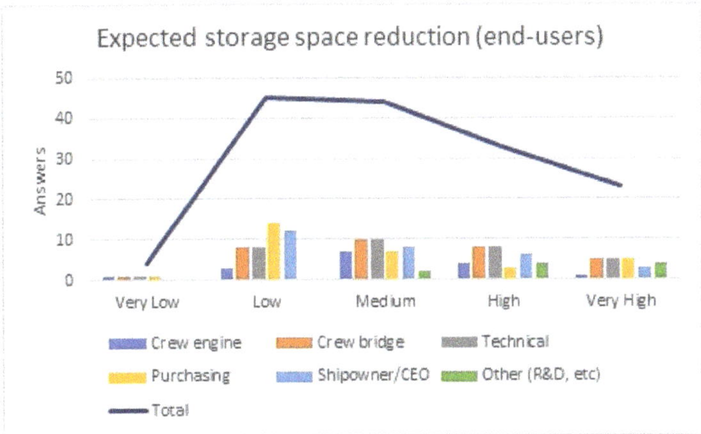

Figure 4. Expected storage space reduction.

Examining the expectations from the participants regardless of the company position, the end-users expected a low to medium space reduction and service improvement and high cost reductions. These are shown in Figure 5.

4.2. Barriers

In reference to the barriers of using parts made by 3D printing, end-user participants thought that the most important was the quality assurance of the spare parts made by the new technology. Technology know-how and staff skills followed. The survey results on these barriers are shown in Figure 6.

The importance of quality came to the surface from the answer to the question for the use of a part made in a different way than the usual procedure (3D printed) followed in the industry.

The majority of the participants answered that they would fit a 3D printed part on the equipment only if it had a quality assurance (Figure 7).

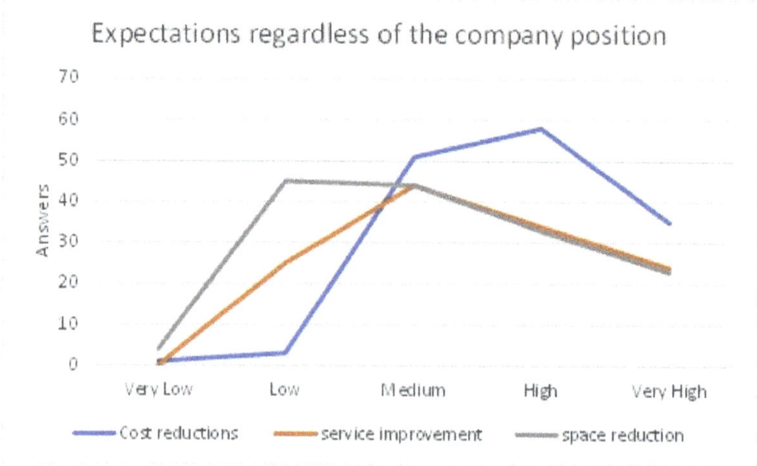

Figure 5. Expectations from the participants regardless of the company position.

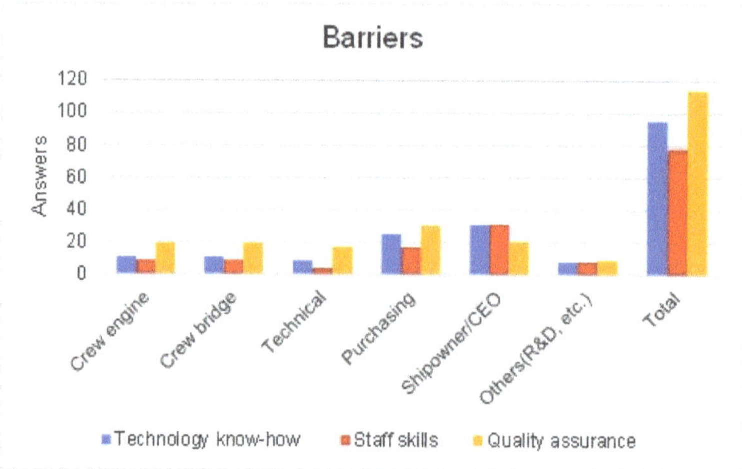

Figure 6. Barriers of using parts made by 3D printing.

4.3. 3D Printing for Low-Cost Items

A great percentage of the survey responders answered that if 3D printing was applied, it would make repairs easier. One could note that all the answers from the land office personnel were positive as they had knowledge of what the cost of a low-value spare could reach if transportation cost was included in addition to the difficulties in finding them (Figure 8). Academics were also positive.

4.4. Seedbed Ground

The use or awareness of 3D printing was an indication for seedbed ground. When asked if they examined or thought about the use of 3D printing, most answered that they did not examine nor did they think about it; many had thought about it, others did not know and a few were concerned about it at the moment. Shipowners/CEOs showed a positive attitude and that was regarded to be more than promising as they are mainly responsible for the acceptance of companies that may lead to the adoption of new technology (Figure 9).

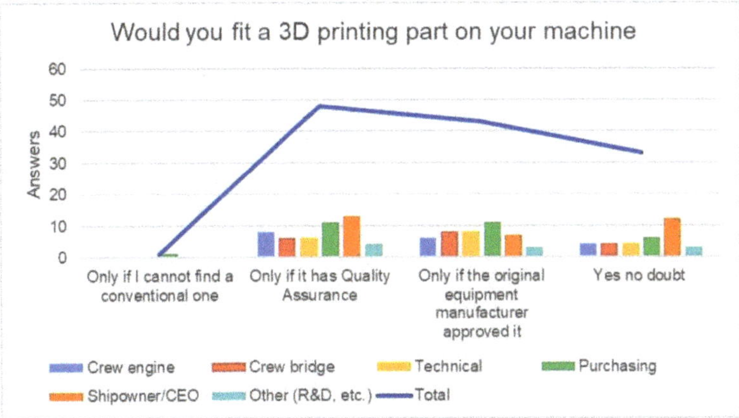

Figure 7. Response to the question "Would you fit a 3D printing part on your machine".

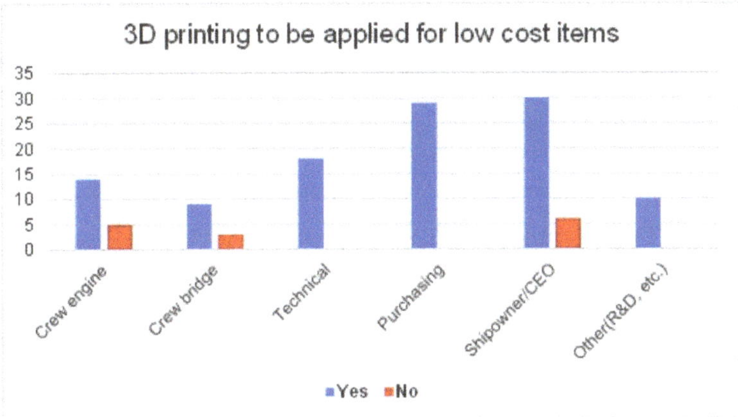

Figure 8. Response to whether 3D printing could be applied for low-cost items.

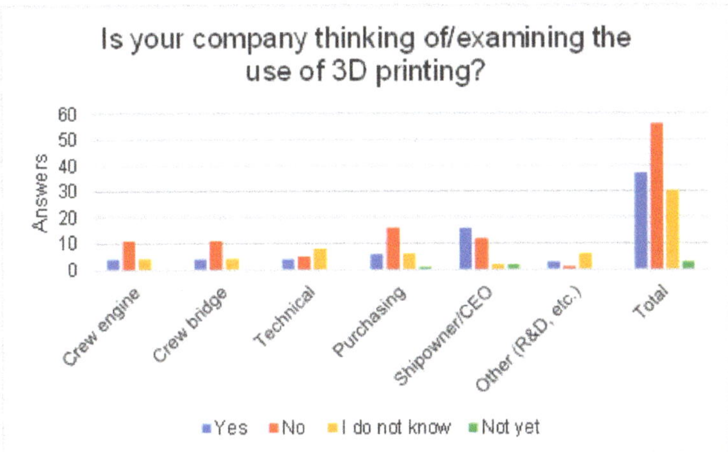

Figure 9. Results for the examination of the use of 3D printing.

The use of CNC technology that would make the transition to 3D printing easier because of the similarities between the two technologies seemed to be absent from the vessels. The positive answers from the staff of the technical department may be due to making parts on shore in local workshops and needs further study. These results are presented in Figure 10.

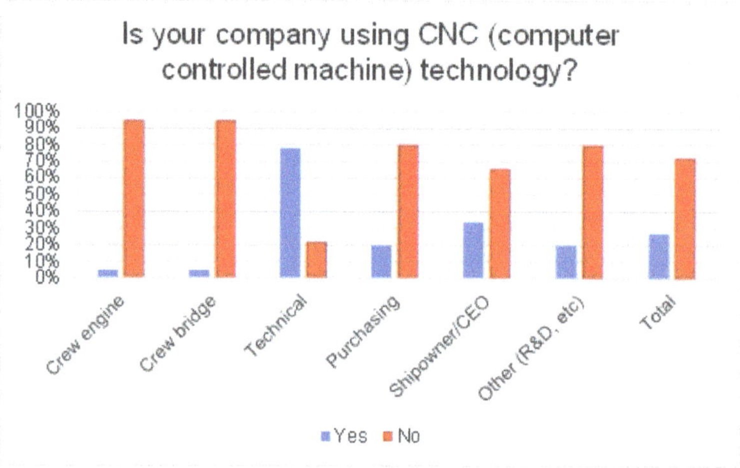

Figure 10. Use of CNC technology.

Spare parts could be made or refurbished on board but only simple and necessary ones. Several participants answered that they preferred to have a stock (Figure 11).

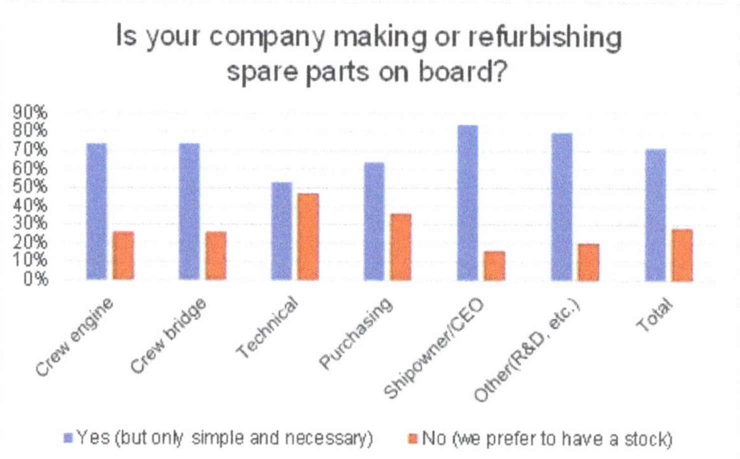

Figure 11. Making or refurbishing spare parts on board.

For making in house (at the company premises or on board) detailed design drawings, the necessary material must be available. However, from the answers, these were not available (Figure 12).

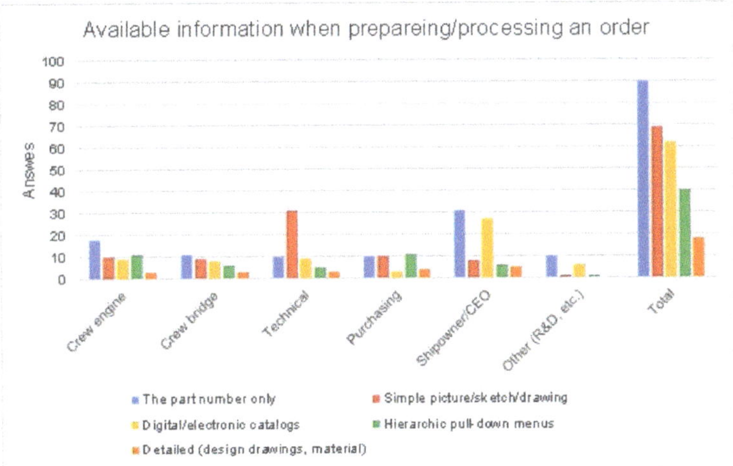

Figure 12. Available information when preparing/processing an order.

5. Conclusions

The digitization process of the spare part supply chain with the introduction of AM involves factors that will push it forward and others that are likely to be an obstacle to it. Our research concentrated on the exploration of the end part of the chain in the maritime industry in order to obtain an understanding of the enablers and inhibitors and by capturing the perceptions of the involved end-user stakeholders. The questionnaire method was selected for capturing the views of the selected target group interviewees: people on board a vessel and on the land office: ship owners, seafarers, superintendents and engineers, for example.

The expected benefits from the introduction of 3D printing (cost reductions, improvement of services and storage space reduction) as ranked by the stakeholders at the end of the SPSC (end-users) were found to vary even within a company according to the role. Cost reduction expectations were medium to very high, the improvement of services was medium and storage space reductions were low to medium. The importance of quality came to the surface also from the answer that the majority of the participants would fit a 3D printed part on the equipment only if it had a quality assurance.

One of the proposed characteristics in the literature for printing spare parts is high value. Due to maritime nature, it makes a low-cost item rise to high value when huge transportation costs to reach the vessels are included. A great percentage of the survey respondents answered that if 3D printing was applied to low-cost items it would make repairs easier.

End-users have the skills to make or refurbish spare parts on board, could make simple and necessary spare parts and preferred to have a stock. The previous use of CNC technology would make the transition easier but was not found on vessels. The majority of the participants were found to be positive in terms of thinking/examining the use of 3D printing of their companies.

The positive attitudes of shipowners/CEOs were regarded to be more than promising as they are mainly responsible for the acceptance of the companies to adopt new technology.

The most important barrier according to the end-user participants was quality assurance of the spare parts made by the new technology followed by the know-how and staff skills. Moreover, material availability, equipment (the printer) cost, staff skills and material cost were to be shown to be of medium importance. In terms of barriers, quality assurance was by far the most vital barrier as all participants were concerned about how well their parts and the materials that were used for are made.

As for the parts that are suitable for 3D printing, the participants' choices seemed to follow the characteristics as gathered from the literature, specifically more specific parts that are infrequently demanded, of high value, of small size, that can fit in the process chamber, are of a high supply risk, a small order quantity percentage, items requiring short delivery times and lastly of a high design complexity. Different from the literature was the finding about low-cost items, which received a positive response from participants.

Critical for the successful implementation of AM in the maritime SPSC is to have clearly set goals (that could be overall cost reduction, better service or whatever is pre-decided). The spare component produced must fit the intended purpose (certification/quality assurance). The overall cost reduction and certification/quality assurance are crucial in order for the venture to gain approval and support at a high level (so the CEO or owner must be convinced first in order to authorize the project and provide the means). Necessary to the decision of the printing is material availability, know-how and skills that should be ensured. Having the aforementioned decisions, approvals and resources, an established project management methodology must be followed for the successful implementation of AM in the SPSC.

The authors have attempted to provide the views of the stakeholders at the end of the chain on board the vessels and at the land office. A similar examination for the supply part of the chain would widen the picture of the SPSC and allow comparisons.

Author Contributions: Conceptualization of the idea, title and layout of the paper, methodology, validation, formal analysis, writing—original draft preparation—review and editing-final manuscript, E.K.; Supervision, direction on the layout and structure of the paper and a review of the draft and final manuscript, N.N.; Review-final manuscript, I.P. All authors have read and agreed to the published version of the manuscript.

Funding: This research received no external funding.

Institutional Review Board Statement: Not applicable.

Informed Consent Statement: Not applicable.

Data Availability Statement: Data are available upon request from the lead author.

Conflicts of Interest: The authors declare no conflict of interest.

References

1. Eruguz, A.S.; Tan, T.; van Houtum, G.-J. A survey of maintenance and service logistics management: Classification and research agenda from a maritime sector perspective. *Comput. Oper. Res.* **2017**, *85*, 184–205. [CrossRef]
2. ISO:ASTM 52900:2015 (en), Additive Manufacturing—General Principles—Terminology. Available online: https://www.iso.org/obp/ui/#iso:std:iso-astm:52900:dis:ed-2:v1:en (accessed on 3 January 2021).
3. Standard A.S. Standard Terminology for Additive Manufacturing Technologies. ASTM International F2792-12a. 2012. Available online: https://www.astm.org/Standards/F2792.htm (accessed on 19 August 2021).
4. Chekurov, S.; Metsä-Kortelainen, S.; Salmi, M.; Roda, I.; Jussila, A. The perceived value of additively manufactured digital spare parts in industry: An empirical investigation. *Int. J. Prod. Econ.* **2018**, *205*, 87–97. [CrossRef]
5. Knofius, N. Towards a Digital Spare Parts Supply Chain. AMQ Services. Available online: https://topsectorlogistiek.nl/wptop/wp-content/uploads/2020/09/White_paper.pdf (accessed on 19 August 2021).
6. Salmi, M.; Partanen, J.; Tuomi, J.; Chekurov, S.; Björkstrand, R.; Huotilainen, E.; Kukko, K.; Kretzschmar, N.; Akmal, J.; Jalava, K.; et al. *Digital Spare Parts*; Aalto University; VTT Technical Research Centre: Espoo, Finland, 2018; pp. 3–4.
7. Final Report Pilot Project 3D Printing of Marine Spares. Port of Rotterdam. 2016. Available online: https://www.portofrotterdam.com/en/files/final-report-pilot-project-3d-printing-of-marine-spares (accessed on 4 April 2020).
8. Green Ship of the Future. Can 3D Print Technology Be Used for Repair and Reconditioning, Reducing the Number of Scrapped Maritime Parts? Available online: https://greenship.org/can-3d-print-technology-be-used-for-repair-and-reconditioning-reducing-the-number-of-scrapped-maritime-parts/ (accessed on 16 February 2020).
9. Kandukuri, S. Additive Manufacturing for Marine Parts—A Market Feasibility Study with Singapore Perspective. Available online: https://www.mpa.gov.sg/web/wcm/connect/www/99a3720f-abfc-4b07-9c9b-467220c1000a/Additive+Manufacturing+Market+Feasibility+Study_Public+Version.pdf?MOD=AJPERES&id=1572312102868 (accessed on 12 June 2020).
10. Rubino, F.; Nisticò, A.; Tucci, F.; Carlone, P. Marine application of fiber reinforced composites: A review. *J. Mar. Sci. Eng.* **2020**, *8*, 26. [CrossRef]

11. Parmar, H.; Khan, T.; Tucci, F.; Umer, R.; Carlone, P. Advanced robotics and additive manufacturing of composites: Towards a new era in Industry 4.0. *Mater. Manuf. Process.* **2021**, 1–35. [CrossRef]
12. Dalaklis, D.; Katsoulis, G.; Kitada, M.; Schröder-Hinrichs, J.-U.; Ölcer, A.I. A "Net-Centric" conduct of navigation and ship management. *Marit. Technol. Res.* **2020**, *2*, 90–107. [CrossRef]
13. Zanoni, S.; Ashourpour, M.; Bacchetti, A.; Zanardini, M.; Perona, M. Supply chain implications of additive manufacturing: A holistic synopsis through a collection of case studies. *Int. J. Adv. Manuf. Technol.* **2019**, *102*, 3325–3340. [CrossRef]
14. Mashhadi, R.; Esmaeilian, B.; Behdad, S. Impact of additive manufacturing adoption on future of supply chains. In Proceedings of the International Manufacturing Science and Engineering Conference, Guangzhou, China, 28–29 November 2015; Volume 56826, p. V001T02A064.
15. Khajavi, S.H.; Partanen, J.; Holmström, J. Additive manufacturing in the spare parts supply chain. *Comput. Ind.* **2014**, *65*, 50–63. [CrossRef]
16. Liu, P.; Huang, S.H.; Mokasdar, A.; Zhou, H.; Hou, L. The impact of additive manufacturing in the aircraft spare parts supply chain: Supply chain operation reference (scor) model-based analysis. *Prod. Plan. Control.* **2014**, *25*, 1169–1181. [CrossRef]
17. Li, Y.; Jia, G.; Cheng, Y.; Hu, Y. Additive manufacturing technology in spare parts supply chain: A comparative study. *Int. J. Prod. Res.* **2017**, *55*, 1498–1515. [CrossRef]
18. Ghadge, A.; Karantoni, G.; Chaudhuri, A.; Srinivasan, A. Impact of additive manufacturing on aircraft supply chain performance. *J. Manuf. Technol. Manag.* **2018**, *29*, 846–865. [CrossRef]
19. Khajavi, S.H.; Holmström, J.; Partanen, J. Additive manufacturing in the spare parts supply chain: Hub configuration and technology maturity. *Rapid Prototyp. J.* **2018**, *24*, 1178–1192. [CrossRef]
20. Westerweel, B.; Basten, R.; den Boer, J.; van Houtum, G.J. Printing spare parts at remote locations: Fulfilling the promise of additive manufacturing. *Prod. Oper. Manag.* **2021**, *30*, 161–532. [CrossRef]
21. Delic, M.; Eyers, D.R. The effect of additive manufacturing adoption on supply chain flexibility and performance: An empirical analysis from the automotive industry. *Int. J. Prod. Econ.* **2020**, *228*, 107689. [CrossRef]
22. Ratnayake, R.M.C. Making Sense of 3D Printing/Additive Layer Manufacturing in Offshore Petroleum Industry: State of the Art. *Ocean Eng.* **2016**, *7*, 2016–54537.
23. Verboeket, V.; Khajavi, S.H.; Krikke, H.; Salmi, M.; Holmstrom, J. Additive Manufacturing for Localized Medical Parts Production: A Case Study. *IEEE Access* **2021**, *9*, 25818–25834. [CrossRef]
24. Turrini, L.; Meissner, J. Spare parts inventory management: New evidence from distribution fitting. *Eur. J. Oper. Res.* **2019**, *273*, 118–130. [CrossRef]
25. Costantino, F.; di Gravio, G.; Patriarca, R.; Petrella, L. Spare parts management for irregular demand items. *Omega* **2018**, *81*, 57–66. [CrossRef]
26. Van der Auweraer, S.; Boute, R.N.; Syntetos, A.A. Forecasting spare part demand with installed base information: A review. *Int. J. Forecast.* **2019**, *35*, 181–196. [CrossRef]
27. Zhu, S.; Dekker, R.; van Jaarsveld, W.; Renjie, R.W.; Koning, A.J. An improved method for forecasting spare parts demand using extreme value theory. *Eur. J. Oper. Res.* **2017**, *261*, 169–181. [CrossRef]
28. Kostidi, E.; Nikitakos, N. Additive manufacturing of Spare parts in the Maritime Industry in the digital era. In Proceedings of the Annual Conference of the International Association of Maritime Economists (IAME), Athens, Greece, 25–28 June 2019; Volume 1.
29. Holmström, J.; Partanen, J.; Tuomi, J.; Walter, M. Rapid manufacturing in the spare parts supply chain: Alternative approaches to capacity deployment. *J. Manuf. Technol. Manag.* **2010**, *21*, 687–697. [CrossRef]
30. Vogelsang, K.; Liere-Netheler, K.; Packmohr, S.; Hoppe, U. Barriers to Digital Transformation in Manufacturing: Development of a Research Agenda. In Proceedings of the 52nd Hawaii International Conference on System Sciences, Maui, HI, USA, 8 January 2019; pp. 4937–4946.
31. Osborn, L. 3D Printing and Intellectual Property. In *Research Handbook on Digital Transformations*; Xavier Olleros, F., Zhegu, M., Eds.; Edward Elgar: Cheltenham, UK, 2016; pp. 254–271. Available online: https://ssrn.com/abstract=2533673 (accessed on 19 August 2021).
32. Kretzschmar, N.; Chekurov, S.; Salmi, M.; Tuomi, J. Evaluating the Readiness Level of Additively Manufactured Digital Spare Parts: An Industrial Perspective. *Appl. Sci.* **2018**, *8*, 1837. [CrossRef]
33. Ballardini, R.M.; Ituarte, I.F.; Pei, E. Printing spare parts through additive manufacturing: Legal and digital business challenges. *J. Manuf. Technol. Manag.* **2018**, *29*, 958–982. [CrossRef]
34. IEEE. What Are Standards? Why Are They Important? SA Beyond Standards. Available online: https://beyondstandards.ieee.org/general-news/what-are-standards-why-are-they-important/ (accessed on 6 August 2020).
35. ISO:IEC 17065:2012 (en), Conformity Assessment—Requirements for Bodies Certifying Products, Processes and Services. 2012. Available online: https://www.iso.org/obp/ui/#iso:std:iso-iec:17065:ed-1:v1:en (accessed on 3 June 2020).
36. Lloyd's Register. Available online: https://www.lr.org/th/additive-manufacturing/additive-manufacturing-resources/guidance-notes-for-the-certification-of-metallic-parts-made-by-additive-manufacturing/ (accessed on 3 May 2020).
37. DNV GL Additive Manufacturing Certification. Available online: https://www.dnvgl.com/services/additive-manufacturing-certification-104684 (accessed on 3 May 2020).

38. ABS Guidance Notes on Additive Manufacturing. Available online: https://ww2.eagle.org/content/dam/eagle/rules-and-guides/current/other/322_guide_for_additive_manufacturing_2021/additive-manufacturing-guide-apr21.pdf (accessed on 17 August 2021).
39. Bureau Veritas Marine and Offshore. Additive Manufacturing—Guidelines for Certification of Product Made Using Wire Arc Additive Manufacturing (WAAM) Process NI662 R00. Available online: https://erules.veristar.com/dy/data/bv/pdf/662-NI_2019-05.pdf (accessed on 17 August 2021).
40. Despeisse, M.; Minshall, T. Skills and Education for Additive Manufacturing: A Review of Emerging Issues. In Proceedings of the IFIP International Conference on Advances in Production Management Systems (APMS), Hamburg, Germany, 31 August 2017; pp. 289–297.
41. Gupta, N.; Tiwari, A.; Bukkapatnam, S.T.; Karri, R. Additive Manufacturing Cyber-Physical System: Supply Chain Cybersecurity and Risks. *IEEE Access* **2020**, *8*, 47322–47333. [CrossRef]
42. Knofius, N.; Van Der Heijden, M.C.; Zijm, W.H. Selecting parts for additive manufacturing in service logistics. *J. Manuf. Technol. Manag.* **2016**, *27*, 915–931. [CrossRef]
43. Lindemann, C.; Reiher, T.; Jahnke, U.; Koch, R. Towards a sustainable and economic selection of part candidates for additive manufacturing. *Rapid Prototyp. J.* **2015**, *21*, 216–227. [CrossRef]
44. Yang, S.; Page, T.; Zhang, Y.; Zhao, Y.F. Towards an automated decision support system for the identification of additive manufacturing part candidates. *J. Intell. Manuf.* **2020**, *31*, 1917–1933. [CrossRef]

Journal of
Marine Science and Engineering

MDPI

Article

Maritime Robotics and Autonomous Systems Operations: Exploring Pathways for Overcoming International Techno-Regulatory Data Barriers

Tafsir Matin Johansson [1,*], Dimitrios Dalaklis [2] and Aspasia Pastra [1]

1 World Maritime University-Sasakawa Global Ocean Institute, World Maritime University, 21118 Malmö, Sweden; asp@wmu.se
2 Maritime Safety and Environmental Administration, World Maritime University, 21118 Malmö, Sweden; dd@wmu.se
* Correspondence: tm@wmu.se; Tel.: +46-764-088-238

Abstract: The current regulatory landscape that applies to maritime service robotics, aptly termed as robotics and autonomous systems (RAS), is quite complex. When it comes to patents, there are multifarious considerations in relation to vessel survey, inspection, and maintenance processes under national and international law. Adherence is challenging, given that the traditional delivery methods are viewed as unsafe, strenuous, and laborious. Service robotics, namely micro aerial vehicles (MAVs) or drones, magnetic-wheeled crawlers (crawlers), and remotely operated vehicles (ROVs), function by relying on the architecture of the Internet of Robotic Things. The aforementioned are being introduced as time-saving apparatuses, accompanied by the promise to acquire concrete and sufficient data for the identification of vessel structural weaknesses with the highest level of accuracy to facilitate decision-making processes upon which temporary and permanent measures are contingent. Nonetheless, a noticeable critical issue associated with RAS effective deployment revolves around non-personal data governance, which comprises the main analytical focus of this research effort. The impetus behind this study stems from the need to enquire whether "data" provisions within the realm of international technological regulatory (techno-regulatory) framework is sufficient, well organized, and harmonized so that there are no current or future conflicts with promulgated theoretical dimensions of data that drive all subject matter-oriented actions. As is noted from the relevant expository research, the challenges are many. Engineering RAS to perfection is not the end-all and be-all. Collateral impediments must be avoided. A safety net needs to be devised to protect non-personal data. The results here indicate that established data decision dimensions call for data security and protection, as well as a consideration of ownership and liability details. An analysis of the state-of-the-art and the comparative results assert that the abovementioned remain neglected in the current international setting. The findings reveal specific data barriers within the existing international framework. The ways forward include strategic actions to remove data barriers towards overall efficacy of maritime RAS operations. The overall findings indicate that an effective transition to RAS operations requires optimizing the international regulatory framework for opening the pathways for effective RAS operations. Conclusions were drawn based on the premise that policy reform is inevitable in order to push the RAS agenda forward before the emanation of 6G and the era of the Internet of Everything, with harmonization and further standardization being very high priority issues.

Keywords: robotic and autonomous systems; drones; crawlers; remotely operated vehicles; Internet of Robotic Things; data governance; user interface; international standards

1. Introduction

Digital data, commonly referred to as data, have a ubiquitous influence in the contemporary information-intensive age, as vast quantities of data are created every second, if not microsecond. The dynamic benefits of data acquisition and subsequent usage have led end-users to the general understanding that data have an unfeigned nexus with decision-making processes, especially ones that concern business-as-usual.

Over the years, the term data has found itself being defined in a myriad of ways. It is further observed that each individual definition of data can be characterized as being discipline-oriented. Whether natural science, economy, or law, scholars have put forth respective definitions from the prism, and to the extent, the term data has interacted with the subject matter of respective disciplines [1]. The quintessential definition of data, however, is found in the common lexicon, i.e., the Cambridge English Dictionary which defines the term as "information, especially facts or numbers, collected to be examined and considered and used to help decision-making, or information in an electronic form that can be stored and used by a computer" [2]. All-embracing in content and ambit, the above definition found in the common lexicon reflects the essence of the varying definitions, to all intents, construction, and purposes, propounded by various subject matter experts, e.g., Senn (1982), Clare and Loucopoulos (1987), and Avison and Fitzgerald (1995) [3–5].

Despite the abundance of discipline-led definitions, experts view modern-day dependency on data with a positive outlook. What began as a mere byproduct of business processes, is now incontrovertibly a valuable asset [6,7]. Similar to land-based industrial counterparts, this so-called dependency on data was observed in the maritime and ocean domain, for example, in marine science concentrated on species distribution modelling, geospatial data in maritime boundary delimitation, submarine communications, nautical charts, and marine scientific research of areas beyond national jurisdiction [8–10]. In short, data are invaluable and reside at the crux of work undertaken by stakeholders that participate in effecting a sustainable transformation in relation to maritime and ocean governance.

There is, of course, the shipping industry's constant reliance on "timely and accurate data to feed its logistical plan" [11]. To take but one example, the voyage data recorder (VDR), which is fitted into passenger ships and other ships of 3000 gross tonnage and above, is considered an important compliance tool in line with the International Maritime Organization's (IMO) International Convention for the Safety of Life at Sea). Not only has VDR facilitated forensic investigations in post incidents and casualties in the past, but also the data gathered by investigators have remained of interest to stakeholders, including academics and researchers that support "green operations" through performance efficiency to ameliorate overall safety while reducing operational costs. Research on data continues with efforts focusing on exploring the means and methods to process "big data" and "data analytics" from an ethical–socio–technical setting [12,13]. In this journey of exploration, the acquisition and translation of "big data" to promote safety culture is a promising direction taken by the shipping industry, following the illustrative development/precedent in the "Big Data Traffic Information Service" system by the automotive company Toyota in 2013 [14].

Technology is undoubtedly a catalyst of change [15]. This statement itself is well grounded in facts, especially considering the manner in which RAS are currently being integrated and adapted into the operational framework of business activities to obtain data. The shipping industry is certainly no exception. Clean and green shipping, one of the clearest manifestations of RAS integration, comes off as a timely response in mitigating the adverse effects of shipping on climate change where data analysis has an invaluable purpose. Second to this crucial aspect is the integration of RAS to carry out statutory and classification tasks, e.g., annual survey, periodic survey, and special survey coupled with damage repairs, satisfied via conventional methods with a view to obtain real-time data for observation of structural weaknesses, thus altogether contributing to maintaining international vessel performance standards [13]. Once fully integrated, RAS would eventually replace traditional human-led survey, inspection, and maintenance. The end-objective of all RAS-driven operations is to gain optimum accuracy in the process of data acquisition during inspection and to improve post-inspection decisions based on data acquired. For example, outer hull is an area prone to damages caused by biofouling. Determining biofouling status for maintaining vessel speed and increasing maneuverability is contingent on accurate data in relation to a number of integral structures. In principle, the processes are governed by international laws and requirements, adopted at the national level and adhered to by flag state authorities [13].

The paradigm shift of RAS reliance on shipping has already begun [15]. In 2020, Bureau Veritas recorded the test results of using drones for close-up inspections and thickness measurements, setting a milestone in this RAS-deployment continuum [16]. Drones, also known as unmanned aerial vehicles (UAVs), are aircrafts that "can be remotely controlled or programmed to fly a predetermined route using information on a specific asset's condition to target known areas of concern" [17]. The other two existing RAS applications include crawlers and ROVs. Very briefly, the former is designed to "crawl along a structure ... to operate on a vertical surface or hull structures in air or underwater" [17]. The latter is exclusively used in underwater operations to "collect visual data, perform Nondestructive Testing (NDT), and measure plate thickness in difficult-to-reach areas" [17].

RAS-interventions today are very data-centric, as they are translated into activities such as data acquisition, data transfer, data storage, and finally data analyses, which remain the principal foci. Although RAS applications—as they exist today—are best described as systems with "human-in-the-loop", their relevant capabilities are promising, leading to the projection that drones, crawlers, and ROVs will achieve full autonomy in the near future [13,15]. As RAS advances towards full autonomy, there is a clear need to spear through the current data governance and corresponding management practices to assess whether procedural details, rules, and requirements follow a water-tight approach securing much needed precautions that would otherwise give rise to complications from the get-go stage.

In furtherance of the foregoing, this paper discusses the results gathered from the European Union Horizon 2020 funded project titled Autonomous Robotic Inspection and Maintenance on Ship Hulls (BUGWRIGHT2) while addressing two pragmatic questions: what are the thorny issues that could invoke data liability in an RAS-led vessel survey and inspection operation, and what are the pathways through which those issues could be mitigated? The motivation behind seeking answers to the above emanates from the current vacuum that exists in the regulatory setting that could call on liability implications that are purely legal in nature. It is important that service robotics engineers, as they move from "human-in-the-loop" technology to "full autonomy" technology, remain aware that technology and law reside in the same continuum and are therefore interrelated. Technology should progress, but in that process, it must be ensured that such progress does not breach or violate any contemporary legal provisions that are globally sensitive in nature. Data acquisition is one subject matter of crucial importance to both engineers as well as global, regional, and national regulators. Simply put, a solution is required before RAS is in mass deployment and before the topic of data protection reaches the contentious stage.

With a view to finding a solution to the questions posed above, this article focuses on identifying the principal data barriers existing in the international techno-regulatory framework (which is followed by ship owners, classification societies, as well as the operational folks engaged under the title of service providers), as well as potential solutions for consideration based on the theoretical construct of data governance. International norms and standards set by the International Association of Classification Societies (IACS) are the ones that regulate all RAS vessel survey activities and are sporadically revised as needed to ensure that the provisions are fit-for-purpose. Based on the current setting as it exists, it is correct to state that shipowners lack a techno-regulatory safety net to protect the non-personal data linked to their asset. The growing needs to eliminate liability drawbacks from such an absence are being discussed in different platforms at the European

Union level, including the European Commission funded project, ROBotics technology for Inspection of Ships (ROBINS).

With the above in mind, the paper commences with an examination of the characteristic peculiarities of the two main classes of data, and this evidently supports the placement of RAS-acquired data under the category of non-personal data, and further allows the authors to explore non-personal data decision domains and building blocks. This is followed by an analytical highlight of the types of stakeholders involved in the non-personal data business model with specific reference to human and non-human actors. Subsequently, this paper provides an expository review of the state-of-the-art international legal framework that maritime RAS falls within. This is done with reference to current international data-related rules and requirements, highlighting the entities typically engaged in vessel survey and inspection operations involving RAS data acquisition, analysis, and validation.

Additionally, an attempt is made to provide an incisive examination of the architecture of Internet of Robotic Things (IoRT) in the context of maritime RAS and through the prism of data. Thereafter, a comparative examination is conducted to highlight best practices offered by selected IMO-recognized organizations, referred to as classification societies that are proactively engaged in regulating vessel survey and inspection processes, including the main subject matter of this article, i.e., data. The major drawbacks of the current international framework are extracted to underline noteworthy challenges before carving out ways forward that provide a first-hand synoptic insight into the regulatory blueprint developed by researchers of the World Maritime University (WMU) under the ongoing BUGWRIGHT2 project. The paper concludes with takeaway notes synthesizing important points gathered from previous discussions, and ones that will allow for digital serenity as maritime RAS engineering progresses.

2. Materials and Methods

To satisfy the aims and objectives of this paper, a combination of the doctrinal and comparative methods was employed. The doctrinal methodology is basically concerned with doctrinal research, which in turn is a combination of "legal theory research" and "expository research" [18]. Given that the paper focuses on technologies engineered to collect data, "legal theory research" (which underpins the detailed study of legal doctrines and principles, jurisprudence, and legal philosophy) remains outside the scope of the work [18].

To promote the results of this article, the authors relied on the technological regulatory (techno-regulatory) international framework (as a part of the "legal theory research") that governs all RAS affairs under IMO's rule of reference. Reliance on the international framework was unavoidable given that they are followed by the RAS community at large, including engineers, manufacturers, technology specialists of RAS, human–machine interface psychologists, shipowners, flag states, as well as service providers. The primary method is therefore expository research that is concerned with the study of technology-based legislation, international treaties, and scholarly literature that touch upon data-related laws, rules, and requirements with respect to RAS—applications of which are in force. This methodology is used to analyze the extant law (i.e., *de lege lata*), pointing out its drawbacks and deficiencies. The subject must be thoroughly understood in order to determine what the law should be in the future (i.e., de lege ferenda). Needless to say, this approach highlights the continuum of past, present, and future in terms of the progress of the law aimed at technologies.

It is important to note that expository research mainly comprises detailed the examination of legal texts that govern the usage of RAS, including international rules and requirements found in legislative material and international legal instruments. In the world of the science–policy interface, these are known as "black-letter law". The exposition of legal texts in the research process includes non-treaty instruments, relevant scholarly literature such as textbooks, academic and professional journals containing legal, management and governance opinions and expert commentaries, standard form international contracts, and the like.

The comparative methodology—essentially comprising comparative research—is used to compare best practices offered. It extends to the examination of law and legal practices in light of current developments, the object of which is to carry out a comparative analysis. The information so gleaned can lead to obtaining an insight into best international practices. Comparative legal research is deemed to be an important tool that can help the researcher to review and juxtapose best practices through rational comparisons. Within the framework of comparative research, interviews were also conducted with relevant stakeholders of the data management process, including an executive of a service supplier company, a legal expert, an IT shipping consultant and a drone team of a maritime technical service company.

It is envisaged that the final outcome of the examination and critical analysis of techno-law carried out through the use of the doctrinal and comparative methods of legal research leads to strategic proposals and pragmatic alternatives being presented by the authors. These should take into account the specific aspects of RAS data governance and management that need changing or introduced. The proposals for ongoing regulatory development should be aimed at making future laws more coherent, precise, and practical (de lege feranda), thereby providing adequate security and protection to beneficiaries.

3. Results

3.1. Theoretical Construct: Dissecting the Dimensions of Data

Despite contemporary high usage, the term "data governance" continues to remain elusive [19,20]. With reference to the definitions implemented by the Data Management Association (DAMA), the term "data governance" is viewed as "the allocation of authority and control and shared decision making over the management of data assets" [21]. With this definition in mind, secondary sources have carved out the end-objectives of data governance: enhancing data quality, data value, and reducing data-related cost and risk [22–25].

At the outset, it is important to note that the principal concept behind data governance brings to the forefront the need for the comprehension of the inherent dichotomy that lies between personal and non-personal data, since any given data could be a combination of different datasets and transformed into "personal data" in cases where there is processing power and data availability (Mattoo and Meltzer, 2018; Chatzimichali and Chyrostomou, 2019) [26,27]. The definition of personal data, which is of practical significance and far from being of theoretical interest, is intrinsically related to data that directly or indirectly relates to an identified or identifiable natural person (Finck and Pallas, 2020) [28].

A concrete definition of "personal data" is found in the EU's Regulation 2016/679 on the General Data Protection Regulation (GDRP). According to Article 4(1) GDPR, the term personal data is defined as "any information relating to an identified or identifiable natural person ('data subject'); an identifiable natural person is one who can be identified, directly or indirectly, in particular by reference to an identifier such as a name, an identification number, location data, an online identifier or to one or more factors specific to the physical, physiological, genetic, mental, economic, cultural or social identity of that natural person" [29] (Article 4.1). Non-personal data, on the other hand, venture outside the scope of identifiable natural person, as they relate to industrial and anonymized raw machine-generated data [30].

Legal and business-environment facets pertaining to non-personal data are regulated by the European Union's (EU) Regulation 2018/1807, as outlined in the framework for the free flow of non-personal data in the European Union—an exemplary governance tool that aims to achieve data-driven growth and innovation between the EU Member States. In the above Regulation, non-personal data are conceptualized as "data other than personal data", as defined in Paragraph 9 of the GDPR and include anonymized datasets used for big data analytics and data on maintenance needs for industrial machines [31]. The Regulation moves unjustified barriers for the free movement of non-personal data and applies to EU natural or legal persons that provide data processing services (e.g., collection, recording, storage, use, disclosure by transmission, dissemination of data), giving individuals and organizations the opportunity to collect and disseminate non-personal data and to use data centers or cloud services located within the EU while protecting individual privacy.

In the context of terminology differentiation, it is also important to note the distinction between data governance and data management, since "governance" is the de facto overarching term within which data management resides. Data governance is related to the upper-level planning and the decisions about the allocation of responsibilities, access, control, and use of data, as opposed to data management, which is linked to the implementation and monitoring of governance-related decisions [32]. Recently published literature indicated that "data management" encompasses the essential sequence of the following activities: collection, storage, processing, using, sharing, and destroying of data [33]. Conversely, data governance ensures appropriate data management and provides the relevant processes and structures for formally managing information and for protecting data as a strategic asset [33–35].

There is also a narrower focus that merits further observation, namely that organizational and technical perspectives should be jointly included in a proper data governance framework [36]. Recent efforts include the work by Abraham et al. [25], who set out to conduct an in-depth literature review that eloquently identified the following data decision domains and building blocks with regards to non-personal data:

Governance mechanisms:

- Identification of the data owner/data manager/data consumer
- allocation of responsibilities
- decision-making authority
- organizational policies and standards
- coordination with the stakeholders;

Application of governance mechanisms on the organizational scope, data scope, and domain scope:

- Quality and integrity of project-related organizational data
- data governance between firms
- machine-generated data
- privacy requirements
- data architecture
- data lifecycle
- data storage;

Antecedents of data governance:

- Internal and external factors that precede data governance practices, such as organizational culture and industry;

Outcomes of data governance:

- Measurement of data governance effectiveness on organizational strategy and business performance [25].

Most of today's data governance programs address goals in two or three data decision domains. Within the settings of the above cited data governance framework, several interrelated decision domains remain dormant: (a) data architecture; (b) data quality; (c) data security and data access; (d) data lifecycle; (e) metadata; and (f) data storage and infrastructure [25,32,37].

The first decision domain, i.e., "data architecture", is a term that refers to the blueprint that aligns data assets with the organizational strategy and sets designs to meet data requirements [21]. In short, the architectural policies and standards denote how data are processed, stored, distributed, and used. Data architecture helps to facilitate the flow of data within an organizational ecosystem by developing interrelated and bidirectional data pipelines. The second domain known as "data quality" focuses on the ability of data to fulfill their intended use through precision, timeliness, completeness, and credibility [32]. In this regard, specific organizational metrics are utilized to safeguard data quality. As for the decision domain that is explicitly tied with safeguarding measures, "data security" as the third domain touches on many issues, including privacy, confidentiality, intellectual property policies, data access, and third-party access [38]. The third domain does not ignore the possibility that effective organizational security mechanisms and tools could help protect data from unauthorized access and corruption throughout their lifecycle.

Similar to biorhythm, the lifecycle of data under the fourth decision domain titled "data lifecycle" captures the intricate timespan of the life of data from its generation and maintenance, up to the very point of their deletion. Data lifecycle represents the approach of archiving and retiring big data until it no longer makes sense to retain them, leading to the assumption that overmanaging information could lead to the waste of capital resources [39]. In terms of data generation, "metadata", whether created manually or automatically, is the fifth category conceptualized as "data-about-data" or as "information stored in IT tools that improves both the business and technical understanding of data and data-related assets" [38]. Metadata are stored in a location different from the original data and are generally invisible to the end-user. There are different types of data that can be considered as "metadata": title, author, keywords, permissions, geolocation of an image, copyright information, date and instruction for the users, among others. Finally, "data storage and infrastructure", the sixth decision domain, is about maintaining or archiving digital data on different types of media for usage by computer devices [40]. Technological storage options are ample whereby technology provides economical and reliable solutions for storage arrangements.

3.2. *Carving Out the Stakeholders of Non-Personal Data*

The need to identify the types of stakeholders that directly or indirectly affect or are directly or indirectly impacted by data decision domains (described in the previous section) with regards to non-personal data is unavoidable in discussions related to data governance. As the first step, one must be cognizant of the key human and non-human actors both within and outside the organization [41]. The interactions between multiple actors have their roots in the actor–network theory (ANT), which is quite often utilized to analyze systems in the fields of social sciences, international relations, and information technology [42].

ANT explores the relational ties within a network in which knowledge is perceived as the product of a network of heterogeneous actors [42]. ANT examines the construction and transformation of heterogeneous networks and collaborative designs, taking social and technical aspects into account. Each network has a generalized symmetry, and each actor has an equal role to play. Therefore, ANT sets the framework for alliance among humans, machines, tools, computers, and other non-human agents. Actions are a conglomerate of many agency groupings that must be slowly untangled [43,44]. An actor–network relationship is never stabilized, owing to the fact that agents are not isolated from each other and are constantly involved in the process of actions and reactions [42,45].

Human actors in a data framework that concerns "machines" are the stakeholders who are involved in and are affected by data-based decisions and actions. The stakeholders that belong to the data governance framework are the participants that are involved in information sharing throughout the data supply chain management [46].Potential stakeholders range from manufacturers, software developers, service providers, business clients, information architects, data governance practitioners, metadata analysts, operations staff, and end-users [47]. Then there are non-human actors that interact with the above: policies and regulations, market demands, economic circumstances, semiotics, technological tools and infrastructure, inter alia [41].

It would seem that the notion of a "governance network" that introduces human and non-human actors' interrelations is of paramount importance in order to mark out the stakeholders that belong to the service robotics non-personal data framework. In doing so, one will come across an important detail: human agents are not unique when it comes to shaping and influencing reality, and for this reason, non-human actors' role in an actor–network domain should not be diminished [44]. Any given system contains diverse entities whereby each entity is a sum of other smaller actors. Inevitably, international standards embedded within the scope of policies and regulations, viewed as a non-human factor, influence the manner in which tasks are completed, covering the essentials that are required. In doing so, international standards cement the close and enduring partnership between and among the stakeholders in the fulfilment of non-personal data acquisition, analysis, and validation objectives.

3.3. RAS and the Internet of Robotic Things: Through the Prism of Data

In its Recommendation titled ITU-T Y.2060, the International Telecommunication Union (ITU) describes the Internet of Things (IoT) as an all-embracing international infrastructure for the informatics and communication society (International Telecommunication Union, 2012; Yousif, 2018) [48,49]. Consequently, IoRT comes forth as a concept that amalgamates IoT and cloud robotics, where cloud robotics is a combination of robotics and cloud computing, with the latter opening the digital doorway for sharing, processing, and storing interoperable information [49]. For maritime RAS, this corresponds to close-up information on vessel structures and thickness measurements of suspect areas.

The fabric of the IoRT architecture anticipates communication of information via five interdependent layers. Localized movements with the help of actuators and the conversion of signals or stimuli (light, motion, sound, and heat) to electrical domains via a wireless sensor network (WSN) form a part of the "hardware layer" that leverages information to the "network layer", which encompasses cellular connectivity, short and medium-range communication technologies, Worldwide Interoperability for Microwave Access (WiMAX), Z-Wave, ZigBee, and low power wide area network (e.g., LoRa) [50]. Within the architecture of IoRT, the "Internet layer" plays a central role, in so far as it supports lightweight information processing, inter alia, through transport protocols including data distributed services (DDS) in the robotics system for carrying out a number of principal tasks, listed as follows: "publish/subscribe messaging, multicast support, real-time instant messaging, packet switched networking, alternative to Transport Control Protocol, disseminating of networked embed system, providing privacy to datagram protocol, message queuing for middleware environment, lightweight local automation, and directly addressing publish/subscribe based communication for real-time embedded systems" [50]. The most invaluable layer of the architecture is the "infrastructure layer", which in turn comprises five distinct and interrelated components, namely a robotic cloud platform, M2M2A cloud platform support, IoT business cloud services, big data services, and an IoT cloud robotics infrastructure [50]. User interaction and interface through monitoring and visualization is realized in the last IoRT architecture element known as the "application layer"—the technical potentials of which are expanding [49,50]. Finally, the aforementioned individual but interconnected layers facilitate data acquisition from the physical world, which are fed to the application layer for observation. The advancements through machine learning (ML), a branch of artificial intelligence, are seen to have revolutionized the IoRT architecture with fast and smooth fault diagnosis, making "ships safer, easier to use and more efficient" [51,52].

Turning to maritime RAS for vessel survey, inspection, and maintenance, it was observed that the key parameters in the implementation and acceptance of RAS platforms in shipping industry are related to operational safety [15]. The so-called operational safety concerns ease of operation during intervention by "human-in-the-loop" during data acquisition, communication and data management, data security, energy autonomy, and power source. Although communication and data management tasks today are conducted using both paper and digital systems, data integrity is nevertheless ensured to a certain extent through the introduction of some form of a blockchain system in the data management process [15]. As the focus remains on integrating RAS into the current manual system, data security and the effectiveness of data collection, data processing, and distribution of analysis outputs (e.g., defects autodetection) need to be demonstrated, ensured, and proven in an effective manner. For each of the three types of maritime RAS, the data communication system in maintaining the data decision dimension needs to be thoroughly revisited in light of user interface (see Table 1). Otherwise, RAS platforms are likely to lack trustworthiness among the stakeholders of the business model. Software developers need to provide details of international standards used and apply those standards to satisfy the data security requirements. For a clear understanding of the different stages involved in RAS in survey and inspection, a diagram (see Figure 1) is presented to familiarize the reader with a real live scenario.

Table 1. State of autonomy, class, and user interface of maritime drones, crawlers, and ROVs.

Maritime Drones: Aviation/Air			
Titles in usage	**State of autonomy**	**Classes**	**User interface**
micro aerial vehicle, unmanned aerial vehicle (UAV), uncrewed aerial vehicle, remote inspection technology, remote inspection vehicle, multidrone system, multi-UAV, drone	Supervised autonomy (machine–human); remotely controlled	A1 (<250 g to <500 g): fly over people but not assemblies; A2 (<2 kg): fly close to people; A3 (<25 kg): fly far from people; CO (<250 g): can fly in subcategory A1 and A3; C1 (<900 g): can fly in subcategory A1 and A3; C2 (<4 kg): can fly in subcategory A1 and A2; C3 and C4 (<25 kg): can fly in subcategory A3). * **Note:** European Union Aviation Safety requires completion of training and passing of exams defined by the national authority for operating drones under category A1, A2, A3, C1, C2, C3, and C4 [53].	High-performance data processing task can be achieved using computer graphical user interface (GUI) or natural user interface (NUI). GUI comprises a control panel, parameter viewer, camera viewer, and requested behaviors viewer. On the other hand, NUI types include visual body interaction, visual marker interaction, hand gesture interaction, and speech command interaction [54]
Crawlers: Climbers on Structures (Predominantly Above Water):			
Titles in usage	**State of autonomy**	**Classes**	**User interface**
Magnetic crawler, vertical surface magnetic crawler robot, remote inspection technology, remote inspection vehicle, magnetic utility crawler, magnetic crawler solutions, crawler	Supervised autonomy (machine–human); remotely controlled	Vacuum adhesion type: pneumatic climbing robot, articulated-limb crawler using rubber suction cups; magnetic adhesion type: electromagnetic wheeled crawlers; propeller type: see [55]	High-performance data processing task can be achieved using a 3D tracking concept and robot operating system. Specific location information is transmitted by tracking unit and visuals are communicated by built-in cameras [56].
ROVs: Underwater (Liquid Environment):			
Titles in usage	**State of autonomy**	**Classes**	**User interface**
ROV, remotely operated underwater vehicle submersible, autonomous underwater vehicle, remote inspection technology, remote inspection vehicle	Supervised autonomy, supervised autonomy (machine–human), remotely controlled using cable and tether	Observation class (up to 100 kg, 300 m typical depth rating): DC voltage, 110/220 V DC/AC voltage); mid-sized class (from 100 kg to 1000 kg, 2000 m typical depth rating): 440/480 V, AC voltage; work-class (>1000 kg, >3000 m typical depth rating): 440/480V [57].	The electronic architecture is fitted with computers—one is located at the surface and is connected to a monitor through which the user can view images taken by the camera and other sensors, and the second is located in the ROV's electronic contained. Both computers control the thruster and the underwater arm to read and process the sensor signal whereby all tasks are performed using data acquisition cards and National Instruments hardware. Communication is materialized via Ethernet protocol [58].

Sources: Adapted by authors from [53–58].

Figure 1. *Cont.*

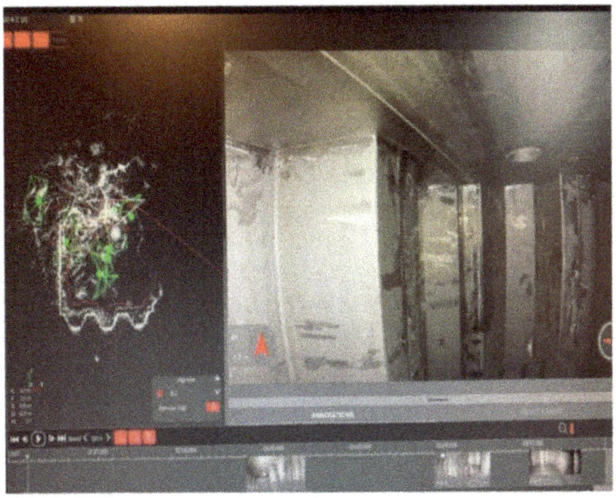

(g)

Figure 1. Overview of the operational sequences considering a drone as a case-study: a typical RAS calibration situation (**a**) and supervised-control during classification survey (**b**); a drone before "accessing" an enclosed space (**c**); drone during survey and inspection (**d**); visual image captured through drone (**left**) (**e**,**f**); data acquisition (**g**). Data can be live-streamed through video conference to facilitate remote surveys if required.

3.4. International Maritime Data Governance Standards

The data obtained and used via RAS clearly belong to the non-personal data domain, given that they concern a vessel and not a natural person. Usage of RAS for gathering non-personal data in shipping, as promulgated in the texts of IACS, stems from two main objectives: (1) to inspect the biofouling status of a ship and (2) to address elevated risks in other structural areas that could result in safety and environmental concerns [59,60]. For statutory surveys, the guidelines in IMO's "Guidelines for the Control and Management of Ship's Biofouling to Minimize the Transfer of Invasive Aquatic Species" issued in 2011 provides two options for conducting in-water inspection: human divers and remotely operated vehicles [60]. In other words, conducting inspection via ROV, i.e., a form of remote inspection technology (RIT), is acceptable and could lead to the issuance of statutory certificates by flag states [13].

In this discussion, the role of classification societies cannot be ignored. More than fifty individual classification society members have been quite influential actors within the shipping industry as far as the history of global maritime trade is concerned. Consequently, classification societies play a crucial role in meeting standards, with the focus recently shifting towards goal-based standards (GBS) through the development of standard rules and regulations in direct conformity with United Nations Convention on the Law of the Sea of 1982 (UNCLOS). It is not a coincidence that according to the official IMO website:

> "Goal-based standards (GBS) are high-level standards and procedures that are to be met through regulations, rules and standards for ships. GBS are comprised of at least one goal, functional requirement(s) associated with that goal, and verification of conformity that rules/regulations meet the functional requirements including goals. In order to meet the goals and functional requirements, classification societies acting as recognized organizations (ROs) and/or national Administrations will develop rules and regulations accordingly. These detailed requirements become a part of a GBS framework when they have been verified, by independent auditors and/or appropriate IMO organs, as conforming to the GBS".

The title under which classification societies operate in the process of developing GBS is known as Recognized Organizations (RO), a title spread across a significant number of

IMO conventions covering environmental and safety facets. Of the many international organizations that contribute to the creation of global maritime technology-based regulatory standards on classification and statutory rules, the IACS is viewed as the leading non-profit international body. Although the International Organization for Standardization has since 1947 has played the traditional role of developing technological standards for robotics and robotic products, the IACS's parallel efforts to date is observed as being the leader in the creation of a techno-regulatory environment for the maritime industry. In the evolution continuum, it is noted that the IACS has come a long way, in so far as its promulgated standards cover more than 90% of the global cargo carrying tonnage. In short, the IACS governs all policy matters that govern the usage of service robotics in the maritime and ocean industry [60]. Some scholars such as Lindøe and Baram purported that standard(s) are private industry driven and serve as an external non-governmental force that supports the national regulatory regime [61] (see Figure 2). Standards containing classification rules developed by the IACS are certainly no exception [60]. Classification surveys following classification rules lead to the issuance of classification certificates, which is an attestation of compliance [60].

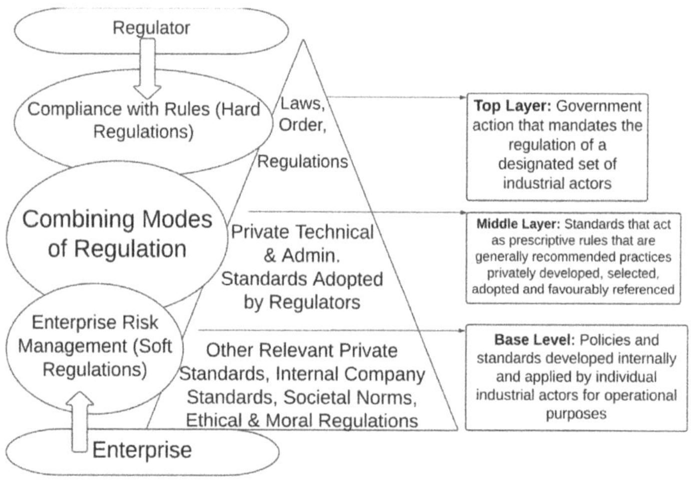

Figure 2. Standards in the regulatory governance regime. Source: Adapted by authors from [61].

Currently, the state-of-the-art survey and certification provisions reside among the various Unified Requirements (UR) developed by the IACS, such as UR Z3, UR Z7.1, UR Z7.2, UR Z10.1, and UR Z10.2. Notwithstanding the plethora of rules and requirements, the IACS has duly considered the role that these emerging technologies could play in the survey process through supervised autonomy. IACS UR Z17, titled "Procedural Requirements for Service Suppliers", creates a regulatory regime that permits the usage of RAS within the ambit of classification survey and is a strong reminder of the IACS's progressive and forward-thinking nature in this technological savvy era. Among the many important provisions, "control of data" as found in s. 5.2.6 of IACS UR Z17 is pertinent to the current discussion on data governance [62]. S. 5.2.6 provides that:

> "When computers are used for the acquisition, processing, recording, reporting, storage, measurement assessment and monitoring of data, the ability of computer software to satisfy the intended application shall be documented and confirmed by the service supplier. This shall be undertaken prior to initial use and reconfirmed as necessary." [62]

Demonstrating adequate control over data as noted in the above section is a precondition to the authorization and subsequent certification of service suppliers by the concerned classification society. Under IACS UR Z17, RAS are permissible pursuant to Sections 3 and 16 of Annex I [62]. The former section applies to firms or service providers that have the capacity to carry out in-water survey on ships (and mobile offshore units) with the help of remotely operated vehicle (ROVs). The latter section applies to firms or service suppliers using RIT (ranging from UAV, ROVs, climbers, unmanned robot arm, drones, etc.) as an alternative to close-up surveys. A close-up survey is an examination "where the details of structural components are within the close visual inspection range of the surveyor" and is an integral formality satisfied, if required, during annual, intermediate and special (or renewal) surveys for detecting fractures, buckling, substantial corrosion, and other types of structural deterioration [63].

A further investigation into the two relevant Annexes of IACS UR Z17 reveals that a number of actors are associated with what could be termed as the data governance and management regime. The stakeholders called upon to remain engaged at various stages include the concerned classification society, flag state administration, as well as service supplier including supervisor, operator, and surveyor. While the classification society and flag state administration deals with approval and certification aspects, the core actors of the service robotic data regime involve the service supplier, who is under obligation to have documented data collection and storage procedures and guidelines in place with the verification of data acquisition tasks, which remain at the discretion of the attending surveyor. All in all, the supplier bears the main responsibilities of ensuring that platforms with data capture/collection/recording devices are readily available and of observing documented procedures that stipulate requirements for location attribution, validation, and storage of data [62].

3.5. Comparative Insight: RAS Data Governance Best Practices

The authors conducted a cross comparative examination with a view of reviewing data governance procedural requirements developed by selected individual member societies (see Appendix A, "Selection Criteria: Class Society Primary Sources on Drones, Crawlers and ROV"). Respondents (including service suppliers, selected classification societies, RAS experts) were also interviewed to confirm that the techno-regulatory provisions were interpreted appropriately. It is noteworthy that in the conduct of the proposed analysis, the IACS UR Z17 data requirement (examined previously) serves as the international evaluative standard against which individual member society rules were benchmarked. Existing rules in the form of best practices aimed at service providers are enshrined by nine individual member societies, i.e., Lloyds Register (LR), Bureau Veritas (BV), Det Norske Veritas (DNV), Registro Italiano Navale (RINA), American Bureau of Shipping (ABS), Russian Maritime Register of Shipping (RS), China Classification Society (CCS), Korean Register (KR), Nippon Kaiji Kyokai (NK); these were the principal materials that were examined out of the twelve IACS member societies (see Appendix A, "Selection Criteria: Class Society Primary Sources on Drones, Crawlers and ROV"). The rationale behind the above selection was guided by the objective of satisfying adequate regional coverage, with LR, BV, DNV, and RINA covering the European Union landscape; ABS representing the Americas; RS covering the Russian region in their own rights; and CCS, KR and NK covering the Asian region. All collected materials were incisively examined through the lens of unique developments.

The results of the cross-comparative examination revealed that while member societies acknowledge IACS UR Z17 as the foundation of respective rules developed in silos, not all selected member societies have explicit provisions on data, let alone innovative out-of-the-box rules. On a positive notion, the term remote inspection technology (RIT) is in principal usage in all classification society guidance documents. We also confirmed that the RIT referred to within guidance documents is synonymous with the term RAS.

In the quest for extracting unique developments, it is apparent that the document titled "Approval of Service Supplier Scheme" captures DNV's diligent effort in regulating the subject matter [64]. Developed in 2016, the above document covers "data storage" rules in s. 16.1.4 of UR Z17's Appendix A. The section covers data storage in the context of RIT-survey reporting and furthers the requirement that all data files should be named after the structure surveyed, and should be stored by the service supplier and readily available at request from DNV for a duration of 5 years [64]. It was also noted that the said provision directly covers the "data storage and infrastructure" decision domain of the data governance framework.

A different approach was observed in the requirements tabled by LR. In the document titled "Procedures for Approval of Service Suppliers", LR upholds the general requirement on "data control" in s. 1.3.7 as found in IACS UR Z17 [65]. The same observation applies to RINA, given that the document titled "Rules for the Certification of Service Suppliers" follows IACS UR Z17 verbatim. LR has nevertheless developed separate guidelines for RIT as well as for unmanned aircraft systems. In terms of RIT, the document titled "Remote Inspection Technique Systems (RITS) Assessment Standard for use on LR Class Surveys of Steel Structure" covers data calibration and analytics, whereas the document titled "Guidance Notes for Inspection using Unmanned Aircraft Systems" provides niche guidance on drone operational as well as data capture and treatment considerations [66,67]. Although the latter document only covers a single RIT, i.e., drones, noteworthy provisions on data can be found in s. 8 entitled "Inspection Data" [67]. Axiomatic from the title, s. 8 branches out into three distinct data-relevant recommendations, with the most important one being s. 8.3, which highlights considerations on data security—an explicit decision domain category on the overarching data governance framework, as discussed earlier [67]. S. 8.3 stresses having appropriate "data security principles, standards and methods" with a view of ensuring that all data acquired receive security and protection against "manipulation or unwanted distribution" [67]. Within the texts of s. 8.3, express reference to ISO/IEC 15408 (on Information Technology–Security Techniques–Evaluation Criteria for IT Security) and ISO/IEC 27001: 2013 (on Information Technology–Security Techniques–Information Security Management Systems) has been provided.

On the American front, best practices are offered in the texts of "Guidance Notes on the Use of Remote Inspection Technologies" developed by the ABS in 2019 [17]. As stated in its "Foreword" note, the document is a holistic approach to governing UAVs, ROVs and robotic crawlers, taking into account rules and requirements as found in IACS Recommendations 42 and 76 and IACS UR Z17 [17]. A remarkable feature of the above document is the manner in which it governs remote inspection vehicle (RIV) post-operation data review and processing tasks, where RIV here is synonymous with both RIT and RAS. Sections 4.9 and 4.11 import detailed requirements that are absent in IACS UR Z17, and prescribe the to-dos in the stages of pre-inspection, during inspection, and post-inspection, touching upon all elements from the data decision domain directly or indirectly. It is important to note that while there is observed adequate emphasis on "data security policies and procedures" in Section 4.11.1.1(h), the paragraph notes that those policies and procedures are to be developed and maintained by the concerned service provider [17]. Whether or not ISO/IEC 15408 and ISO/IEC 27001: 2013 would have been appropriate references is a matter that requires further consideration by concerned members of the ABS.

Among the Asian counterparts, CCS utilizes two distinct documents: one on remote surveys and the other one on the usage of UAVs [68,69]. Of the two, the latter contains provisions on data and information in Section 3, which covers data acquisition, data processing, and data security aspects in a fashion similar to the ABS Guidance Note [68]. Indeed, Section 3 touches on the three stages of inspection (as mentioned above) and prescribes adherence with "statutory or regulatory requirements, company regulations and contractual agreements requirements" [68]. As for NK, the "Rules for Approval of Manufacturers and Service Suppliers" issued in 2020 incorporates IACS UR Z17 rule to the letter in Section 1.4.1, and hence the focus remains confined to the "data control" aspect [70]. The same is observed in the case of KR when observing the texts of Section 5(2)(f) (Annexes 1–11) as found in the document titled "Guidance Relating to the Rules for the Classification of Steel Ships" [71].

Finally, a thorough investigation into the Annexes of the "Guidelines on Technical Supervision of Ships in Service" published by the RS provided interesting insight into the domain of inspection using RIT, inter alia [72]. Unfortunately, neither the general provisions nor the main Annexes of the 2019 Guidelines prescribes data considerations for ship owners and service suppliers, other than prioritizing the basic conditions and procedures for the conduct of such operations.

3.6. Hovering over RAS Governance Gray Areas

The importance of good data governance is internationally recognized. Regions from all across the world have endeavored to regulate data through the adoption of protection and privacy legislation. According to the United Nations Conference on Trade and Development (UNCTAD), a total of 204 countries currently have either national "legislation" or national "draft legislation" (184 countries enacted legislation; 20 countries tabled draft legislation) in place [73]. In other words, statistics show that 66% countries developed legislation, 10% countries developed draft legislation, 19% countries have no legislation, and 5% countries are yet to provide information on the subject matter [73]. A cursory look at the titles of existing national legislation indicated that all status quo efforts remain within the parameters of personal data or personal information protection.

National emphasis to protect personal data is justified. Regardless of the ensuing tension created by the requirements, the big picture renders the international influence or "global reach" of GDPR as the substantiated ground on which this justification rests [74]. As for holistic efforts to safeguard data covering both personal and non-personal categories, six countries (Russia, USA, Australia, Norway, China, and the Netherlands) to date have managed to device cyber strategies [75]. With this in view, the question then is, what progress have the international actors made in the maritime sector?

In an effort to answer the preceding question, it becomes clear that emphasis on both personal and non-personal data security is omnipresent at the international level. Markedly, in the context of non-personal data, "cyber security" is the current buzzword in both the maritime and ocean domain [76]. Cyber security breaches through electronic virus and malware have disrupted commerce on more than one occasion, exposing the vulnerable face of the shipping industry [76]. Even the IMO mainframe has not been spared as it has already encountered a cyber-attack in 2020.

In retrospect, technology applications such as the automatic identification system (AIS), the global positioning system (GPS), the satellite-based long-range identification and tracking system (LRITS), as well as the vessel monitoring system (VMS), while collectively supporting the notion of safety and environmental protection, have also paved the way for discussions on illegal electronic interferences and disruptions, in turn prompting the feasibility of the above systems to be questioned. The inadequacy of the current international regime has called for the need to undertake further action by IMO Member States (MS) that adopted IMO Resolution MSC.428(98) titled "Maritime Cyber Risk Management in Safety Management Systems" in 2017. Bearing in mind the recommendations on maritime cyber risk management embedded within the IMO's "Guidelines on Maritime Cyber Risk Management", the 2017 Guidelines require the incorporation of safety management systems in a company's Document of Compliance by 1 January 2021 (International Maritime Organization, 2017). The 2017 Guidelines are apparently aligned with the overall objectives of International Ship and Port Facility Security (ISPS) Code, developed at the IMO under the International Safety of Life at Sea Convention as well as the International Management Code for the Safe Operation of Ships and for Pollution Prevention (ISM Code). Awareness remains at the heart of all international initiatives. However, we assert that awareness while serving as the stepping stone towards good data governance nevertheless requires in-tandem acknowledgment of a number of emerging challenges. The aforementioned statement thus begs the question: what are the existing challenges?

Self-evident from the above discussion, ongoing efforts are apparently concentrated on developing digital awareness among organizations and companies that are vessel-centric. The IMO's current focus remains on maritime autonomous surface ships (MASS), which leaves the bulk of work on RIT regulatory standard development tasks at the hands of the IACS (Johansson, 2021). Although the IACS is closely tied to the IMO through a Memorandum of Agreement (MoA), there are currently no hints that international cooperation on the subject matter will be underway any time soon.

It is axiomatic that the current maritime service robotics data requirements are driven by IACS UR Z17. Prima facie, s. 5.2.6 is terse, rendering the IACS standards insufficient, unsettled, and incomplete. Data security and data storage considerations are overlooked, which are likely to segue into more complex challenges in the future once service robotics are in mass deployment after they are met with acceptance by all flag states and classification societies. This bleak situation is further exacerbated by the lack of a horizontal initiative by member societies to flesh out a much-needed data governance framework. In addition, the existing initiatives are disharmonized, given that there are no restraints that inhibit individual member societies integral to the IACS's "big 12" or the societies outside of the "big 12" from crafting their own class rules [60]. As was observed in the comparative insight segment (previously discussed), this practice is already constant. Some individual member states have promulgated standards that touch upon all the major issues pertinent to the data governance framework, and individual progress is commendable. The nature and scope of those standards are indeed commendable, but it is difficult to single out a document from any member society as containing the data governance "gold standards" for maritime RAS. Again, the member societies that have provided guidance on the subject matter do not cover all types of RITs. This in turn raises the question, do individual types of RIT deployed in vessel inspection and survey require separate attention?

The answer to the question posed above requires an assessment of two different strands of analysis: firstly, the difference in data generated by individual RITs, and secondly and distinctly, the sufficiency of a single ambient data governance framework that ties all important decision domain threads.

With a view of assessing the first strand, the end-product of RITs requires a separate focus. What is observed is that the current permissible limit of RIT deployment is limited to a close-up survey (and thickness measurement where required) of ships' structures and in-water survey (in lieu of docking survey) of ships' underwater structures. Despite the various types of RITs that are available to complete the tasks that remain within the above regulatory limitation, the principal objective is to acquire data for verification and validation by the principal surveyor. RITs have built-in image sensors that transform photons into electrical signals that are then viewed on high-definition display screens, recorded and analyzed by operators engaged in commercial inspection activities. Despite the various types of RITs, the principal objective of all technological deployment remains unchanged. Data, whether generated from a drone, magnetic crawler, or an ROV, still need to adhere to the IACS UR Z17 prescribed form, i.e., high-quality video images and still images, and therefore the format of the end-product will not vary.

Insight into the second strand unveils that a common data governance framework is sufficient to cover all types of RIT relevant actions. Examples of this are seen in the ABS's 2019 Guidance document, which offers a functional and consolidated approach. A close look at its s. 4.9 and s. 4.11 confirms that all "data review" and "data post-processing" provisions are one-size-fits-all and apply to the three ABS-approved remote inspection vehicles (RIVs), i.e., drones, crawlers, and ROVs. Whether or not new technologies with hybrid applications emerge in the future and require separate data governance framework is still a "wait and see" game.

In terms of data preservation, a topic that also needs attention is the duration aspect. As is tradition, once the data acquisition tasks are complete, the operator is under obligation to provide those videos and still images and data in a format acceptable to the attending surveyor for review and examination, after which survey reports are developed and submitted to the classification society for approval and certification. The operational procedures are shepherded by a contractual agreement entered into between ship owners and service providers or ship owners and a classification society. What is apparent is that the IACS, as well as a majority of the individual classification societies that belong to the "big 12", are rather silent on post data acquisition steps, which might indirectly affect the data storage and security dimension. The current practice gives copyright ownership to the service providers, which is also coupled with the right to retain data for a limited duration within which data need to be communicated subject to request from the concerned classification society.

In reality, RITs are programmed to acquire vessel structural information that forms a part of the vessel-history. The vessel itself is a business asset, and from that standpoint, adequate protection should be given to safeguard the information so gathered in the operational process. While charter parties are only interested in assurance of vessel seaworthiness, should the data containing structural defects fall into the wrong hands, unforeseen negative effects could ensue. Shipping is a competitive industry, and that is why asset-related information should be treated with utmost confidentiality. Although not a contentious issue at this point, nonetheless a number of topics require further clarification: Who should retain the copyright ownership of data gathered from RITs? What is the duration of preservation of data and image from close-up and in-water surveys? Should there be any safeguard mechanisms for service providers against third-party liability? These are a few outstanding matters that require consideration in order to create a level playing field for all stakeholders involved in the RIT business model.

4. Discussion Based on Priori Synthesis: Ways Forward

The findings from BUGWRIGHT2 identified a set of twenty-six action items that comprise the international regulatory blueprint (see Table 2). The blueprint accommodates cardinal elements while bearing in mind the mandate of relevant international actors with law-making, technical, and operational capabilities when maneuvering in the so-called business model landscape. All elements address the inconsistencies and barriers that have the potential to impede the successful deployment of emerging service robotic technologies for taking evidence-based decisions in the process of vessel inspection, survey, and maintenance. In exploring niche areas, the propositions tabled in eight out of the twenty-seven items of the blueprint is predicated on the view that data are the nucleus around which the business-model rotates. Therefore, the role of the all actors involved in data acquisition, data processing and analysis, and data validation requires acknowledgment followed by appropriate review and revision.

Table 2. Overview of the regulatory blueprint action items for harmonization of international arrangements (synthesis derived from BUGWRIGHT2 Deliverable 1.4.1).

Underlying Element	Action Item
Interorganizational consultation	**Action Item 1 in conjunction with Action Item 2:** Creation of a forum to take part in revisions and reforms undertaken by IACS in relation to Unified Requirements.
Categorization based on capacity	**Action Item 3:** Classify based on capacity and determine whether MAVs, AUVs, and crawlers fall under the scope of "mobile robots".
Standalone definitions	**Action Item 4:** Consider standalone definitions (in the following manner) for MAVs, AUVs, and crawlers rather than referring to all technologies under the overarching term "remote inspection technologies".
Classification pursuant to degree of autonomy	**Action Item 5:** In order for procedural rules and requirements to keep pace with technological innovation (towards full autonomy), service robots require a form of categorization along the lines of degree of autonomy. A potential way forward is to follow closely the degrees rendered to vessels and how the different stages were set by IMO's MASS (as the first step in scoping exercise).
Operational limitations and conditions for service robotics	**Action Item 6:** Consider operational limitations for MAVs, AUVs, and crawlers that help ensure effective completion of survey process. **Action Item 7:** Consider pre-operation, in-operation, and post-operation conditions for service robots.

Table 2. *Cont.*

Underlying Element	Action Item
Expand existing provisions on "alterations"	**Action Item 8:** Revise provisions in relation to survey and inspection planning, and consider all potential risk assessment options. **Action Item 9:** Indicate in detail the procedures in cases where the service supplier alters the certified system, which in turn affects the quality system. **Action Item 10:** Consider revision to include provisions on certification of multisite organizations under the rules concerning certification. **Action item 11:** Consider revision of existing provisions related to survey procedures) to include a client service system (CSM), live-streaming during remote survey, and real-time collection of survey process information, solutions in case of problem with live-streaming, recording of process and conclusion in the ship log, and conditions for certification. **Action item 12:** Consider technological platforms for facilitation collection and delivery of survey-related information such as computers, intelligent remote glasses, and digital cameras for live-streaming purposes. **Action Item 13**: Consider using more a factual and objective post-reporting system.
Safety management system	**Action Item 14:** Consider incorporating valuable provisions related to a safety management system, with explicit reference to safety policy, safety risk management, safety assurance, and safety promotion.
Liability clause	**Action Item 15:** Consider incorporating a liability clause in UR Z17 for maintaining third-party liability insurance in case of accidents or deaths.
Data governance and management	**Action Item 16:** Hull inspection data should be kept confidential as this may constitute a trade secret for the shipowner. Legitimate practices should be in place for the collection, storage, and use of unpublished data of economic significance. For overcoming barriers to data governance and data management, explicit provisions are needed in the form of a contract that specifies the allocation of responsibilities and roles for the ownership, storage, security, and sharing of information between service suppliers, classification societies, and shipowners. Sound data governance principles are essential to help minimize risks and keep external cyber securities threats out of their networks. The contract should be signed during the planning stage of hull inspection. **Action Item 17:** IACS guidelines should elaborate on their provisions about data management from the use of remote techniques. **Action Item 18:** Data ownership, which is one of the most critical parts of the data governance process, defines the rightful owner of the data elements, sharing policy, and access rights to third parties granted by the data owner. During the planning stage of hull inspection, a clear understanding between service suppliers, classification societies, and ship owners/managers should be maintained about data ownership. **Action Item 19:** Digital data preservation gives reliable protection to information and systems needed to ensure the long-term usability of data and metadata. Clear allocation of responsibility should be given to the party that is responsible for data and image preservation. **Action Item 20:** Distributed data between the different stakeholders intensifies data security efforts between participants in the data process. Cloud environments encounter increased security threats due to inadequate access management and system vulnerabilities. Measures should be in place for the security and confidentiality of remote inspection technology data by all the relevant stakeholders to ensure a sound data governance process. **Action Item 21:** During the planning stage of hull inspection, it should be specified how the data are shared between the different stakeholders to ensure secure data transfer between data owners and users. Provisions should exist about the sharing of data in the formal agreement. A secure industry platform could be utilized for secure data transfer between data owners and users, when saving and sharing the video stream from the remote survey. **Action Item 22:** The current copyright regime does not protect computer-generated data; thus, explicit provisions in the contract should be made to safeguard computer-generated works. **Action Item 23:** Potential liability issues that stem from the use of data should be underlined in the contract. Input material supplied by the asset owner to the service supplier before the hull inspection (i.e., images, drawings, and designs) should not infringe the copyright or other rights of a third party.

Table 2. *Cont.*

Underlying Element	Action Item
Harmonizing statutory and class rules with reference to close-up survey	**Action Item 24:** Consider aligning the definition of a "close-up survey" found in IMO's Enhanced Survey Programme, given that the current definition of close-up survey is inadequate as the IACS created the possibility to use RITs for remote inspection, allowing the surveyor to conduct close-up surveys through sensors.
Controlling variety for optimum quality performance by regulating technical and operational standards	**Action Item 25:** Develop a methodology to establish standards based on product categorization with the aim of reducing variety to identify the best product from all categories of RITs.
Creating a remote inspection technology "trustworthy ecosystem"	**Action Item 26:** There should not only be efforts to market quality products so that smooth integration is possible but also efforts to create lawful, ethical, and robust service robots that can render end-users trust in the products deployed for survey and maintenance tasks.

The first data-themed action item furthered in the regulatory blueprint is a recommendation calling for the security and protection of all data acquired in relation to the structural elements of a vessel. The recommendation tied to the first action item anticipates dialogue and discussion among ship owners and all other actors noted within IACS rules and requirements. Based on constructive exchanges, it should be confirmed whether legitimate practices should be in place for the collection, storage, and use of unpublished data of economic significance. The platform should support recommendations to integrate sound data governance principles in the context of the data decision domains to safeguard information from cyber security threats—a common phenomenon that could lead to unforeseen damages.

The second and third thematic items are founded on a simple notion: IACS is the principal actor that governs sound data governance system for service robotics. However, a consideration of the requirements implemented by other international organizations is important to address incidental loopholes. LR, for example, has referenced ISO/IEC 15408 (Information Technology–Security Techniques–Valuation Criteria for IT Security) and ISO/IEC 27001:2013 (Information Technology–Security Techniques–Information Security Management Systems) in the procedural document that highlight valuable tools for stakeholders in the development of secure IT systems. This invokes the thought whether IACS, as the lead international organization on the topic, should adhere to the same. Through the second item, the authors assert that harmonization can be achieved through an explicit reference of ISO in both IACS Guidelines No. 42 and IACS UR Z17, especially considering the fact that a majority of the countries are concerned with GDPR and personal data, which leaves non-personal data unprotected and vulnerable.

The third item then focuses on finding a pathway to determine data ownership—an important "detail" that should be considered and integrated into the general provisions of IACS UR Z17 procedural requirements. It is through this third action item recommendation where we submit that there should be a clear understanding of the concept "ownership" among service suppliers, classification societies, and ship owners/managers. Generally speaking, "data ownership" is set by the enterprise's upper echelons and is related to the allocation of responsibilities over data [77]. The ownership decisions encompass a wide range of issues such as domains, data availability, accessibility, and frequency of updates [78]. The outcome of the third action item, according to our views, should help determine whether the fourth action item (i.e., incorporation of a separate section on data security and protection) is required within Section 3 and Section 16 of Annex I of IACS URZ17. We hold the position that clear provisions on data control and data security should exist; otherwise, the data governance framework may suffer. Determining which stakeholder organization should retain data ownership when working within a dynamic business model certainly has positive implications. A quality system under ISO 900 series, referred to in documents developed by individual member societies, prescribes accountability as a part of the engagement process under the quality management principles. Provisions on data control and data protection would certainly help monitor data-centric activities by holding the organization responsible for data control and protection accountable, which is a sound way forward.

Incorporating the much-needed provisions in the texts of IACS UR Z17 would also create the right set of circumstances to review the current contractual practices (see Figure 3). This comes as a timely recommendation and serves as the fifth action item. The lack of adequate coverage of the specifics has paved the opportunity for service suppliers to develop contracts based on convenience and as seen fit. The roles and responsibilities for data ownership, storage, security, and sharing of information remain not catered for to say the least and require also an in-depth review of all private contracts developed by service suppliers. It is important to enquire how data are preserved and whether the service suppliers have adequate and reliable mechanisms in place to ensure the long-term usability of data and metadata gathered from RITs and ROVs.

Connected to the fifth action item is the need to verify whether organizations in the business model, especially service suppliers have a well-documented organizational "information systems security policy" and "backup strategy", both of which comprise the sixth action item, followed by feasible recommendations for consideration. For the former, a viable way forward could be that the entities of the business model implement tools and techniques to prohibit unauthorized access to programs and information resources. Users should have a unique user identifier (through passwords or other authentication mechanisms) to keep track of users for establishing accountability. As for the latter, i.e., a backup strategy, organizations could consider a digital infrastructure containing schedules for routine backup check and fast recovery tools. Again, one should come across ISO/IEC 15408 and ISO/IEC 27001:2013, which articulates general principles that could inspire a panoply of thoughts leading to concrete solutions post-verification and in cases where both security policy and backup strategy is absent.

The seventh listed action item in the regulatory blueprint corresponds to data transfer and secured sharing. Prior to the commencement of the operational phase, the IACS prescribes a planning stage with a number of items that are discussed at length by the stakeholders involved in the business model. We believe that specific discussions should be aimed at the best ways to share data among the stakeholders with a view to ensuring secure data transfer between the "owner of the data" and the others including end-users. Whether or not this item could encapsulate the incorporation of provisions alongside data control and data protection, or whether this should be covered under the formal contract between service suppliers and ship owners remain at the discretion of the international actors. Notwithstanding, we opine that this is an important step because this also concerns data security, and therefore a secured platform for data transfer between data owners and users could be the best way forward when saving and sharing the video stream from the remote survey. In doing so, the party providing access should implement and administer access restrictions to ensure that only authorized individuals have the ability to access or use information resources. The use of the universal serial bus (USB) for data sharing, according to the authors, should be avoided.

The final action item of the data segment relates to liability and contains a precautionary approach in order to avoid legal consequences. Caution should be exercised by the asset owners when sharing "input material" with the service supplier before the survey, e.g., images, drawings, and designs, lest this infringes the copyright or other rights of a third party. In case of infringement, the service supplier should be held unaccountable against any loss, damage, or other claims arising from such violation. On the other hand, service suppliers should be discouraged from using survey data for marketing reasons without the prior approval of the asset owner. It is understood that the specifics are currently absent in IACS UR Z17, paving the opportunity for service suppliers to develop contracts based on convenience and as seen fit. The roles and responsibilities for data ownership, storage, security, and sharing of information remains ambiguous to say the least. A possible way forward, should IACS deem it appropriate, is to mark out ways to strengthen due diligence on the part of service providers when holding the copyright or ownership of data obtained through the usage of RITs or ROVs.

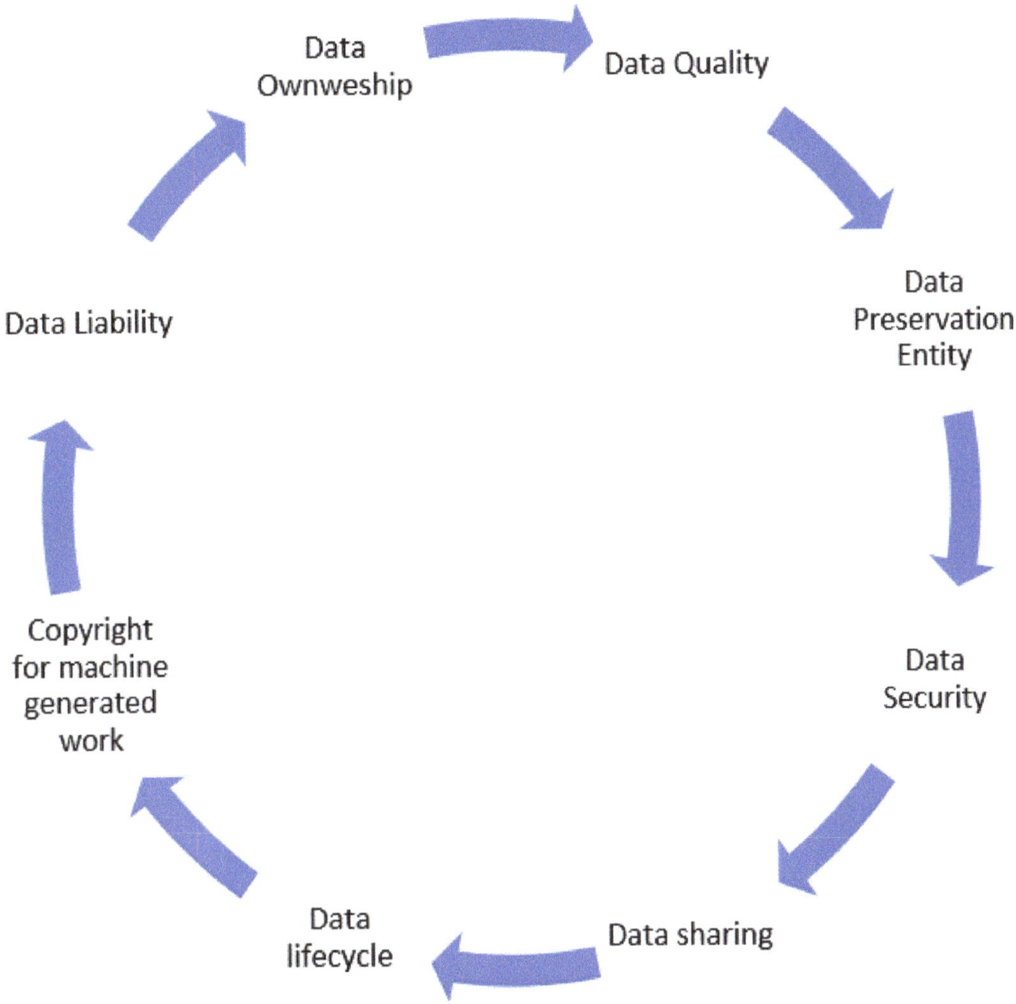

Figure 3. Visual-cycle of the data elements to be included in the contract between service suppliers, classification societies, and asset owners/operators.

5. Conclusions

Approximately 100,000 commercial vessels with a displacement of more than 100 tons navigate the oceans and the seas worldwide [79]. Altogether, there are more than 9000 large ships and more than 4000 very large ships that are above the age of five years [80]. Ship owners need to ensure that these vessels are fit for service, which comes off as a requirement under the UNCLOS [80]. Obtaining certification for all class and statutory surveys provides evidence that all survey and inspection related procedures are complete. However, the procedures to ensure fitness depend on the type of survey (annual surveys need around 1–2 days, intermediate surveys take around 3–4 days, specials surveys go around 1–2 weeks), and there is a need to also factor in post-inspection actions to mitigate defects and deterioration. Maritime RAS could just well be the time savior in a world where shipping is conducted in a just-in-time fashion.

Regardless of the many opportunities that surface with maritime RAS technological applications, a well-implemented check and balance system should continue to operate in tandem with classification and statutory tasks. The topic of maritime non-personal data needs separate attention, broadly owing to the fact that a majority of the world's national strategies are struggling to protect data. Alongside personal data, non-personal data protection issues have already emerged, prompting the adequacy of the current data governance framework to be questioned. Questions are directed at international organizations that have an obligation to ensure that legal provisions aimed at governing technological products are free from imbalance and asymmetry. The success of the science–policy interface for maritime non-personal data requires a review of all the building blocks to ensure that asset related information is protected throughout the entire lifecycle. Generic problems and questions on non-personal data protection are likely to recur in other instances once fully autonomous ships using e-navigation and other forms of autonomous communication to establish contact between and among vessels are in place. This could invoke bring in other questions, such as the feasibility and effectiveness of current big data analytical techniques for an increased rate and volume of data—an issue of significant importance in offshore environments that needs to be considered as the industry moves towards the Internet of Shipping [80–85].

It is also necessary to clearly highlight that there are more desired capabilities in the pipeline, such as autonomous maritime RAS that connect automatically with autonomous vessels through machine-to-machine communication and are guided by the principles of AI. As scientists, engineers, and technologists follow the AI route and usher in the advent of the sixth-generation wireless (6G) communication (i.e., the successor of the fifth-generation wireless) for discharging "unprecedented capacity and latency", there needs to be a solid protection regime to avoid unwanted exposure of data and liability issues [81]. It can be said that 6G communication envisages a tech-savvy environment that has the potential to blur the understanding between human-to-machine and machine-to-machine interactions. We speculate that 6G itself combined with the groundbreaking concept of Internet of Everything (IoE) is likely to be labyrinthine in nature and is expected to contain hidden layers of communication that may require countermeasures to protect the data decision framework as well as the necessary building blocks for smooth user interface. However, until the introduction of 6G and IoE, engineers and policy makers need to establish a settled discourse of data integrity in maritime and ocean domain by observing the current IoRT architecture.

In reality, a sound data governance approach goes beyond just the protection factor simply because the data decision dimension and building blocks secure a "trustworthy" environment. A trustworthy business model entails trust in the product that is likely to have a positive repercussion among the stakeholders that are a part of the business model of non-personal data [82,83]. The current system is characterized as supervised autonomy (see Table 1). Detecting inherent surreptitious vices is a pre-condition in order to help engineers and manufactures of maritime RAS pass the design bottleneck and to ensure market success and subsequent successful integration by flag states and classification societies into mandatory survey processes. The key action as envisaged by authors is to flag out data-related liability provisions in international standards, and when doing so, consider the theoretical constructs of a sound data governance framework before harmonizing the international RAS data management system with other available best practices. Should the action materialize, the benefits are likely to be reaped by the actors involved in the business model, thus creating an environment where each and all are fully aware of the responsibilities and liabilities, whether strict, absolute, or vicarious

Here, collaboration among different international stakeholders of the maritime RAS ecosystem is climacteric. This could include discussions among all international bodies, including organizations that have RAS and maritime RAS mandate, e.g., IMO, IACS, ISO, engineers, policy makers, flag state authorities, manufacturers service suppliers, classification societies, and ship owners and operators to flesh-out solutions on this topic. All in all, what must be hindered is the development of maritime RAS technology-based laws, rules, and requirements in silo. Harmonization has always been at the epicenter of international efforts. Therefore, cultivating international harmonization through optimal solutions that answers the current burning questions leading to a regime that safeguards non-personal data is in order. After all, it is not only that the world contemplates a future where there exists maritime RAS-to-autonomous vessel communication, but also there is a silent aspiration that works to support safe and liability-free maritime RAS-to-autonomous vessel communication. Still much work lies ahead to keep the implications positive in mankind's epic era of combating to establish digital serenity and its unfortunately already inevitable impacts.

Author Contributions: T.M.J., as the WMU Principal Investigator of BUGWRIGHT2 was responsible for developing the main draft, which was based on raw data provided by A.P. Significant revisions were made by D.D. in the development of this article. All authors have read and agreed to the published version of the manuscript.

Funding: This paper is derived from research conducted under the European Union (EU) Horizon 2020 funded project titled Autonomous Robotic Inspection and Maintenance on Ship Hulls (BUGWRIGHT2) under grant agreement No. 871260.

Institutional Review Board Statement: Not applicable.

Informed Consent Statement: Not applicable.

Data Availability Statement: Not applicable.

Acknowledgments: The authors would like to thank Ronán Long and Clive Schofield of the World Maritime University–Sasakawa Global Ocean Institute and the Nippon Foundation for their generous support. The authors are grateful to Anastasios Tsalavoutas and David Knukkel for providing first-hand information incorporated in the section titled "RAS and the Internet of Robotic Things: Through the Prism of Data".

Conflicts of Interest: The authors declare no conflict of interest.

Appendix A. Tabular Amalgamation of Published Literature and Primary Sources Based on Selection Criteria

Selection Criteria: Literature on Data and Information	
Author(s)	**Title (Year of Publication)**
Zins, C.	What is the meaning of "data", "information", and "knowledge"?
Selection Criteria: Literature on Data and Asset	
Author(s)	**Title (Year of Publication)**
Lake, P.; Crowther, P.	Data an Organizational Asset. In *Concise Guide to Databases* (2013)
Selection Criteria: Literature on Big Data	
Author(s)	**Title (Year of Publication)**
Boyd, D.; Crawford, K.	Critical Questions for Big Data: Critical Questions for Big Data (2012)
Zhang, S.; Gao, H.; Yang, L. and Song, J.	Research on big data governance based on actor-network theory and Petri nets. In IEEE 21st International Conference on Computer Supported Cooperative Work in Design (CSCWD) (2017)
Johansson, T.; Long, R.; Dalaklis, D.	The Role of WMU-Sasakawa Global Ocean Institute in the Era of Big Data (2019)

	Selection Criteria: Data Governance
Author(s)	**Title (Year of Publication)**
McGilvray, D.	Data governance: The Missing Link for Data Quality Success. San Francisco Bay Area (2006)
Khatri, V.; Brown, C.V.	Designing data governance. *Communications of the ACM* (2010)
Al-Badi, A; Tarhini, A.; Khan, A.I.	Exploring Big Data Governance Frameworks (2018)
Al-Ruithe, M.; Benkhelifa, E.; Hameed, K.	Data Governance Taxonomy: Cloud versus Non-Cloud. *MDPI, Open Access Journal* (2018)
Al-Ruithe, M.; Benkhelifa, E.; Hameed, K.	A systematic literature review of data governance and cloud data governance. Personal and Ubiquitous Computing (2019)
Benfeldt, O.; Persson, J.S.; Madsen, S;	Data Governance as a Collective Action Problem. *Information Systems Frontier* (2020)
Janssen, M.; Brous, P.; Estevez, E., Barbosa, L. S and Janowski, T.	Data governance: Organizing data for trustworthy Artificial Intelligence (2020)
Chakravorty, R.	Common challenges of data governance (2020)
Thuraisingham, B.	Cloud Governance. In *IEEE 13th International Conference on Cloud Computing (CLOUD)* (2020)
	Selection Criteria: Data Privacy and User Rights
Author(s)	**Title (Year of Publication)**
Mattoo A.; Meltzer, J. P.	International data flows and privacy: the conflict and its resolution. *Journal of International Economic Law* (2018)
Chatzimichali, A.; Chrysostomou, D.	Human-data interaction and user rights at the personal robot era. In 4th International Conference on Robot Ethics and Standards: ICRES 2019, (2019)
	Selection Criteria: Personal and Non-Personal Data
Author(s)	**Title (Year of Publication)**
Finck, M.; Pallas F.	They who must not be identified-distinguishing personal from non-personal data under the GDPR (2020)
Somaini, L.	Regulating the Dynamic Concept of Non-Personal Data in the EU: From Ownership to Portability (2020)
	Selection Criteria: Actor–Network Theory
Author(s)	**Title (Year of Publication)**
Montenegro, L. M.; Bulgacov, S.	Reflections on actor-network theory, governance networks, and strategic outcomes (2004)
Latour, B.	*Reassembling the Social: An Introduction to Actor-Network Theory* (2005)
Dincer, D.	The Act-Shifts Between Humans and Nonhumans in Architecture: A Reading of Bruno Latour's Actor-Network Theory (2020)
	Selection Criteria: Internet of Robotic Things
Author(s)	**Title (Year of Publication)**
Ray, P.P.	Internet of Robotic Things: Concept, Technologies and Challenges (2017)
Yousif, R.	A Practical Approach of an Internet of Robotic Things Platform, Master of Science Thesis, KTH Royal Institute of Technology (2018)
	Selection Criteria: IACS Primary Sources
The International Association of Classification Societies	UR Z17, Procedural Requirements for Service Suppliers (1997)
The International Association of Classification Societies	Guidelines for Surveys, Assessment and Repair of Hull Structure–Corr. 1 (Recommendation 76) (2008)

Selection Criteria: Class Society Primary Sources on Drones, Crawlers, and ROV	
Author(s)	**Title (Year of Publication)**
Lloyds Register	Guidance Notes for Inspection Using Unmanned Aircraft Systems (2016)
Lloyds Register	Remote Inspection Technique Systems (RITS): Assessment Standard for Use on LR Class Surveys of Steel Structure (2018)
American Bureau of Shipping	Guidance Notes on the Use of Remote Inspection (2019)
Bureau Veritas	Approval of Service Suppliers (2020)
China Classification Society	Guidelines for Use of Unmanned Aerial Vehicles (2018)
China Classification Society	Rules for Classification of Sea-going Steel Ships (2015)
Det Norske Veritas	Approval of Service Supplier Scheme (2016)
Korean Register	Rules for the Classification of Steel Ships (2017)
Nippon Kaiji Kyokai	Rules for Approval of Manufacturers and Service Suppliers (2020) (Part 1 Chapter 1)
Registro Italiano Navale	Rules for the Certification of Service Suppliers (2020)
Russian Maritime Register of Shipping	RIT Requirements: Annex 39 (Guidelines for the Use of Remote Inspection Techniques for a Survey of Ships and Marine Structures)
Russian Maritime Register of Shipping	ROV Requirements: Annex I (Procedure for In-water Survey of Ships and Offshore Installations) of Guidelines on Technical Supervision of Ships in Service, Russian Maritime Register of Shipping
Selection Criteria: Data Management	
Authors(s)	**Title (Year of Publication)**
Loshin, D.	*Master data management* (2008)
Vilminko-Heikkinen, P.; Pekkola, S.	Changes in roles, responsibilities and ownership in organizing master data management (2019)
Earley, S.; Henderson, D.; Data Management Association	*DAMA-DMBOK: Data Management Body of Knowledge* (2017)
Brous P.; Janssen M.; Vilminko-Heikkinen R.	Coordinating Decision-Making in Data Management Activities: A Systematic Review of Data Governance Principles (2016)

References

1. Zins, C. What Is the Meaning of "Data", "Information", and "Knowledge"? Available online: https://pciucr.files.wordpress.com/2011/03/what-is-the-meaning-of-data.pdf (accessed on 1 April 2021).
2. Cambridge English Dictionary. Available online: https://dictionary.cambridge.org/dictionary/english/data (accessed on 1 April 2021).
3. Senn, J.A. *Information Systems in Management*, 2nd ed.; Cengage South-Western: Boston, MA, USA, 1982.
4. Clare, C.P.; Loucopoulos, P. *Business Information Systems*; Paradigm: London, UK, 1987.
5. Avison, D.E.; Fitzgerald, G. *Information Systems Development: Methodologies, Techniques and Tools*, 2nd ed.; McGraw-Hill: Maidenhead, UK, 1995.
6. Bughin, J.; Chui, M.; Manyika, J. Clouds, Big Data, and Smart Assets: Ten Tech-Enabled Business Trends to Watch. *Mckinsey Quarterly* **2010**, *56*, 75–86.
7. Lake, P.; Crowther, P. Data an Organizational Asset. In *Concise Guide to Databases*; Springer: London, UK, 2013.
8. Arsana, A.; Rizos, C.; Schofield, C. The Application of GIS in Maritime boundary Delimitation, Innovations in 3D Geo Information Systems. In Proceedings of the First International Workshop on 3D Geoinformation, Kuala Lumpur, Malaysia, 7–8 August 2006. [CrossRef]
9. Graham, C.H.; Elith, J.; Hijmans, R.J.; Guisan, A.; Peterson, A.T.; Loiselle, B.A. Nceas Predicting Species Distributions Working Group. The influence of spatial errors in species occurrence data used in distribution models. *J. Appl. Ecol.* **2008**, *45*, 239–247. [CrossRef]
10. Davenport, T. Submarine Communications Cables and Science: A New Frontier in Ocean Governance? In *Science, Technology and New Challenges to Ocean Law*; Scheiber, H.N., Kraska, J., Kwon, M.-S., Eds.; Brill-Nijhoff: Leiden, The Netherlands, 2015; pp. 209–252, ISBN 9789004299610.

11. Martjin, N.; Hulstjin, J.; De Bruijne, M.; Yao-Hua, T. Determining the Effects of Data Governance on the Performance and Compliance of Enterprises in the Logistics and Retail Sector. In Proceedings of the 14th IFIP Conference on e-Business, e-Services, and e-Society, Delft, The Netherlands, 13–15 October 2015; Janssen, M., Mäntymäki, M., Hidders, J., Klievink, B., Lamersdorf, W., Van Loenen, B., Zuiderwijk, A., Eds.; Springer: Delft, The Netherlands, 2015; pp. 454–466. [CrossRef]
12. Boyd, D.; Crawford, K. Critical Questions for Big Data: Critical Questions for Big Data. *Inf. Commun. Soc.* **2012**, *15*, 662–679. [CrossRef]
13. Johansson, T.; Long, R.; Dalaklis, D. The Role of WMU-Sasakawa Global Ocean Institute in the Era of Big Data. *J. Ocean Technol.* **2019**, *14*, 22–29.
14. Toyota. TMC Announces New Big Data Traffic Information Service. 2013. Available online: https://global.toyota/en/detail/70065 (accessed on 12 March 2021).
15. Dalaklis, D. Unleashing the Power of Data to Optimise Vessel Performance. In Proceedings of the Vessel Optimisation Webinar Week (Virtual Event), London, UK, 8 September 2020. [CrossRef]
16. Bureau Veritas-Seeing Remotely—In Safety: Bureau Veritas Performs First Survey by Drone. Available online: https://marine-offshore.bureauveritas.com/newsroom/seeing-remotely-safety-bureau-veritas-performs-first-survey-drone (accessed on 10 March 2021).
17. American Bureau of Shipping. Guidance Notes on the Use of Remote Inspection Technologies. 2019. Available online: https://ww2.eagle.org/content/dam/eagle/rules-and-guides/current/other/242-gn-remote-inspection-tech/rit-gn-feb19.pdf (accessed on 1 April 2021).
18. Chynoweth, P. Legal Research. In *Advanced Research Methods in the Built Environment*; Knight, A., Ruddock, L., Eds.; Wiley: Blackwell, UK, 2009; pp. 28–38.
19. Al-Ruithe, M.; Benkhelifa, E.; Hameed, K. A systematic literature review of data governance and cloud data governance. *Pers. Ubiquitous Comput.* **2019**, *23*, 839–859. [CrossRef]
20. Benfeldt, O.; Persson, J.S.; Madsen, S. Data Governance as a Collective Action Problem. *Inf. Syst. Front.* **2020**, *22*, 299–313. [CrossRef]
21. Earley, S.; Henderson, D.; Data Management Association. *DAMA-DMBOK: Data Management Body of Knowledge*, 2nd ed.; Technics Publications, LLC: Bradley Beach, NJ, USA, 2017.
22. Alhassan, I.; Sammon, D.; Daly, M. Data governance activities: An analysis of the literature. *J. Decis. Syst.* **2016**, *25*, 64–75. [CrossRef]
23. Ransbotham, S.; Kiron, D.; Analytics as a Source of Business Innovation. MIT Sloan Management Review. 2017. Available online: https://sloanreview.mit.edu/projects/analytics-as-a-source-of-business-innovation/ (accessed on 26 March 2021).
24. Al-Ruithe, M.; Benkhelifa, E.; Hameed, K. Data Governance Taxonomy: Cloud versus Non-Cloud. *Sustainability* **2018**, *10*, 95. [CrossRef]
25. Abraham, R.; Schneider, J.; Vom Brocke, J. Data governance: A conceptual framework, structured review, and research agenda. *Int. J. Inf. Manag.* **2019**, *49*, 424–438. [CrossRef]
26. Mattoo, A.; Meltzer, J.P. International data flows and privacy: The conflict and its resolution. *J. Int. Econ. Law* **2018**, *21*, 769–789. [CrossRef]
27. Chatzimichali, A.; Chrysostomou, D. Human-data interaction and user rights at the personal robot era. In Proceedings of the 4th International Conference on Robot Ethics and Standards: ICRES 2019, London, UK, 20 October 2019. [CrossRef]
28. Finck, M.; Pallas, F. They who must not be identified—Distinguishing personal from non-personal data under the GDPR. *Int. Data Priv. Law* **2020**, *10*, 11–36. [CrossRef]
29. GDPR; European Union. Regulation (EU) 2016/679 of the European Parliament and of the Council of 27 April 2016 on the Protection of Natural Persons with Regard to the Processing of Personal Data and on the Free Movement of Such Data, and Repealing Directive 95/46/EC (General Data Protection Regulation). Available online: https://eur-lex.europa.eu/eli/reg/2016/679/oj (accessed on 1 April 2021).
30. Somaini, L. Regulating the Dynamic Concept of Non-Personal Data in the EU: From Ownership to Portability. *Eur. Data Prot. Law Rev.* **2020**, *6*, 84–93. [CrossRef]
31. European Union. Regulation (EU) 2018/1807 of the European Parliament and of the Council of 14 November 2018 on a Framework for the Free Flow of Non-Personal Data in the European Union European Union's (EU) Regulation 2018/1807 on a Framework for the Free Flow of Non-Personal Data in the European Union. Available online: https://eur-lex.europa.eu/legal-content/EN/TXT/?uri=CELEX%3A32018R1807 (accessed on 1 April 2021).
32. Khatri, V.; Brown, C.V. Designing data governance. *Commun. ACM* **2010**, *53*, 148–152. [CrossRef]
33. Janssen, M.; Brous, P.; Estevez, E.; Barbosa, L.S.; Janowski, T. Data governance: Organizing data for trustworthy Artificial Intelligence. *Gov. Inf. Q.* **2020**, *37*, 1–8. [CrossRef]
34. McGilvray, D. *Data Governance: The Missing Link for Data Quality Success*; Data Management Association Presentation for DAMA Day; Granite Falls Consulting Inc.: San Francisco, CA, USA, 2006.
35. Brous, P.; Janssen, M.; Vilminko-Heikkinen, R. Coordinating Decision-Making in Data Management Activities: A Systematic Review of Data Governance Principles. *Lect. Notes Comput. Sci.* **2016**, 115–125. [CrossRef]
36. Lucas, A. Towards Corporate Data Quality Management. *Port. J. Manag. Stud.* **2010**, *15*, 173–196.

37. Al-Badi, A.; Tarhini, A.; Khan, A.I. Exploring Big Data Governance Frameworks. *Procedia Comput. Sci.* **2018**, *141*, 271–277. [CrossRef]
38. Chakravorty, R. Common challenges of data governance. *J. Secur. Oper. Custody* **2020**, *13*, 23–43.
39. Morabito, V. *Big Data and Analytics Strategic and Organizational Impacts*, 1st ed.; Springer: Switzerland, Zurich, 2005. [CrossRef]
40. Dell Technologies. 2020. Available online: https://www.delltechnologies.com/sv-se/storage/data-storage.htm (accessed on 12 March 2021).
41. Zhang, S.; Gao, H.; Yang, L.; Song, J. Research on big data governance based on actor-network theory and Petri nets. In Proceedings of the IEEE 21st International Conference on Computer Supported Cooperative Work in Design (CSCWD), Wellington, New Zealand, 26–28 April 2017. [CrossRef]
42. Latour, B. *Science in Action: How to Follow Scientists and Engineers through Society*; Harvard University Press: Cambridge, MA, USA, 1988.
43. Latour, B. *Reassembling the Social: An Introduction to Actor-Network Theory*; Oxford University Press: Oxford, UK, 2005.
44. Montenegro, L.M.; Bulgacov, S. Reflections on actor-network theory, governance networks, and strategic outcomes. *Braz. Adm. Rev.* **2014**, *11*, 107–124. [CrossRef]
45. Dincer, D. The Act-Shifts between Humans and Nonhumans in Architecture: A Reading of Bruno Latour's Actor-Network Theory. In *Contemporary Applications of Actor Network Theory*; Williams, I., Ed.; Palgrave Macmillan: Singapore, 2020; pp. 33–50. [CrossRef]
46. Thuraisingham, B. Cloud Governance. In Proceedings of the IEEE 13th International Conference on Cloud Computing (CLOUD), Beijing, China, 18–24 October 2020; pp. 86–90. [CrossRef]
47. Loshin, D. *Master Data Management*, 1st ed.; Kaufmann: Burlington, MA, USA, 2008.
48. International Telecommunication Union. Series Y: Global Information Infrastructure, Internet Protocol Aspects and Next-Generation Networks. 2012. Available online: https://www.itu.int/rec/T-REC-Y.2060-201206-I (accessed on 20 April 2021).
49. Yousif, R. A Practical Approach of an Internet of Robotic Things Platform. Master's Thesis, KTH Royal Institute of Technology, Stockholm, Sweeden, 2018. Available online: https://www.diva-portal.org/smash/get/diva2:1290401/FULLTEXT01.pdf (accessed on 20 April 2020).
50. Ray, P.P. Internet of Robotic Things: Concept, Technologies and Challenges. *IEEE Access* **2017**, *4*, 9489–9500. [CrossRef]
51. Dalaklis, D. Exploring the issue of technology trends in the "era of digitalisation". In Proceedings of the World Maritime Day Parallel Event (IMO 70: Our Heritage—Better Shipping for a Better Future), Szczecin, Poland, 13 June 2018. [CrossRef]
52. Tsaganos, G.; Nikitakos, N.; Dalaklis, D.; Ölcer, A.I.; Papachristos, D. Machine Learning Algorithms in Shipping: Improving Engine Fault Detection and Diagnosis via Ensemble Methods. *WMU J. Marit. Aff.* **2020**, *19*, 51–72. [CrossRef]
53. Open Category—Civil Drone. Official Homepage of European Union Aviation Safety Agency. Available online: https://www.google.com/search?q=EASA&oq=EASA+&aqs=chrome..69i57j014j69i60j69i65j69i60.1240j0j4&sourceid=chrome&ie=UTF-8 (accessed on 18 April 2020).
54. Fernandez, R.A.S.; Sanchez-Lopez, J.L.; Sampedro, C.; Bavle, H.; Molina, M.; Campoy, P. Natural User Interfaces for Human-drone Multi-modal Interaction. In Proceedings of the 2016 International Conference on Unmanned Aircraft Systems (ICUAS), Arlington, VA, USA, 7–10 June 2016. [CrossRef]
55. Sattar, T.P. Chapter 3.4: Wall Climbing Crawlers for Nondestructive Testing. In *Topics on Nondestructive Evaluation (TONE): Volume 4: Automation, Miniature Robotics and Sensors for Nondestructive Evaluation and Testing*; Bar-Cohen, Y., Ed.; ASNT Publications: Columbus, OH, USA, 2000; Volume 4, pp. 77–100.
56. Fondahl, K.; Eich, M.; Wollenberg, J.; Kirchner, F. A Magnetic Climbing Robot for Marine Inspection Services. In Proceedings of the 11th International Conference on Computer and IT Applications in the Maritime Industries (COMPIT '12), Liege, Belgium, 16–18 April 2012.
57. Remouit, F. Underwater Electrical Connections and Remotely Operated Vehicles. Licensiate Thesis, Uppsala Universitet, Uppsala, Sweden, 2016.
58. Garcia-Valdovinos, L.G.; Salgado-Jimenez, T.; Bandala-Sanchez, M.; Nava-Balanzar, L.; Hernandez-Alvarado, R.; Cruz-Ledesma, J.A. Modelling, Design and Robust Control of a Remotely Operated Underwater Vehicle. *Int. J. Adv. Robot. Syst.* **2014**, *11*, 1. [CrossRef]
59. International Association of Classification Societies. Classification Societies. What, Why and How? Available online: http://www.iacs.org.uk/media/3785/iacs-class-what-why-how.pdf (accessed on 1 April 2021).
60. Johansson, T. International Standards for Hull Inspection and Maintenance of Robotics and Autonomous. In *Emerging Technology and the Law of the Sea*; Kraska, J., Part, Y.-K., Eds.; Cambridge University Press: Cambridge, UK, 2021; in press.
61. Lindøe, P.H.; Michael, S.; Baram, M.S. The Role of Standards in Hard and Soft Approaches. In *Standardization and Risk Governance*; Odd Einar Olsen, O.E., Juhl, K., Lindøe, P.H., Engen, O.A., Eds.; Routledge: London, UK, 2020; pp. 235–254.
62. The International Association of Classification Societies (IACS). UR Z17, Procedural Requirements for Service Suppliers. 1997. Available online: http://www.iacs.org.uk/publications/unified-requirements/ur-z/?page=2 (accessed on 10 March 2021).
63. International Association of Classification Societies. Classification Societies (IACS). Guidelines for Surveys, Assessment and Repair of Hull Structure (1994)—Corr. 1 (Recommendation 76). September 2008. Available online: http://www.iacs.org.uk/publications/recommendations/61-80/rec-76-rev2-corr1-cln/ (accessed on 10 March 2021).
64. DNV, Det Norske Veritas-Germanischer Lloyd. Approval of Service Supplier Scheme. 2019. Available online: http://rules.dnvgl.com/docs/pdf/DNVGL/CP/2019-02/DNVGL-CP-0484.pdf (accessed on 31 March 2021).

65. Lloyds Register. Procedures for Approval of Service Suppliers. 2020. Available online: https://www.lr.org/en/approval-of-service-suppliers/ (accessed on 10 March 2021).
66. Lloyds Register. Remote Inspection Technique Systems (RITS) Assessment Standard for Use on LR Class Surveys of Steel Structure. 2018. Available online: https://www.cdinfo.lr.org/information/Documents/Approvals/Remote%20Inspection%20Techniques/Remote_Inspection_Technique_Systems_RITS_Assessment_Standard_for_use_on_LR_Class_Surveys_of_Steel_Structure.pdf (accessed on 20 March 2021).
67. Lloyds Register. Guidance Notes for Inspection Using Unmanned Aircraft Systems. 2016. Available online: https://www.lr.org/en/latest-news/lloyds-register-releases-guidance-notes-for-inspection-with-uas/ (accessed on 10 March 2021).
68. China Classification Society. Guidelines for Use of Unmanned Aerial Vehicles for Surveys. 2018. Available online: https://www.ccs.org.cn/ccswzen/articleDetail?id=201910000000003817 (accessed on 2 April 2021).
69. China Classification Society Guidelines for Ship Remote Surveys. 2019. Available online: https://www.ccs.org.cn/ccswzen/articleDetail?id=201910000000003625 (accessed on 2 April 2021).
70. Nippon Kaiji Kyokai. Rules for Approval of Manufacturers and Service Suppliers. 2020. Available online: https://www.classnk.or.jp/hp/pdf/Rules_Guidance/publish/371_servicesuppliers_e_2020.pdf (accessed on 10 March 2021).
71. Korean Register. Guidance Relating to the Rules for the Classification of Steel Ships. 2017. Available online: https://eclass.krs.co.kr/KRRules/KRRules2017/KRRulesE.html#%EC%84%A0%EA%B8%89%20%EB%B0%8F%20%EA%B0%95%EC%84%A0%EA%B7%9C%EC%B9%99/%EC%A0%81%EC%9A%A9%EC%A7%80%EC%B9%A8 (accessed on 10 March 2021).
72. Russian Maritime Register of Shipping, Guidelines on Technical Supervision of Ships in Service. 2019. Available online: https://rs-class.org/upload/iblock/1ed/1ed89e7715ca8ae9d2d65f5a990177f3.pdf (accessed on 20 March 2021).
73. UNCTAD. Data Protection and Privacy Legislation Worldwide. Available online: Unctad.org/page/data-protection-and-privacy-legislation-worldwide (accessed on 26 March 2021).
74. de Búrca, G. Introduction to the Symposium on the GDPR and International Law. *AJIL Unbound* **2020**, *114*, 1–4. [CrossRef]
75. Tikk, E. *International Law in Cyberspace: Mind the Gap*; Cyber Policy Institute: Jyväskylä, Finland, 2020. Available online: https://eucyberdirect.eu/wp-content/uploads/2020/03/tikk_rif-forpublication.pdf (accessed on 3 April 2021).
76. Dalaklis, D.; Schröder-Hinrichs, J.U. The Cyber-Security Element of Hybrid Warfare: Is there a Need to "Formalise" Training Requirements? In Proceedings of the 10th NMIOTC Annual Conference (Countering Hybrid Threats: An Emerging Maritime Security Challenge), Chania, Greece, 4 June 2019. [CrossRef]
77. Silvola, R.; Jaaskelainen, O.; Kropsu-Vehkapera, H.; Haapasalo, H. Managing one master data—Challenges and preconditions. *Ind. Manag. Data Syst.* **2011**, *111*, 146–162. [CrossRef]
78. Vilminko-Heikkinen, P.; Pekkola, S. Changes in roles, responsibilities and ownership in organizing master data management. *Int. J. Inf. Manag.* **2019**, *47*, 76–87. [CrossRef]
79. UNCTAD. *Review of Maritime Transport, United Nations Conference on Trade and Development*; UNCTAD/RMT/2019/Corr.1, U.N. Sales No. E.19.II.D.20, U.N. Doc.; UNCTAD: Geneva, Switzerland, 2020.
80. UNCLOS. United Nations Convention on the Law of the Sea pt. XIII, Dec. 10, 1982, 1833 U.N.T.S. 397.
81. Christensen, G. Definition: 6G 2020. Available online: https://searchnetworking.techtarget.com/definition/6G (accessed on 1 April 2021).
82. Pastra, A.; Koufopoulos, D.N.; Samac, N.; Johansson, T. Behavioral integration in the boardroom. *Team Perform. Manag.* **2021**. [CrossRef]
83. Pastra, A.; Koufopoulos, D.N.; Skintzti, V.; Johansson, T.; Samac, N. Exploring trust in the boardroom: The case of Nordic region. *Team Perform. Manag.* **2021**. [CrossRef]
84. Theodoropoulos, P.; Spandonidis, C.C.; Themelis, N.; Giordamlis, C.; Fassois, S. Evaluation of Different Deep-Learning Models for the Prediction of a Ship's Propulsion Power. *J. Mar. Sci. Eng.* **2021**, *9*, 116. [CrossRef]
85. Spandonidis, C.; Theodoropoulos, P.; Giordamlis, C. Combined Multi-layered Big Data and Responsible AI Techniques for Enhanced Decision Support in Shipping. In Proceedings of the International Conference on Decision Aid Sciences and Applications (DASA20), Sakheer, Bahrain, 8–9 November 2020; pp. 669–673.

Journal of
*Marine Science
and Engineering*

MDPI

Article

Prediction of a Ship's Operational Parameters Using Artificial Intelligence Techniques

Kiriakos Alexiou [1], Efthimios G. Pariotis [2], Theodoros C. Zannis [2,*] and Helen C. Leligou [1]

1 Department of Industrial Design and Production Engineering, University of West Attica, 12243 Athens, Greece; kiriakosal2004@yahoo.gr (K.A.); e.leligkou@uniwa.gr (H.C.L.)
2 Naval Architecture and Marine Engineering Section, Hellenic Naval Academy, 18539 Piraeus, Greece; pariotis@hna.gr
* Correspondence: thzannis@hna.gr

Abstract: The maritime industry is one of the most competitive industries today. However, there is a tendency for the profit margins of shipping companies to reduce due to an increase in operational costs, and it does not seem that this trend will change in the near future. The most important reason for the increase in operating costs relates to the increase in fuel prices. To compensate for the increase in operating costs, shipping companies can either renew their fleet or try to make use of new technologies to optimize the performance of their existing one. The software structure in the maritime industry has changed and is now leaning towards the use of Artificial Intelligence (AI) and, more specifically, Machine Learning (ML) for calculating its operational scenarios as a way to compensate the reduction of profit. While AI is a technology for creating intelligent systems that can simulate human intelligence, ML is a subfield of AI, which enables machines to learn from past data without being explicitly programmed. ML has been used in other industries for increasing both availability and profitability, and it seems that there is also great potential for the maritime industry. In this paper the authors compares the performance of multiple regression algorithms like Artificial Neural Network (ANN), Tree Regressor (TRs), Random Forest Regressor (RFR), K-Nearest Neighbor (kNN), Linear Regression, and AdaBoost, in predicting the output power of the Main Engines (M/E) of an ocean going vessel. These regression algorithms are selected because they are commonly used and are well supported by the main software developers in the area of ML. For this scope, measured values that are collected from the onboard Automated Data Logging & Monitoring (ADLM) system of the vessel for a period of six months have been used. The study shows that ML, with the proper processing of the measured parameters based on fundamental knowledge of naval architecture, can achieve remarkable prediction results. With the use of the proposed method there was a vast reduction in both the computational power needed for calculations, and the maximum absolute error value of prediction.

Keywords: Artificial Intelligence (AI); Fuel Oil Consumption (FOC); prediction; regression algorithms; data mining; Machine Learning (ML)

1. Introduction

The prediction of a ship's crucial operational parameters, such as M/E power, can lead to major operation sustainability. Required M/E power is directly related to Fuel Oil Consumption (FOC). Predicting the increase in FOC due to ship's hull degradation can lead shipowners to choose the right time and place for a dry docking, achieving the maximum profitability. On the other hand, the ability to "decode" and compare the information that lies in the comparison of ships' performance values between any given time and the values that were achieved in the sea trials can lead to accurate prediction of performance reduction of a specific component. This constant comparison will lead to a better control of the service life of a ship.

During the last decades, the maritime industry (shipyards, maritime companies, engineers/researchers) has embarked on a process of continuous research and development of new technologies. This process is even more important for forthcoming years, in order to comply with progressively stricter environmental regulations and, at the same time, keep operational costs as low as possible to maintain shipping as one of the most cost-effective means of transportation. To make this possible, the whole process of ship design and operation is based on a balanced mixture of computational and experimental investigations which are highly demanding in terms of time required and capital cost. After a long and heavy computational period which mostly relies on theoretical/deterministic models as much as empirical methods, prototypes have to be constructed and extensive tests have to be carried out. One promising technique which could help to reduce the effort required for ship design and cost effective ship operation is ML, which has already been used successfully in many land-based industrial applications. Building prediction algorithms based on ML can be the answer for improving ship performance in terms of fuel cost and exhaust emissions (Nitrogen Oxides, Sulfur Oxides, etc.). The current study relies on this field, focusing on the prediction of the ship's M/E power required, depending on the operational conditions, which could become valuable tools to optimize a ship's operational profile.

In shipping and maritime transport in general, the penetration of AI is still in its infancy but this seems likely to change. In other transportation sectors, the use of Internet of Things (IoT) and big data are pursued towards augmentation of a system's intelligence [1], while in other sectors like manufacturing and Information and Communication Technology (ICT) the penetration of artificial intelligence is already in a more advanced stage [2,3]. Looking back at the international literature, studies appear more and more often in the general context of the utilization of ML in shipping. In 2005 a hybrid model for ship operational performance monitoring was developed, and the uncertainty framework specific to ship's performance was analyzed [4]. The application of a gray-box modelling approach to the prediction of ship fuel consumption was proposed in 2006, which can be used as a tool for online trim optimization [5]. A fuel consumption model that utilizes data obtained from the noon-reports that are transmitted daily back to shore was developed in 2018 [6]. A ML approach, for predicting FOC on ships, by training an ML algorithm with dynamic data from the ships' logging system and using high time intervals for fuel sums has been presented [7]. In more recent studies, the efficiency of several multiple regression algorithms on the task of ship FOC prediction under different data sampling frequencies was studied [8]. An estimation model of FOC based on a deep-learning ANN was conducted [9]. Finally, an FOC optimization study of a container vessel using regression algorithms, with data selected by correlation analysis, and comparison of the results has also been carried out [10,11].

The aim of this study is to determine a simple procedure for pre-processing the vast amount of data that are collected from modern ships ADLM systems. Although the use of ML in the prediction of operational parameters has been addressed before, this study focuses on the reduction of the computational work needed for this task. The proposed procedure does not depend on the type of ship and uses a short fragment of the original data set, thus limiting the computational demand. The results are verified for actual use. In this direction, we employ different regression algorithms in the prediction of a ship's required main engine power, using data from an ADLM system in their original form and after a specific preprocess procedure. The performance of algorithms is compared in terms of accuracy and required processing power.

The rest of the paper is organized as follows: in Section 2, we present the proposed method of approach. We explain the logical steps that form the data set and the basic working principle of the different regression algorithms that are going to be used. The application of the proposed method in our data set, the benchmark tools for evaluation, and the actual comparison of the results are presented in Section 3. Finally, Section 4 is dedicated to the conclusion of this paper.

2. Methodology

Two different data sets are used for the procedure of predicting a ship's required main engine power. A raw data set is constructed, with the use of all values gathered in a period of 6 months, and a preprocessed one that is produced from the above with a certain proposed methodology. The results are then benchmarked and compared. The flow diagram of the prediction method is shown in Figure 1.

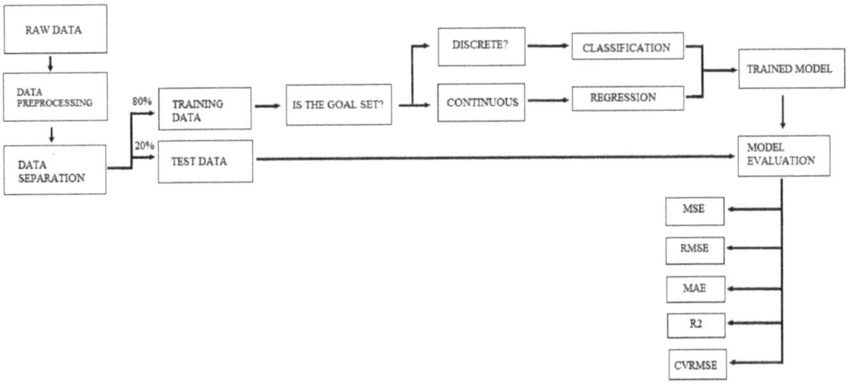

Figure 1. Flow diagram of the prediction method.

The significance of the prediction of a ship behavior is obvious. Until recently, only white box model approaches (that rely on physics) have been adopted for this purpose in the maritime industry. This approach has the disadvantage of relying on experience and demands heavy computational work. These two factors make the use of this approach an expert-only privilege. In this paper we propose the use of ML in addition to common and well-known practices in order to achieve respective results with less computational work and with no need for previous experience. ML is a rapidly growing science with great penetration in human activity. Establishing a framework is essential for the quickly adoption of ML in maritime.

The biggest challenge in ML is the pre-processing of the data set that will be used. For this purpose, many different approaches have been used. Purely computational methods like Principal Component Analysis (PCA) have been adopted for the purpose of formation and restriction of the vast amount of data that form data sets. For an industry that traditionally relies on more conventional approaches we propose a hybrid model of data pre-processing.

The data set is crucial in ML, since choosing the values that will be imported to the ML algorithms is of utmost importance in any ML procedure. The main objective is to be able to achieve acceptable prediction scores with as little values as possible. On the other hand, choosing the multitude of different attributes is equally important so that with the right regression algorithm and tuning, there will be no need for heavy computational power. In most ML applications, the choice of the different attributes that will form the data set is chosen by the value of correlation factor [12]. Different wrapper or filter methods are used for the selection of attributes that increase the computational work. In this paper the above selection is based on the physical principles that affect a ship's movement on the sea:

- Maneuverability
- Stability
- Engineering
- Seakeeping
- Resistance

- Powering

The first step of data formation procedure is Sampling. The data we used were synchronized to refer to specific time periods (1-min interval), which is typical for currently deployed ADLM systems. Then, they were identified and given the appropriate unit of measurement.

Attribute selection. From the original attributes (178 in our test case), we opted for selection based on the possibility of influencing the object of the study (e.g., the speed of the ship). The attributes that were judged not to affect the object of the study at all (e.g., electric oil pressure) were removed, in order to reduce the total volume of data and to properly configure the data set. (From the 178 entries, 42 were selected).

Feature Engineering/data transformation. In the context of data limitation and based on the physical nature of the phenomenon under consideration, it is legitimate to transform attributes in order to either limit the data set, or replace some characteristics with others that better describe the affect on the phenomenon. Based on this, the attributes (Wind Direction, Wind Speed) were replaced by the Head Wind Speed (HWS). Head Wind Speed (the component of the wind speed that is opposite of ships movement) is the value that affects the resistance of a ship due to the wind according to literature [13–15]. The replacement of values was done using Formula (1):

$$\begin{gathered} X = \text{Wind Direction} - \text{True Heading} \\ \text{If } X < -180 \text{ then } X' = 360 + X \\ \text{If } X > 180 \text{ then } X' = 360 - X \\ X'' = \text{ABS}(X') \\ \text{HWS} = \text{COS}(X'') \times \text{Wind Speed} \end{gathered} \tag{1}$$

A schematic presentation of the above formula is presented in Figure 2.

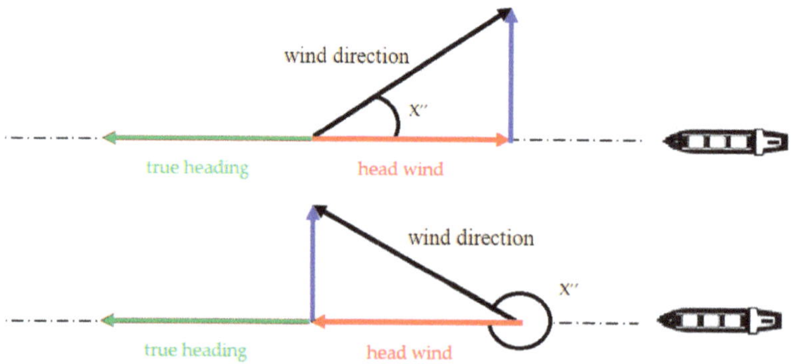

Figure 2. Schematic presentation of HWS calculation.

Also, the attributes that correspond to a ship's relative movement due to weather conditions, such as Inclinometer_X_max (the maximum trim value of ship), Inclinometer_X_min (the minimum trim value of ship), Inclinometer_Y_max (the maximum list of ship), and Inclinometer_Y_min (minimum list of ship), were replaced with the average values, Inclinometer_X_average and Inclinometer_Y_max, respectively.

Instance's selection. There are many methods in literature to deal with missing values (NaN) in the data set [16,17]: deleting the entire line, using a fixed value for all lost values, replacing the lost value with the average value of the column, replacing the lost value with the average value of the class, and loss value prediction (with 1-NN algorithm) are some of the most commonly used. In this paper the first method was selected. All the lines (Instances) which contained even one lost value (NaN) have been deleted. (This way in our sample, from 236,161 lines initially available, a subset of 39,475 was actually used).

We also proceeded to one more improvement of the initial dataset. There is a power limit given by the manufacturer (Hyundai in this case), below which any Diesel engine must not operate for a long time. [18]. This value of power is in most cases approximately the 15–20% of the maximum continuous (L1) for electronically controlled machines or 20–25% for mechanically controlled (camshaft controlled). Assuming that the crew is aware of these limitations, the operation of a ship's M/E in these conditions can be considered a transient phenomenon. Based on the results of the ship sea trials, it appears that, with 25% of the engine power, about 57 revolutions per minute are achieved. The condition of the ship (in terms of hull degradation) could not be identified, so it is possible for 25% of M/E power at the time of the study to correspond to different performances. Having that in mind, all the lines of the data set in which the value of the M/E revolutions per minute (RPM) was less than 50 were excluded and the final data set with 29,067 lines was obtained.

Data cleansing/ "noise" removal. As there is always the possibility of wrong values being captured due to several reasons (e.g., temporal malfunctioning or transmission error in the sensing network), we also processed the available dataset to remove such values. Towards removing the values that refer to transient states (e.g., acceleration process), the data were renumbered (maintaining their chronological order) and divided into equal frequency intervals to alleviate the computational burden. These intervals have the same number of values, which in our case was 5000. At these intervals and with the K-means method [19], the average and dispersion of the values were identified (always in terms of the actual speed of the ship which is a quantity that was considered fundamental in terms of the transience of the phenomenon under consideration). In all the above intervals, the speed values (mean value ± scatter) were between a minimum and a maximum value, i.e., 9.8384 (position A in Figure 3) and 15.5759 (position B in Figure 3) in our case. Lines with speed values outside this field were identified as noise and removed from the data set. The final data set consisted of 39 columns (attributes) and 27,155 rows (instances).

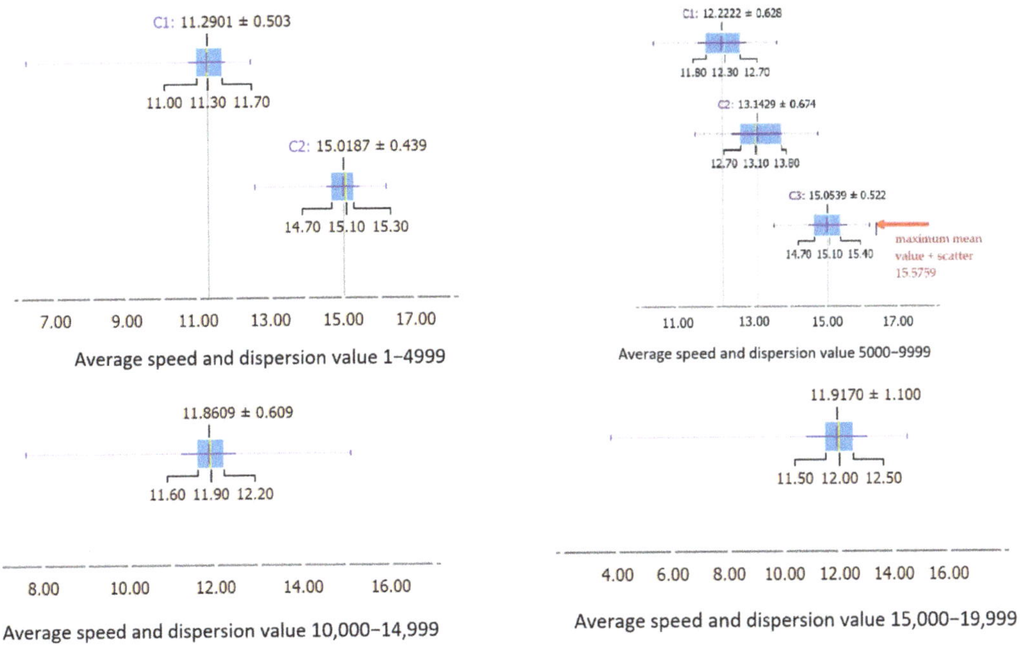

Figure 3. *Cont.*

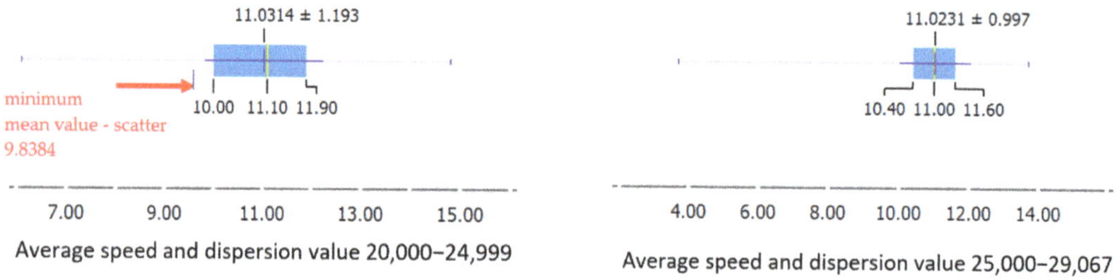

Figure 3. Ship's speed values (mean value ± scatter).

Normalization. The use of neural networks is particularly facilitated if the values of the input data range between 0 and 1. For this reason, a minimum–maximum normalization was performed, and a data set was constructed with the values of all fields in the range of 0.1 using the following Formula (2):

$$X\prime = \frac{X - min}{max - min}(1 - 0) + min \quad (2)$$

After the completion of the pre-process, data were randomly divided into a training data set (consisting of 80% of the lines) which is used for algorithm training, and validation data set (consisting of the remaining 20% of the lines) for result testing. The power of the main engine was set as a given goal.

The following algorithms were used for prediction and are graphically presented in Figure 4:

(1) Linear Regression. The linear regression algorithm locates the linear relationship between the dependent variable we are looking for (Y) and one or more independent (X) variables. Since this relationship is linear, we can easily determine how the dependent variable changes with the value of the independent variable [20]. The graphical representation (Figure 4a) of the linear regression operating model is the draw of a sloping and straight line, representing the correlation between independent and dependent variables. The red line represents the prediction that minimizes the overall error. All the predicted values are part of this line and the difference in Y axis between the line and the actual values is the error.

(2) Decision Tree. This is a tree-like classifier. It is a structure that is constructed by nodes. There are three different types of nodes: chance nodes, decision nodes, and end nodes. Chance nodes represents the probability of a certain result, decision node represent the decision to be made and the end node shows the result [21]. The tree acts like a map of all possible outcomes for the question asked. Decisions are made based on the characteristics of the data set. It is essentially a graph that aims to explore all possible solutions to a problem we have posed under specific circumstances. In the example presented in Figure 4b, a decision has to be made for a kid to be allowed to go out to play. In left node, the combination of cloudy weather with high temperature is a 'go' criteria. The same decision (to allow the kid to play outside) has been made with completely different weather conditions: sunshine with low wind (right node).

(3) kNN. The kNN algorithm stores the available data and categorizes them. It forms data categories based on their common characteristics. The new data given are classified based on their similarity to the basic characteristics of each category that are already stored [22]. This process is performed dynamically. This means that when new data appears can easily be sorted into a category from the available ones. In Figure 4c, the kNN algorithm has to classify the new spot as part of category A or category B.

(4) Random Forest. This is based on the concept of ensemble learning, in which multiple classifiers are combined to improve the performance of the model, in order to solve

a complex problem. It was named forest because it essentially contains a number of decision trees which, using subsets of the original data, make decisions and predictions in parallel. Once the predictions have been made by the decision trees, it takes the average value of them to improve the predictive accuracy of the total and initial data set [23]. In the example of Figure 4d, each of the decision trees A, B . . . N give results to the problem given (red). The result that appeared more often is the end result of Random Forest Algorithm.

(5) Neural Network. The ANN are the mathematical equivalent of the function of the biological neurons that make up the human brain. They consist of a number of simple and internally interconnected processing units, which are organized in layers. The fundamental structure is consisted of one input layer and one output but it is possible that in addition to those there are hidden layers in between. The above layers consist of a number of units or nodes that are connected to each other in such a way that one unit has links with many other units of the same or another level [24].

(6) AdaBoost. The Adaptive Boosting (AdaBoost) algorithm is an enhancement technique that aims to combine multiple weak classifiers to in order to create a strong classifier. An individual (weak) classifier may not be able to accurately predict the class of an object, but when we group many weak classifiers together, which gradually learn from the wrong classifications of each other, we can create a very powerful model.

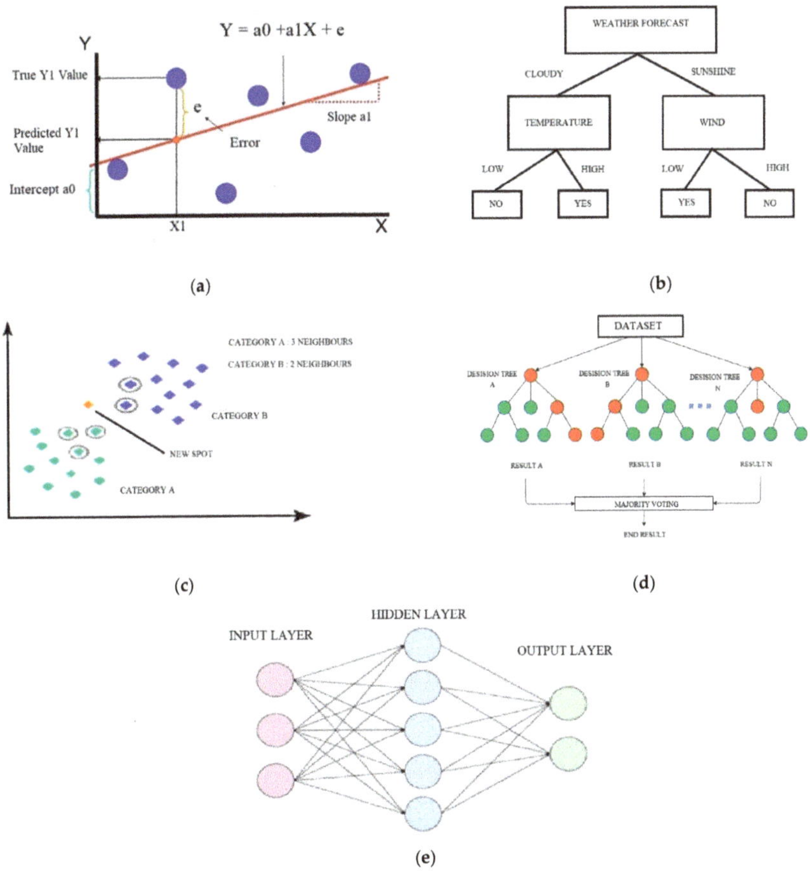

Figure 4. Graphical representation of the ML algorithms. (**a**) Linear Regression graphical representation. (**b**) Decision Tree example. (**c**) Match value (spot) with k to nearest neighbors. (**d**) Random Forest example. (**e**) ANN example with one hidden layer.

The above algorithms used the data selected according to the aforementioned procedure and the performance of the forecasts was measured. The same algorithms with the use of the original data set (after a mild preprocess procedure) were used and the results have been compared. In both cases, all of the algorithm parameters stayed the same. The procedure was repeated 10 times with a different selection of data used for training.

3. Evaluation of the Approach

3.1. Our Case and the Dataset

To evaluate the proposed approach, we used a dataset acquired from the ships ADLM system. The ship used as reference in this paper is a Crude oil Tanker with 165.000 tons displacement. The ship is relatively new since it was commissioned ten years ago. A raw data set was constructed with the use of all values gathered in a period of 6 months. Due to the owner shipping company's strict policy regarding data release, the values describing the ship and routes are slightly modified in this presentation (the accurate values were used for problem formulation). The ship's features are presented in Table 1.

Table 1. The reference ship's main features.

Feature	Value
Ship Type	Crude Oil Tanker
Displacement	165.000 (t)
Main Engine	Hyundai 6S70MC-C 18,660 kW × 91 RPM
Propeller	FPP 4Blades DIA 8200 (mm)
Generator's Engine	Hyundai 5H21/32 905 kW × 900 RPM
Generators	Hyundai HFC7 506-84K 850 kW × 450 V
LBL	265 (m)
B (mld)	50 (m)
D (mld)	23 (m)

The data were extracted from the ship's ADLM system in a period of 6 months, from mid-February of 2020 until end of July 2020. A total of 178 different attributes were collected. The sample frequency that was chosen for forming the data set was one complete set of values every minute, so there was a total number of 236,161 instances. The ship's ADLM system reads values from each different sensor and stores them in a database with a much higher frequency than 1 per minute. The procedure of recording all of the sensor values introduces time deviations but these values (in a ship) don't change so rapidly as to introduce negligent error. In the data set there were many cases of missing values (NaN) due to many factors, sensor anomalies, data entry errors, periods in which the recorded accessory was shutdown, etc. The fields of data that are recorded from the ship's ADLM, which make up the data set attributes, belong to four main categories.

- Kinematics (Inclinometer, Speed etc.)
- Propulsion (Rpm, Torque etc.)
- Weather (Wind, Swell Wave etc.)
- Time-Space (Latitude, Time etc.)

At this period of 6 months, the ship executed many different itineraries. The routes expand through the perimeter of Africa, North Europe, Mediterranean Sea, and the Indian and Pacific Oceans. The significant differences in the geographical area of routes resulted in great differentiation of environmental conditions. The different routes are graphically presented in Figure 5.

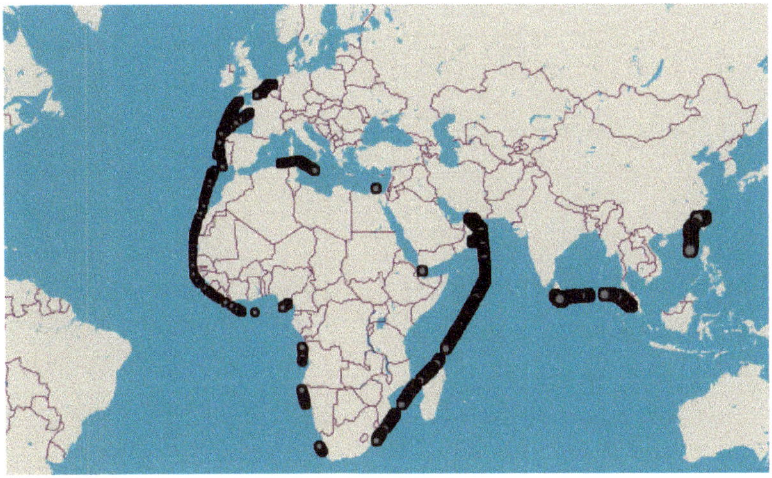

Figure 5. Graphical presentation of executed routes.

The above routes were executed in various cruising speeds but the area around 12 Knots dominates, as depicted by Figure 6, where the distribution and frequency of the different cruising speeds are presented.

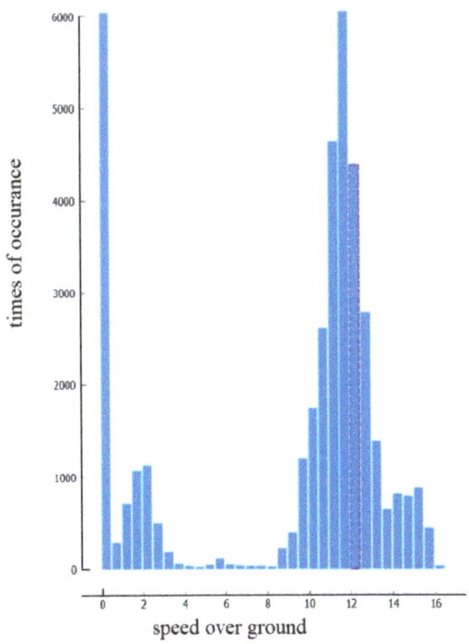

Figure 6. The ship's cruising speed distribution.

Each cruising speed was kept relatively steady through each route by the ship's crew so the speed over ground values have low fluctuation (with the exception for the transitional periods) [25]. Figure 7 presents the actual values of ship speed over the total number of instances used in the hybrid model, in groups of 5000 instances at each sub-figure (due to software limitation on the number of instances that can be interpreted).

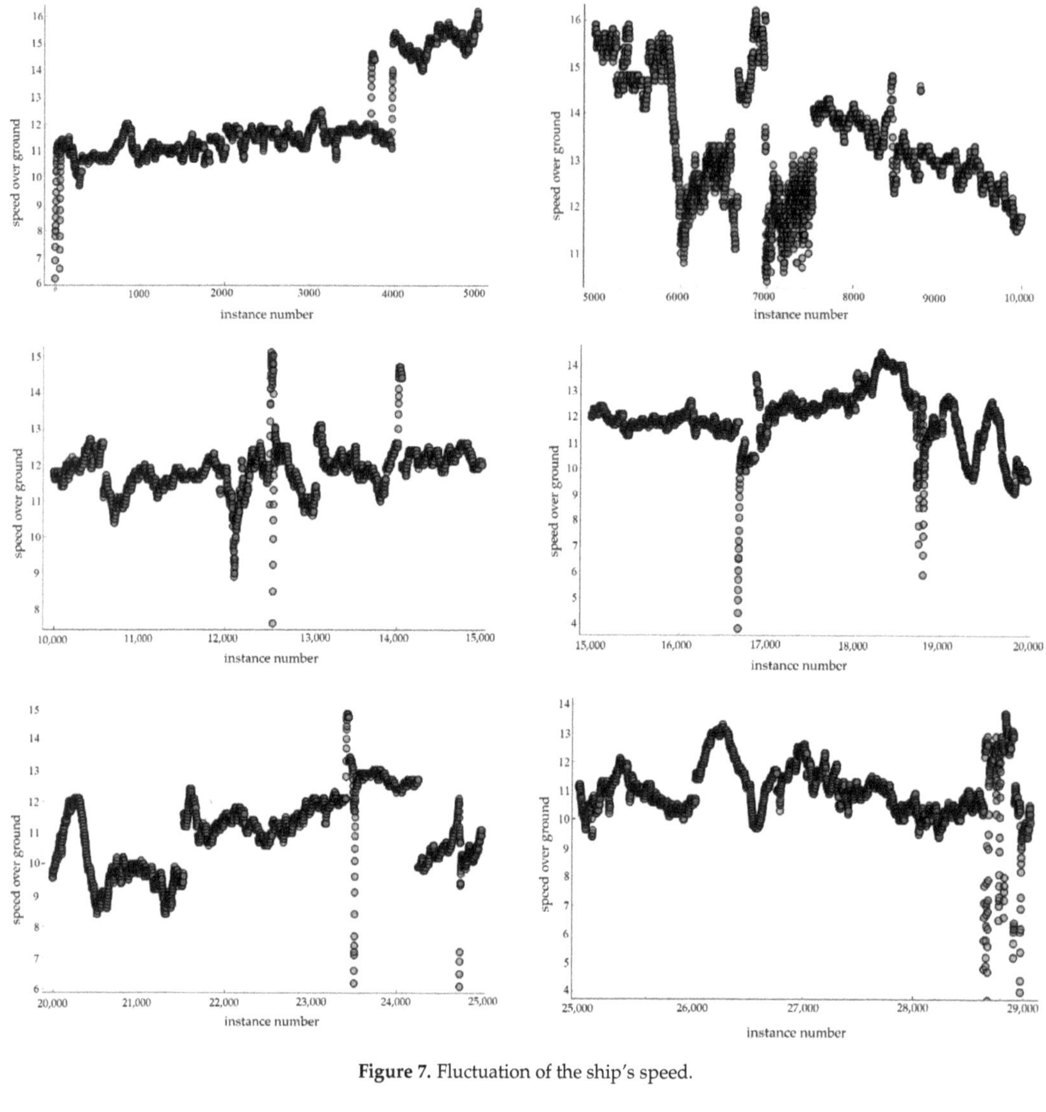

Figure 7. Fluctuation of the ship's speed.

3.2. Evaluation Methodology

The accuracy of a prediction is defined as the total number of correct predictions divided by the total number of predictions made for a data set. However, in a regression problem that seeks to predict the specific value of an attribute, this definition isn't sufficient. We have to define methods that will essentially set forecast deviation values and through them we will be able to compare prediction performance [26]. The most widely used indicators for conducting comparative performance benchmarks are the following:

- Mean Absolute Error (MAE)
- Mean Squared Error (MSE)
- Root Mean Squared Error (RMSE)
- Coefficient of Variation of Root-Mean Squared Error (CVRMSE)
- R Square (R^2)

Figure 8 is showing a graphical representation of the prediction results of a regression algorithm. The blue line represents the actual (true) values and the purple dots the predicted ones.

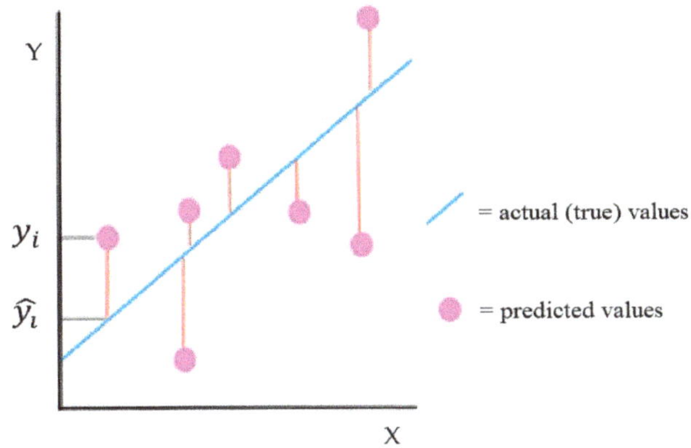

Figure 8. Prediction results of regression algorithm.

MAE is the average of the absolute differences between predicted and actual prices and gives us a measure of how far forecasts are from reality. Mathematically it is defined by Equation (3):

$$MAE = \frac{1}{N}\sum_{i=1}^{N}|y_i - \hat{y}_i| \tag{3}$$

MSE is quite similar to the *MAE*, with the only difference being that the *MSE* takes the average of the square of the difference between the actual values and the predicted values (Equation (4)).

$$MSE = \frac{1}{N}\sum_{i=1}^{N}(y_i - \hat{y}_i)^2 \tag{4}$$

RMSE is the square root of *MSE*, is essentially the same measurement tool as it but has prevailed and is widely used in the literature (Equation (5)).

$$RMSE = \sqrt{\frac{1}{N}\sum_{i=1}^{N}(y_i - \hat{y}_i)^2} \tag{5}$$

CVRMSE is the next step from *RMSE* by normalizing it from the mean dependent variable (Equation (6)).

$$CV(RMSE) = \frac{1}{\bar{Y}}\sqrt{\frac{\sum_{i=1}^{N}(Y_i - \hat{Y}_i)^2}{N}} \tag{6}$$

Finally, R^2 focuses not so much on the results but on the operation of the algorithm itself. It specifies the degree to which the variations in the values of the dependent variable (prediction target) can be explained by changes in the values of the independent variables (data set) (Equation (7)).

$$R^2 = 1 - \frac{\sum_i (y_i - \hat{y}_i)^2}{\sum_i (y_i - \bar{y})^2} \tag{7}$$

In Table 2 the benchmark scores, using the above-mentioned performance indicators for both cases, either using the complete original data set or the modified ones (hybrid model) are presented for all the prediction algorithms examined. In the case of RFR,

Decision Tree, and AdaBoost, using the complete data set resulted in the achievement of better results. The use of the proposed hybrid model for pre-processing ships data increased the error in all of the benchmark scores. On the other hand, in the cases of kNN, Linear Regression, and ANN the results are clearly better with the use of pre-processed selected data (hybrid model). Therefore, it is not possible to determine clearly which methodology is preferable.

Table 2. Benchmark scores of the prediction algorithms using the complete data set and the hybrid model (modified data set).

Model	*MSE*	*RMSE*	*MAE*	R^2	*CVRMSE*	
Random Forest	298.902	17.289	2.421	1.000	0.293	Use of complete Data
	4542.351	67.397	41.162	0.999	0.966	Hybrid Model
kNN	31,590.858	177.738	72.709	0.998	3.010	
	6126.236	78.270	43.311	0.999	1.258	
Linear Regression	Na	Na	Na	Na	Na	
	12,666.868	112.547	84.931	0.998	1.614	
Neural Network	111,588.711	334.049	233.929	0.992	5.657	
	6096.33	78.079	84.931	0.999	1.120	
Decision Tree	639.464	25.288	4.957	1.000	0.428	
	7702.947	87.766	53.181	0.999	1.122	
Adaboost	356.969	18.894	4.105	1.000	0.320	
	4116.474	64.16	39.196	0.999	0.920	

The second comparison criteria used was based in real life applications where a deviation of 3% on predicted M/E power output is considered acceptable. Therefore, a threshold of 3% error was set to the predicted values so the results were evaluated accordingly. The percentage of the predicted values in which the difference from the actual ones is less than 3% was recorded. The prediction results with the use of the complete data set along with the proposed hybrid model were compared in terms of computational work and error size values. The use of the proposed method resulted in most cases in a minor deterioration of the percentage of values with an error below 3%, as shown in Figure 9. However, in the case of the kNN algorithm, using the hybrid model (modified data set) leads to slight improvement in the predictive performance, while in the case of the ANN algorithm the improvement is considerably higher, which is a clear evidence of the better performance achieved using the hybrid model.

The maximum difference between the predicted and the actual value of M/E power, is very important because it represents the worst performance (spike) that we can expect from the use of each method. Figure 10 presents the maximum deviation between the actual and predicted value of M/E power, using either the complete data set or the hybrid model (modified data set). As observed, using the hybrid model, for all algorithms, a considerable reduction on the maximum deviation between actual and predicted value is obtained, with the best improvement obtained in case of Linear regression and the best performance observed for the case of the Decision Tree algorithm, with the maximum deviation being equal to 10.16%. This is a significant performance improvement, however, for real life applications an even better predictive accuracy is required. To this end, further improvements on the hybrid model are required.

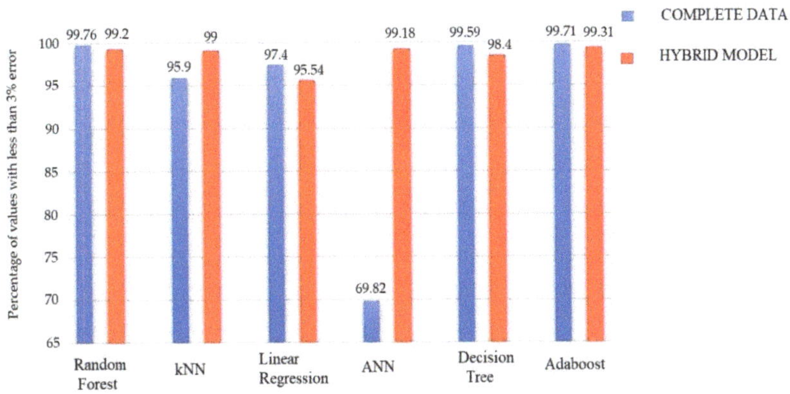

Figure 9. Comparison of percentage in predicted M/E power values (with an error of less than 3%) between the complete data set and the hybrid model.

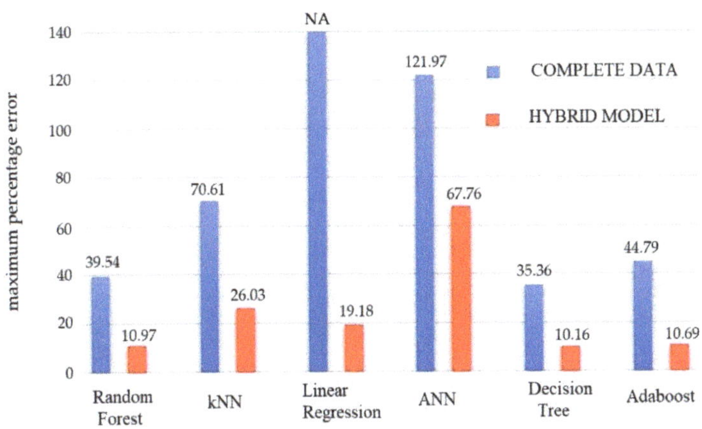

Figure 10. Comparison of the maximum percentage error in the prediction of M/E power with the complete data set and the hybrid model.

Finally, a comparison of the time needed for training and testing the algorithms was recorded. Figure 11 presents the comparison of the reduced time needed (on a percentage basis) for the algorithms to complete the calculations for training, using the hybrid model. All calculations in this study were conducted in the same personal computer, under the same conditions to make the comparison reliable.

The time needed for training was dramatically reduced in all cases using the hybrid model (in the range of 93.37% to 99.77%), with the exception of the ANN algorithm where an increase of 14.35% is observed. Finally, all algorithms performed better in terms of computational power needed for testing, as shown in Figure 12.

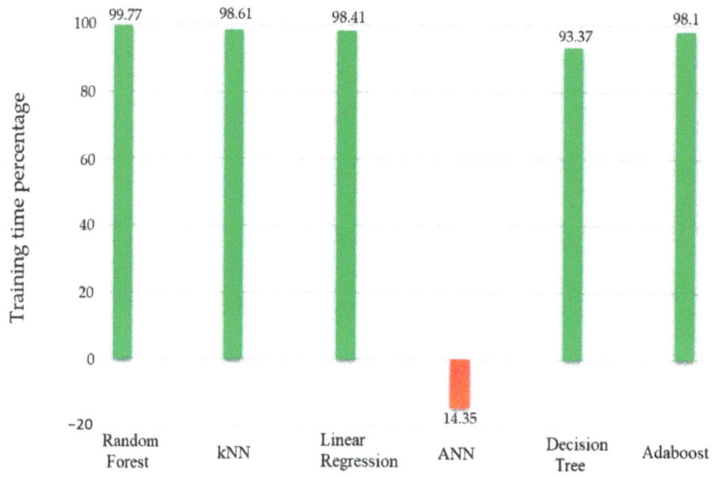

Figure 11. Comparison of the predictive algorithms training times required on a percentage basis using the hybrid model (modified data set).

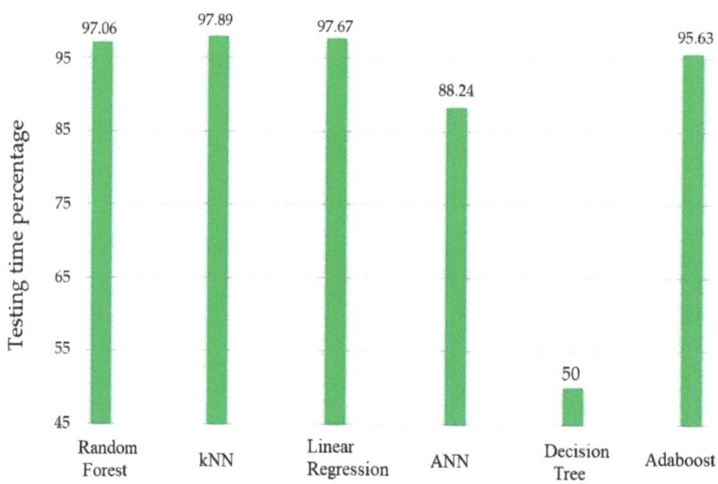

Figure 12. Comparison of the predictive algorithms testing times required on a percentage basis using the hybrid model (modified data set).

As observed, there is a significant using the hybrid model for the prediction of M/E power. The reduction in CPU time demand ranges from 50% to 97.89%. It has been noticed that although, as shown in Figure 11, the ANN algorithm need more time for the training using the hybrid model, the situation is altered when using the trained algorithm for M/E power prediction. Therefore, it is obvious that the hybrid model offers great potential in the reduction of the required overall computational time for algorithm training and application for the M/E power prediction. This is very important if we also take into account that, in real life applications, the computational power demand of the prediction algorithms is often an equally significant parameter as the accuracy of the predictions.

Figure 13 shows, in the same diagram, the actual (true) values of M/E power output (black color) in conjunction with the predicted values (green color) using the hybrid model (modified data set) for all algorithms.

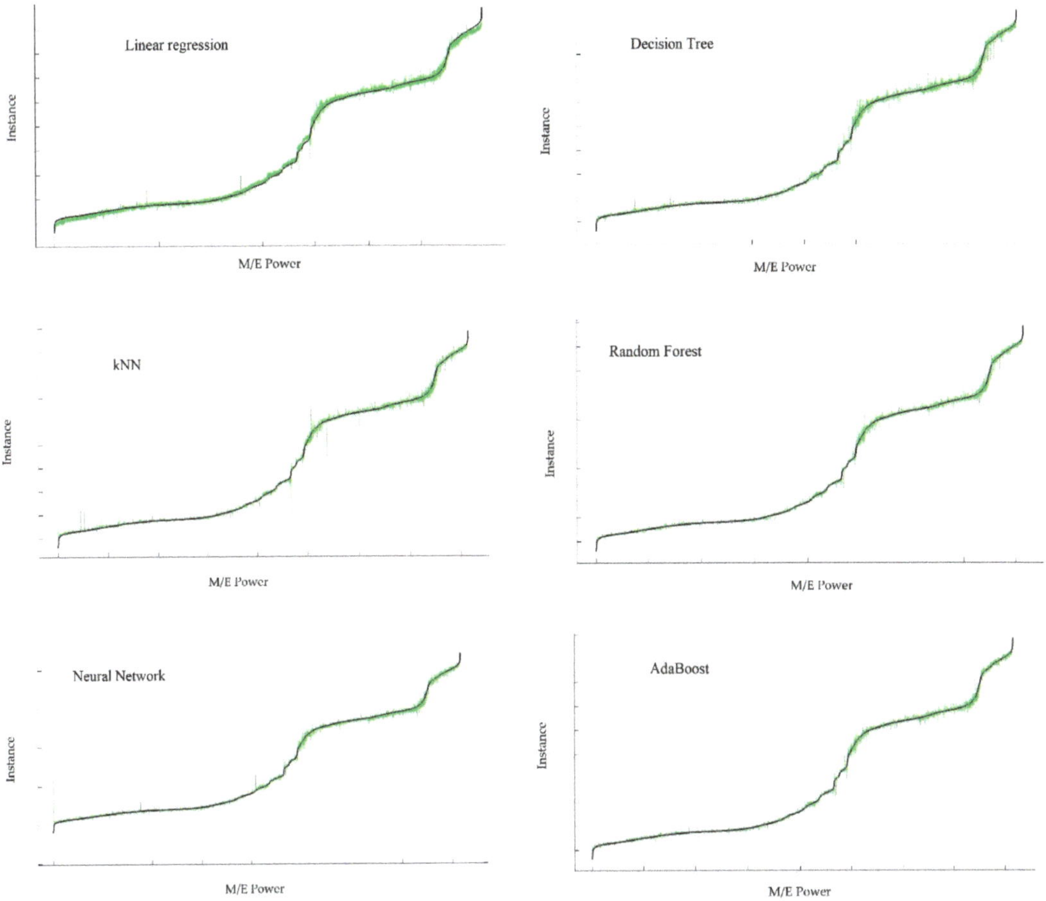

Figure 13. Conjunction between actual and predicted values.

As observed, in all cases (prediction algorithms), there is a good coincidence between actual and predicted values of M/E power, during the whole range of the data set used. However, there are many instances where high discrepancies are observed (spikes) for all prediction algorithms examined, while it seems that the linear regression algorithm is the least accurate, with the discrepancies between actual and predicted M/E power being relatively high at all the data points examined.

4. Conclusions

The results which are presented in this paper indicate that ML can become a valuable tool for the marine industry. With the use of regression algorithms and the data from ship's ADLM systems we can predict with good accuracy a very crucial operational parameter, such as the engine power, almost in real time. This knowledge can be used to lower the operational costs in shipping companies. The use of a straightforward pre-processing procedure proposed by the authors, for selecting and/or transforming the vast data that ADLM systems collects, significantly improves the performance of the prediction AI algorithms in terms of computational power demand. Actually, there is a reduction in the range of 50% to 99.97% of the required computational time, while keeping the accuracy at almost the same level, i.e., there is a negligible reduction of the predicted values with an error less than 3%, as shown in Figure 9. At the same time, the hybrid model reduces considerably the maximum deviation of the predicted values of M/E power (spikes) compared to the performance of the same algorithms using the complete original data set. This deviations (spikes) reach relatively high values (in the range of 10%), which are not acceptable in real life applications, although this is observed for a very low percentage of the data points examined, i.e., below 4.46% of the data in the worst case, as shown in Figure 9, for the linear regression algorithm. Therefore, further research is needed to overcome these deficiencies of the proposed model and determine the most suitable prediction algorithm for the estimation of M/E power demand using real-time data obtained by the ADLM system of the ship. In this way, it will become possible to introduce ML in real life applications (like the prediction of M/E power demand of the vessel according to the operating conditions), where predictive accuracy and computational power demand are crucial parameters.

Author Contributions: Conceptualization, K.A., E.G.P., H.C.L.; Data curation, K.A.; Formal analysis, K.A., E.G.P.; Investigation, K.A., E.G.P., H.C.L.; Methodology, K.A., E.G.P.; Project administration, H.C.L., E.G.P.; Software, K.A.; Validation, T.C.Z.; Writing—original draft, K.A.; Writing—review & editing, K.A., E.G.P., T.C.Z. and H.C.L. All authors have read and agreed to the published version of the manuscript.

Funding: This research received no external funding.

Institutional Review Board Statement: Not applicable.

Informed Consent Statement: Not applicable.

Data Availability Statement: Not applicable.

Conflicts of Interest: The authors declare no conflict of interest.

References

1. Orphanoudakis, T.; Fanariotis, A.; Politis, I.; Karachalios, T.; Demestichas, K.; Xynogalas, S.; Loumiotis, I.; Sykas, E. Exploiting IoT and Big-Data for Building Multiservice Capable Intelligent Transportation Systems. In Proceedings of the 2019 4th South-East Europe Design Automation, Computer Engineering, Computer Networks and Social Media Conference (SEEDA-CECNSM), Piraeus, Greece, 20–22 September 2019; pp. 1–7. [CrossRef]
2. Zafeiropoulos, A.; Fotopoulou, E.; Peuster, M.; Schneider, S.; Gouvas, P.; Behnke, D.; Muller, M.; Bok, P.-B.; Trakadas, P.; Karkazis, P.; et al. Benchmarking and Profiling 5G Verticals' Applications: An Industrial IoT Use Case. In Proceedings of the 2020 6th IEEE Conference on Network Softwarization (NetSoft), Ghent, Belgium, 29 June–3 July 2020; pp. 310–318. [CrossRef]
3. Trakadas, P.; Simoens, P.; Gkonis, P.; Sarakis, L.; Angelopoulos, A.; Ramallo-González, A.P.; Skarmeta, A.; Trochoutsos, C.; Calvo, D.; Pariente, T.; et al. An Artificial Intelligence-Based Collaboration Approach in Industrial IoT Manufacturing: Key Concepts, Architectural Extensions and Potential Applications. *Sensors* **2020**, *20*, 5480. [CrossRef] [PubMed]
4. Aldous, L.G. Ship Operational Efficiency: Performance Models and Uncertainty Analysis. Thesis submitted in ful-filment of the requirements for the degree of Doctor of Philosophy. Ph.D. Thesis, University College London, London, UK, 2015.
5. Coraddua, A.; Oneto, L.; Baldi, F.; Anguita, D. *Vessels Fuel Consumption Forecast and Trim Optimisation: A Data Analytics Perspective*; Elsevier Ltd.: Amsterdam, The Netherlands, 2016.
6. Gkerekos, C.; Lazakis, I.; Papageorgiou, S. Leveraging Big Data for Fuel Oil Consumption Modelling. In Proceedings of the 17th Conference on Computer and IT Applications in the Maritime Industries, Pavone Canavese, Italy, 14–16 May 2018; Technische Universität Hamburg-Harburg: Hamburg, Germany, 2018; pp. 144–152.
7. Ahlgren, F.; Mondejar, M.E.; Thern, M. *Predicting Dynamic Fuel Oil Consumption on Ships with Automated Machine Learning*; Elsevier Ltd.: Amsterdam, The Netherlands, 2018.
8. Gkerekos, C.; Lazakis, I.; Theotokatos, G. *Machine Learning Models for Predicting Ship Main Engine Fuel Oil Consumption: A Comparative Study*; Elsevier Ltd.: Amsterdam, The Netherlands, 2019.
9. Bui-Duy, L.; Vu-Thi-Minh, N. *Utilization of a Deep Learning-Based Fuel Consumption Model in Choosing a Liner Shipping Route for Container Ships in Asia*; Elsevier Ltd.: Amsterdam, The Netherlands, 2020.
10. Uyanık, T.; Karatuğ, Ç.; Arslanoğlu, Y. *Machine Learning Approach to Ship Fuel Consumption: A Case of Container Vessel*; Elsevier Ltd.: Amsterdam, The Netherlands, 2020.
11. Yan, R.; Wang, S.; Du, Y. *Development of a Two-Stage Ship Fuel Consumption Prediction and Reduction Model for a Dry Bulk Ship*; Elsevier Ltd.: Amsterdam, The Netherlands, 2020.
12. Olson, D.L.; Wu, D. *Predictive Data Mining Models*, 2nd ed.; Springer Nature Singapore Pte Ltd.: Singapore, 2020.

.3. Molland, A.F.; Turnock, S.R.; Hudson, D.A. *Ship Resistance and Propulsion: Practical Estimation of Ship Propulsive Power*; Cambridge University Press: Cambridge, UK, 2011.
.4. Politis, G.K. *Ship Resistance and Promotion*, 5th ed.; National Technical University of Athens: Athens, Greece, 2018.
.5. Birk, L. *Fundamentals of Ship Hydrodynamics: Fluid Mechanics, Ship Resistance and Propulsion*; John Wiley & Sons: Hoboken, NJ, USA, 2019.
.6. Refaat, M. *Data Preparation for Data Mining Using SAS (The Morgan Kaufmann Series in Data Management Systems)*; libgen.lc: San Francisco, CA, USA, 2006.
.7. Pyle, D. *Data Preparation for Data Mining*; Morgan Kaufmann Publishers, Inc.: Burlington, MA, USA, 1999.
.8. Woodyard, D. *Pounder's Marine Diesel Engines and Gas Turbines*; Butterworth-Heinemann: Oxford, UK, 2004.
.9. Kantardzic, M. *Data Mining Concepts, Models, Methods, and Algorithms*; Wiley-IEEE Press: Hoboken, NJ, USA, 2020.
20. Weisberg, S. *Applied Linear Regression*; Wiley-Interscience: Hoboken, NJ, USA, 2005.
21. Barros, R.C.; De Carvalho, A.C.P.L.F.; Freitas, A.A. *Automatic Design of Decision-Tree Induction Algorithms*; Springer International Publishing: Berlin/Heidelberg, Germany, 2015.
22. Yang, X.-S. *Introduction to Algorithms for Data Mining and Machine Learning*; Academic Press: Cambridge, MA, USA, 2019.
23. Sullivan, W. *Machine Learning For Beginners: Algorithms, Decision Tree & Random Forest Introduction*; Publishdrive: Budapest, Hungary, 2017.
24. Parkes, A.I.; Sobey, A.J.; Hudson, D.A. *Physics-Based Shaft Power Prediction for Large Merchant Ships Using Neural Networks*; Elsevier Ltd.: Amsterdam, The Netherlands, 2018.
25. Maoa, W.; Rychlik, I.; Wallin, J.; Storhaug, G. *Statistical Models for the Speed Prediction of a Container Ship*; Elsevier Ltd.: Amsterdam, The Netherlands, 2016.
26. Binieli, M. Machine Learning: An Introduction to Mean Squared Error and Regression Lines. 2018. Available online: www.freecodecamp.org (accessed on 21 June 2021).

Journal of
Marine Science and Engineering

MDPI

Article

Real-Time Management of Vessel Carbon Dioxide Emissions Based on Automatic Identification System Database Using Deep Learning

Yongpeng Wang [1], Daisuke Watanabe [2,*], Enna Hirata [3] and Shigeki Toriumi [4]

1 Graduate School of Marine Science and Technology, Tokyo University of Marine Science and Technology, Tokyo 135-8533, Japan; m185018@edu.kaiyodai.ac.jp
2 Department of Logistics and Information Engineering, Tokyo University of Marine Science and Technology, Tokyo 135-8533, Japan
3 Center for Mathematical and Data Sciences, Kobe University, Kobe 657-8501, Japan; enna.hirata@platinum.kobe-u.ac.jp
4 Department of Information and System Engineering, Chuo University, Tokyo 112-8551, Japan; toriumi@ise.chuo-u.ac.jp
* Correspondence: daisuke@kaiyodai.ac.jp; Tel.: +81-3-5245-7367

Abstract: In this study, we propose an effective method using deep learning to strengthen real-time vessel carbon dioxide emission management. We propose a method to predict real-time carbon dioxide emissions of the vessel in three steps: (1) convert the trajectory data of the fixed time interval into a spatial–temporal sequence, (2) apply a long short-term memory (LSTM) model to predict the future trajectory and vessel status data of the vessel, and (3) predict the carbon dioxide emissions. Automatic identification system (AIS) database of a liquefied natural gas (LNG) vessel were selected as the sample and we reconstructed the trajectory data with a fixed time interval using cubic spline interpolation. Applying the interpolated AIS data, the carbon dioxide emissions of the vessel were calculated based on the International Towing Tank Conference (ITTC) recommended procedures. The experimental results are twofold. First, it reveals that vessel emissions are currently underestimated. This study clearly indicates that the actual carbon dioxide emissions are higher than those reported. The finding offers insight into how to accurately measure the emissions of vessels, and hence, better execute a greenhouse gases (GHGs) reduction strategy. Second, the LSTM model has a better trajectory prediction performance than the recurrent neural network (RNN) model. The errors of the trajectory endpoint and carbon dioxide emissions were small, which shows that the LSTM model is suitable for spatial–temporal data prediction with excellent performance. Therefore, this study offers insights to strengthen the real-time management and control of vessel greenhouse gas emissions and handle those in a more efficient way.

Keywords: vessel trajectory prediction; LSTM; cubic spline interpolation; LNG; CO_2 emissions

1. Introduction

1.1. Greenhouse Gases (GHGs) of Vessels

With global warming progressing worldwide, each government's emission management for greenhouse gases (GHGs) is becoming more meticulous. GHG emissions from vessels are receiving increasing attention, especially in port cities, since vessels have become a major source of polluting emissions. Most governments have introduced emission tax rates to reduce the GHG emissions of vessels, especially emissions of carbon dioxide (CO_2), which is one of the most important greenhouse gases.

Since automatic identification system (AIS) database store the detailed real-time trajectory data of vessels, we can use longitude, latitude, speed over ground (SOG) and course over ground (COG) to accurately, and in real-time, estimate the carbon dioxide emissions while the vessel is sailing. Many studies have been carried out to estimate vessel inventories using AIS data. Mabunda et al. [1] estimated a ship emission inventory near the strait of Gibraltar. Winther et al. [2] implemented emission inventory estimation in the Arctic though a S-AIS (satellite automatic identification system). The International Maritime Organization [3] implemented full-scale vessel emission inventory analysis using AIS data. Coello et al. [4] estimated an emission inventory from a UK fishing fleet, and Yao et al. [5] estimated vessel emission inventories in an estuary of the Yangtze River. Wang et al. [6] used the Ship Traffic Emissions Assessment Model (STEAM2) to estimate the fuel consumption of a vessel, though wave and wind resistance were not considered and the average speed had insufficient accuracy. Kim et al. [7] considered that the accuracy of the calculation could be improved by solving the problem of an unstable AIS data interval.

1.2. *Vessel Trajectory Prediction in Deep Learning*

Generally, vessel trajectory data can be represented as a set of multi-dimensional spatial–temporal sequences $\{(p_1, a_1, t_1), (p_2, a_2, t_2), (p_3, a_3, t_3), \ldots, (p_n, a_n, t_n)\}$, where p_i is the position (longitude, latitude), a_i is the data attribute (SOG, COG) and t_i is the recording time of the AIS data.

Traditional trajectory prediction methods are linear prediction methods, but linear methods have limitations when it comes to constructing the kinematics equations of the vessel. Some traditional methods apply Gaussian-based models. Laxhammar et al. [8] proposed a Gaussian mixture model that combines multiple probability distributions to model vessel trajectories. Its disadvantage, however, was in determining the number and quantity of Gaussian components; few Gaussian components made it difficult to describe the vessel's trajectory, and too many were prone to overfitting. Zhao et al. [9] proposed an improved Kalman filter algorithm with system noise estimation to predict the vessel's trajectory; however, this method of setting parameters was complicated and difficult to reproduce. In addition, the marine environment had a great influence on the trajectory of vessels, which made research difficult.

Considering the inefficiency of linear relationships for different spatial–temporal data, as discussed in previous studies, we opted to use the neural network method to fit the relationship between the historical trajectory data and the next-time data. Deep learning is an algorithm that uses artificial neural networks as a framework to perform characterization learning on data. It can automatically learn the information in the data, and can learn the essential characteristics of the dataset. As a subset of machine learning methods, deep learning automatically extracts data features, which reduces the manual extraction workload and improves prediction efficiency.

Some scholars have applied deep learning to AIS data prediction. Quan et al. [10] established the recurrent neural network-long short-term memory (RNN-LSTM) model to predict vessel trajectory. Nguyen et al. [11] used the sequence-to-sequence model to predict vessel trajectory, while the encoder and decoder used long short-term memory (LSTM). However, the vessel trajectory data were spatial–temporal sequence data that had both temporal and spatial correlation. The SOG of the vessel will affect the distance the vessel travels, and the COG of the vessel will affect the position of the vessel. Therefore, some scholars proposed using the convolutional layer to extract the spatial characteristics of the trajectory data, and then used the LSTM model to analyze the temporal characteristics. This also achieved precise prediction results. For example, Ljunggren et al. [12] used convolutional networks of different sizes to learn spatial–temporal sequences in trajectory features, separate the different features in the data and finally, synthesize a feature to perform trajectory prediction. Wang and Liu [13] proposed a convolutional neural network model combined with LSTM, studying the spatial structure and temporal characteristics of vessel trajectory data simultaneously; the prediction effect was better than using the convolutional neural network alone, but the prediction error was not overwhelming compared to the LSTM model. There are also studies that examine the accuracy of machine learning methodologies on real datasets in vessels [14,15].

Existing studies suggest that the LSTM model is a suitable algorithm for vessel trajectory prediction applications. However, studies predicting emissions applying reconstructed trajectory data in deep-learning models have been scarce. This study aims to fill the research gap.

1.3. The Motivation of the Study

The motivation of this study is to strengthen the real-time management of vessel carbon dioxide emissions and to control vessel GHG emissions in a more efficient way. To achieve that, we propose a method that can accurately grasp the current emissions of the ship and predict the future emissions status. We consider that the incompleteness of the AIS trajectory data will affect the accuracy of the vessel carbon dioxide emissions estimation. As such, we propose using the cubic spline interpolation method to enrich the vessel trajectory. With this method, we can estimate the vessel carbon dioxide emissions every 1 s to improve the estimation accuracy. Thereafter, we use the LSTM model to learn the historical SOG characteristics and trajectory characteristics of vessels, to predict the spatial–temporal distribution of vessels' future carbon dioxide emissions.

Deep learning can learn the basic characteristics of the data from a small number of samples through a deep non-linear network structure, which is suitable for rapid analysis of vessel data in the marine environment, so we chose deep learning. To improve the validity of the prediction data, we emphasized a real-time and short-term nature. Figure 1 outlines the flow of this study.

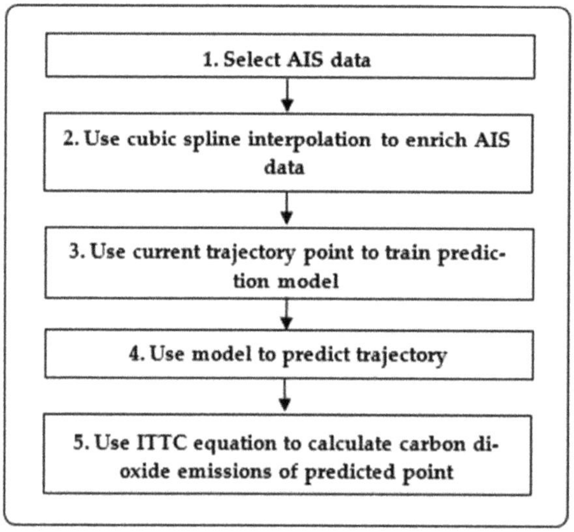

Figure 1. Flow chat of this study.

The rest of the paper is structured as follows. Section 2 describes the formation of the cubic spline interpolation model for reconstructing trajectory data, the LSTM model for prediction and the CO_2 emission estimation model. Section 3 discusses the experiments and results. Section 4 summarizes the work and the prospects for future studies.

2. Equations of the Model

2.1. Cubic Spline Interpolation Model

Due to incorrect operation of the AIS system by shore and vessel personnel, information transmission failure between AIS and the shore base, subjective and objective factors such as the random failure of the AIS system itself or the problem of artificial improper maintenance, we need to select a method of interpolation and repair the trajectory to obtain more accurate and complete vessel trajectory information. Compared with the linear interpolation model, the non-linear interpolation model is more in line with the actual sailing state of the ship; especially near the coast, the ship cannot always keep a straight line sailing, but will adjust the course according to different sea conditions. Therefore, we select the method of cubic spline interpolation to smooth the vessel trajectory.

We suppose that the interval of trajectory data is $[a, b]$, divided $[a, b]$ into n intervals, as $[(x_0, x_1), (x_1, x_2), \cdots, (x_{n-1}, x_n)]$, $x_0 = a$, $x_n = b$, where the function expression for each interval is $S(x)$. Cubic spline means that the curve of each interval is a cubic equation $S_i(x)$ that meets the interpolation conditions $S(x_i) = y_i$ and meets the condition of a smooth curve where $S_i(x)$, $S_i'(x)$, $S_i''(x)$ are a continuous function. The solved equation (Bartels et al. [16]) is as follows:

$$S_i(x) = a_i + b_i(x - x_i) + c_i(x - x_i)^2 + d_i(x - x_i)^3 \tag{1}$$

$$S_i'(x) = b_i + 2c_i(x - x_i) + 3d_i(x - x_i)^2 \tag{2}$$

$$S_i''(x) = 2c_i + 6d_i(x - x_i) \tag{3}$$

where $S_i(x)$ is the cubic spline model expression, and a_i, b_i, c_i, d_i are the parameters to be solved.

Accordingly, $S_i(x)$ must meet interpolation conditions $S(x_i) = y_i$ and Equations (1)–(3). Then, we can obtain the equation as follows:

$$a_i = y_i \tag{4}$$

$$h_i = x_{i+1} - x_i \tag{5}$$

$$a_i + b_i h_i + c_i h_i^2 + d_i h_i^3 = y_{i+1} \tag{6}$$

According to continuous function conditions,

$$S_i''(x_{i+1}) = S_{i+1}''(x_{i+1}) \tag{7}$$

$$S_i'''(x_{i+1}) = S_{i+1}'''(x_{i+1}) \tag{8}$$

We can obtain the equation as follows:

$$b_i + 2h_i c_i + 3h_i^2 d_i = b_{i+1} \tag{9}$$

$$2c_i + 6h_i d_i = 2c_{i+1} \tag{10}$$

$$d_i = \frac{m_{i+1} - m_i}{6h_i} (m_i = 2c_i) \tag{11}$$

Inputting Equations (4), (10) and (11) into Equation (6), we can obtain the equation as follows:

$$b_i = \frac{y_{i+1} - y_i}{h_i} - \frac{h_i}{2} m_i - \frac{h_i}{6}(m_{i+1} - m_i) \tag{12}$$

By inputting Equations (4) and (10)–(12) into Equation (9), we can obtain the equation as follows:

$$h_i m_i + 2(h_i + h_{i+1}) m_{i+1} + h_{i+1} m_{i+2} = 6\left[\frac{y_{i+2} - y_{i+1}}{h_{i+1}} - \frac{y_{i+1} - y_i}{h_i}\right] \tag{13}$$

We build linear equations with m as the unknown ($m_0 = 0$, $m_n = 0$):

$$\begin{bmatrix} 1 & 0 & 0 & 0 & \cdots & 0 \\ h_0 & 2(h_0+h_1) & h_1 & 0 & \cdots & 0 \\ 0 & h_1 & 2(h_1+h_2) & h_2 & \cdots & 0 \\ 0 & 0 & h_2 & 2(h_1+h_2) & h_3 & \vdots \\ \vdots & \vdots & \ddots & \ddots & \ddots & 0 \\ 0 & 0 & 0 & h_{n-2} & 2(h_{n-2}+h_{n-1}) & h_{n-1} \\ 0 & 0 & 0 & 0 & 0 & 1 \end{bmatrix} \begin{bmatrix} m_0 \\ m_1 \\ m_2 \\ m_3 \\ \vdots \\ m_n \end{bmatrix} = 6 \begin{bmatrix} 0 \\ \frac{y_2-y_1}{h_1} - \frac{y_1-y_0}{h_0} \\ \frac{y_3-y_2}{h_2} - \frac{y_2-y_1}{h_1} \\ \vdots \\ \frac{y_n-y_{n-1}}{h_{n-1}} - \frac{y_{n-1}-y_{n-2}}{h_{n-2}} \\ 0 \end{bmatrix} \tag{14}$$

We can calculate $m_0, m_1, \cdots, m_n$ from Equation (14) and use it to calculate a_i, b_i, c_i, d_i and know the function expression for each interval to enrich the vessel trajectory data.

2.2. Long Short-Term Memory Model

The LSTM model (Hochreiter et al. [17]) is a variant of the recurrent neural network (RNN). The RNN cannot learn longer histories' data, resulting in a gradient decline or even disappearance at further time steps. To solve this problem, the LSTM model introduces storage units and unit states to control information transfer based on the RNN.

There are four gates (forget gate, input gate, update gate and output gate) in the storage unit of the LSTM model. The input gate controls the addition of new information. The forget gate can forget information that needs to be discarded and retain the useful information of the past. The update gate can update data. The output gate causes the storage unit to output only information related to the current time step. These four gate structures perform matrix multiplication and non-linear summing in the memory cells so that the memory does not decay in constant iterations.

As shown in Figure 2, the structure of LSTM neural network consists of three layers: input layer, hidden layer (LSTM_Layer_1 and Other Layer) and output layer. Figure 3a shows the structure of the RNN. Its cell contains only one activation function, and the cells are only linked in order. Figure 3b is the structure of LSTM. Its cell is more complex than the RNN. It needs four gate calculations to output to the next cell. Figure 3c is an enlarged view of Figure 3b.

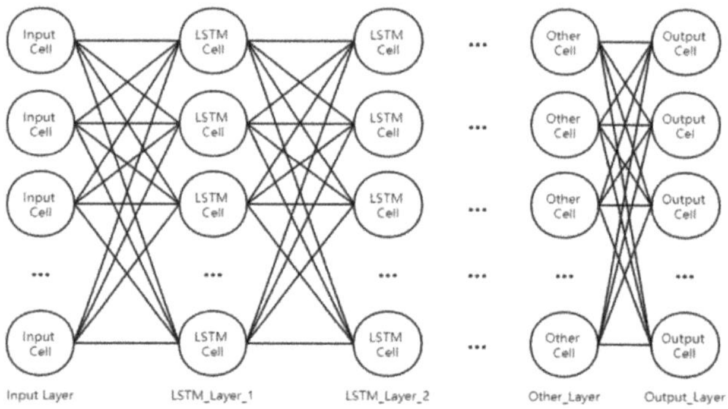

Figure 2. Structure of long short-term memory (LSTM) neural network.

Forget Gate: For time t, the state h_{t-1} at the previous time and the current training data x_t can get f_t through the forget gate; the formula is as follows:

$$f_t = \sigma\left(W_f \cdot [x_t, h_{t-1}] + b_f\right) \tag{15}$$

Input Gate: $\widetilde{C}_t$ decides the amount that can be added to the cell state in the *tanh* network layer, and i_t outputs a number between 0 and 1 through the Sigmoid network layer to decide which status values to update; the formula is as follows:

$$\widetilde{C}_t = tanh(W_c \cdot [x_t, h_{t-1}] + b_c) \tag{16}$$

$$i_t = \sigma(W_i \cdot [x_t, h_{t-1}] + b_i) \tag{17}$$

Update Gate: Updates old state C_{t-1} to new state C_t. This gate retains long-term and short-term memory in different proportions of the cell; the formula is as follows:

$$C_t = i_t * \widetilde{C}_t + f_t * C_{t-1} \tag{18}$$

Output Gate: The third sigmoid network layer determines parts of the output cell state, combined with Equation (20), to obtain the output value of the cell; the formula is as follows:

$$o_t = \sigma(W_o \cdot [x_t, h_{t-1}] + b_o) \tag{19}$$

$$h_t = o_t * \theta_h(C_t) \tag{20}$$

where f_t: forget gate, $\widetilde{C}_t$: input cell, i_t: input gate, C_t: update gate, o_t: output gate, σ: sigmoid activation function, W: weights for different gates, x_t: input value at time t, h_{t-1}: output value at the previous moment, and b: bias term for different gates.

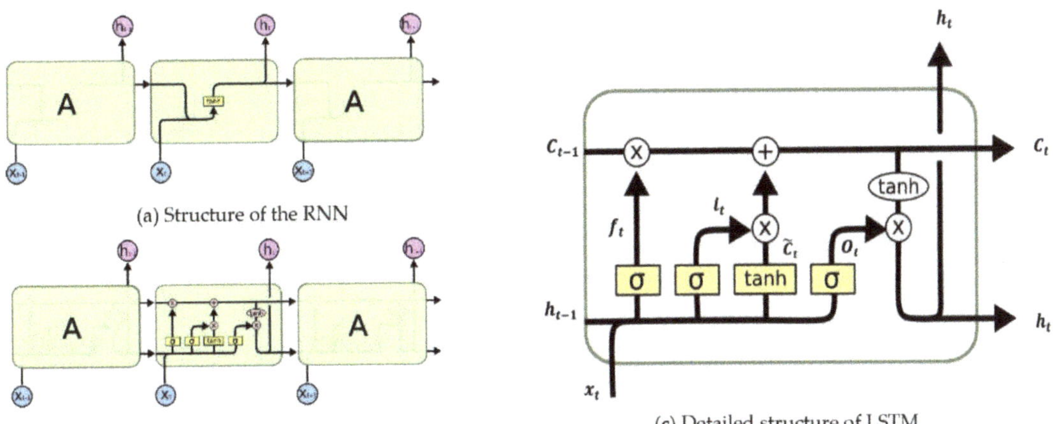

Figure 3. Structure of recurrent neural network (RNN) and LSTM cells. Source: adapted from Olah [18] with permission from Olah, C., 2021.

2.3. CO_2 Emission Estimation Model

In this study, we adopted the ITTC recommended procedure [19,20] to estimate vessel resistance. It is necessary to derive the total resistance first. Total resistance can be denoted as:

$$R_T = \frac{1}{2} C_T \rho S V^2 \tag{21}$$

where R_T: total resistance, C_T: total resistance coefficient, ρ: density of water, S: wetted surface of the hull, V: SOG.

C_T, the total resistance coefficient, can denoted as:

$$C_T = C_F + C_A + C_{AA} + C_R \tag{22}$$

where C_F: frictional resistance coefficient, C_A: incremental resistance coefficient, C_{AA}: air resistance coefficient, C_R: residual resistance coefficient.

Based on calculated total resistance of the vessel, an estimation of the required power for the vessel to sail at speed V in a calm sea condition can be calculated by considering the components of the propulsion efficiencies. Installed power is the power required to tow the vessel with speed V in a calm sea. Service power can be derived from:

$$P_I = \frac{R_T V}{(\eta_T \eta_D)} + m \quad (23)$$

where P_I: installed power, η_T: transmission efficiency,η_D: quasi-propulsive coefficient, m: sea margin.

Fuel oil consumption is calculated by using the specific fuel oil consumption (SFOC) as shown in Table 1 from Smith et al. [21]. Table 1 shows the value of the SFOC for each diesel engine type with engine age. As the vessel engine becomes older, the efficiency of the engine decreases and the advent of technology makes a newer engine more efficient.

Table 1. Specific fuel oil consumption (SFOC) and distribution of engine age of vessels included in automatic identification system (AIS) data.

Engine Age	Slow-Speed Diesel (SSD) Engine	Medium-Speed Diesel (MSD) Engine	High-Speed Diesel (HSD) Engine
Before 1983	205	215	225
1984–2000	185	195	205
Post–2001	175	185	195

The marine liquefied natural gas (LNG) CO_2 emissions factor is 2.75. So, the CO_2 emission estimation model can be denoted as:

$$E_i = \sum P_I \times SFOC \times 2.75 \times T \quad (24)$$

where E_i, the total CO_2 emissions, is calculated by summing the carbon dioxide emissions at each trajectory point; T is the time interval of each track point.

3. Experiments and Results

3.1. Automatic Identification System (AIS) Dataset Analysis

This study used AIS data of an LNG vessel provided by exactEarth, as shown in Figure 4. The time period was from 1 January 2016 to 30 June 2016, and number of messages was about 50 million. It included vessel names, callsigns, maritime mobile service identities (MMSIs), vessel types, vessel cargos, vessel classes, lengths, widths, flag countries, destinations, estimated times of arrival (ETAs), draughts, longitudes, latitudes, SOGs, COGs, rates of turn (ROTs), headings, navigation (nav) statuses, source, times, main vessel types and sub vessel types.

The AIS data used this research took a comma-separated values (CSV) form. Every datum was divided by day based on Greenwich Mean Time (GMT). A total of 182 days were included, so the complete data of the vessel were divided into 182 small datasets. This was inconvenient, so we used MySQL to build a trajectory database combining the 182 small datasets into a large dataset.

This trajectory dataset mainly included: MMSI, longitude, latitude, SOG, COG and time.

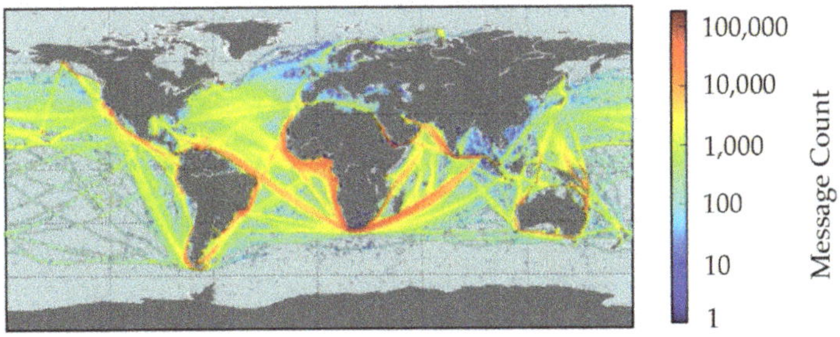

Figure 4. Heat map of data.

In the original dataset, we found that the average time interval of each piece of AIS data in Table 2 was 520.52 s. Furthermore, 25% of the data had an interval time within 6 s, 50% were within 17 s, and 75% were within 42 s. According to Figure 5, we also found that more than 90% of the data had a time interval of more than 30 min. Only about 8% of the data had a time interval of 2 s. This may be because AIS data collected through satellites show longer data collecting intervals when the vessel was sailing in areas with high traffic compared to areas with less traffic.

Table 2. Summary statistics of data interval.

Items	Hour	Minutes	Seconds
Count (number)	9,072,291	9,072,291	9,072,291
Mean	0.14	8.68	520.52
25%	0.00	0.10	6.00
50%	0.00	0.28	17.00
75%	0.01	0.70	42.00

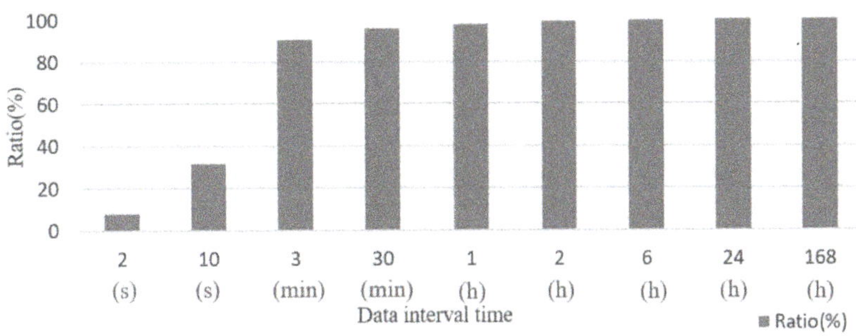

Figure 5. Distribution of data interval time.

3.2. Interpolation Calculation

After database analysis, taking the data integrity and continuity as the principle, we extracted the LNG vessel trajectory data of MMSI 310028000 from the trajectory database as a sample to illustrate the feasibility of the model. The vessel trajectory over six months is shown in Figure 6. The vessel travels mainly between Japan and Australia; the IMO number is "8913174", built in 1992.

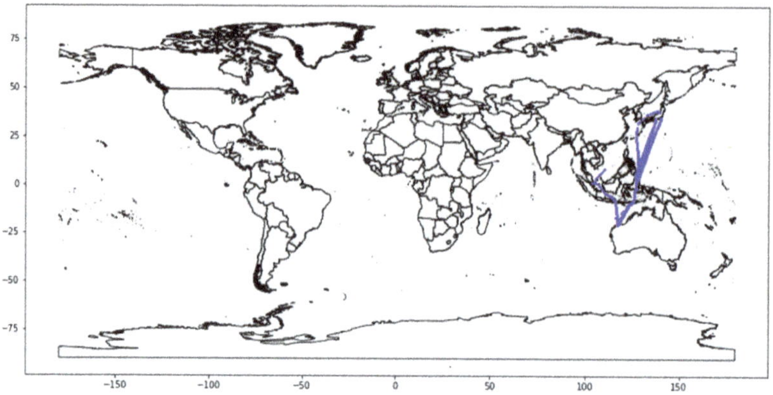

Figure 6. Distribution of sample vessel trajectory.

The focus of this research was to grasp the carbon dioxide emissions of vessels in real-time. We chose to use vessel data from 5 January 2016 over 00:35:10–00:59:40 for analysis. The reason for the selection was that the AIS data has relatively larger data points during the period, which could minimize the workload of trajectory data reconstruction. In addition, the trajectory trend is a straight line instead of a curve, which is more appropriate for simplicity. The total time interval was 26 min, and there were 23 trajectory data points, as shown in Table 3. For example, we found that the data intervals were different: 1 s, 7 s, 11 s, 12 s, 7 s from 00:35:10 to 00:35:48. Therefore, this study proposes using the cubic spline interpolation method to resample the data at one second intervals.

Table 3. Partial sample data.

MMSI	Time	Longitude	Latitude	SOG	COG
310028000	00:35:10	118.6957	−17.0403	15.9	24.7
310028000	00:35:11	118.6961	−17.0395	15.8	25.1
310028000	00:35:18	118.6963	−17.0391	15.9	25.1
310028000	00:35:29	118.6966	−17.0383	15.8	25
310028000	00:35:41	118.697	−17.0375	15.8	24.8
310028000	00:35:48	118.6972	−17.0371	15.8	25.2

We used Equations (1)–(14) to calculate the trajectory interpolation. Taking longitude as an example, $x_i = longitude_i$. The partial calculation results are shown in Figure 7. We found that the line connecting the original data points was no longer a simple straight line (orange color), but had become a curve (blue color), which was more in line with the actual situation. In terms of the actual geographic location, we could see the restored trajectory after interpolation calculation from Figure 8, where the yellow point is the trajectory point of the vessel at 00:59:40, the red point is the trajectory point on 5 January, the orange point is on other days in January and the green points are interpolation points.

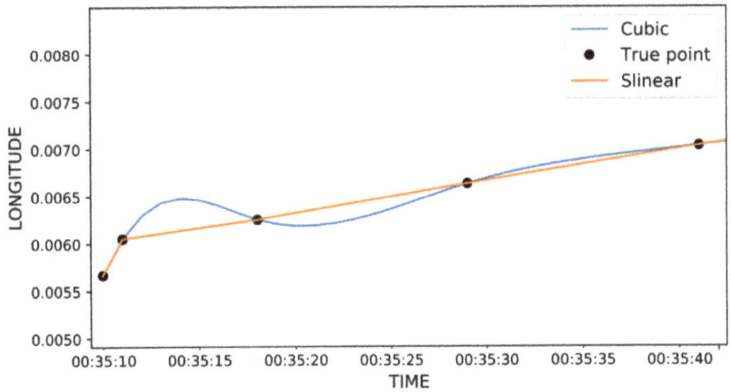

Figure 7. Distribution of longitude interpolation.

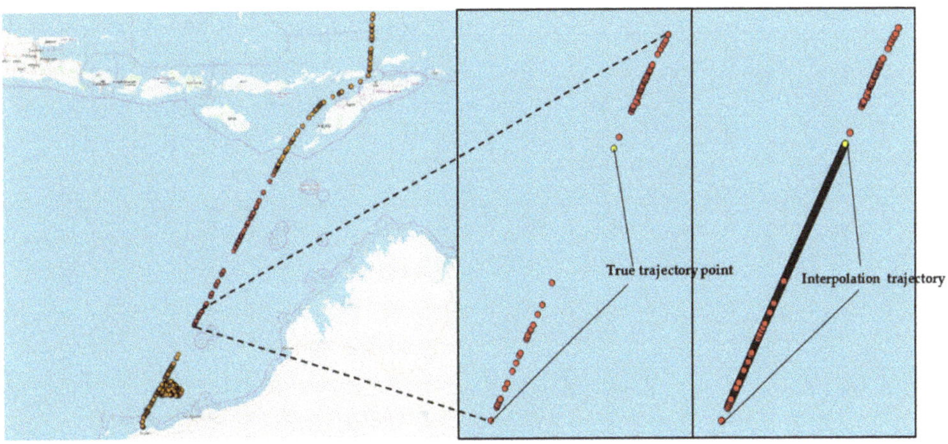

Figure 8. Geographical distribution of trajectory interpolation.

3.3. Vessel Trajectory Prediction

The recurrent neural network is a typical framework for deep learning; it can be used to deal with spatial–temporal sequence problems. The characteristics are that the output of the current moment depends on the calculation result of the previous moment and that the timing is strong. The LSTM model is an improvement on the recurrent neural network as it also analyzes historical calculation results that are much older, automatically removes invalid historical calculation results, and remembers useful historical calculation results by introducing gating functions. The timeliness is stronger than the recurrent neural network.

As described in Section 1.2, for a vessel, its trajectory characteristics at time t could be expressed as $Y_{(t)} = \{(p_1, a_1, t_1),(p_2, a_2, t_2),(p_3, a_3, t_3), \ldots ,(p_n, a_n, t_n)\}$; we could use this as an input value of the LSTM model. The number of data after the interpolation calculation changed from 23 to 1472. The experimental environment for this study was the DELL OptiPlex 7050 desktop computer, the CPU was Intel(R) Core (TM) i7-7700 CPU @3.60GHz, the memory was 16.0 GB, the operating system was Windows10 Pro, the program development environment was PyCharm (JetBrains, s.r.o., Prague, Czech Republic) in python 3.7 and we used an LSTM model provided by Keras (Chollet, F., & others, Retrieved from https://github.com/fchollet/keras (accessed on 20 July 2021)). After experiments and manual adjustment of parameters, the optimal parameters of the LSTM model in this study were finally determined. As shown in Table 4, the time required to run the model once was 195 s.

Table 4. LSTM model parameters.

Items	Parameters	Items	Parameters
Base Learning Rate	0.001	LSTM_layer_1	256
Optimizer	Adaptive moment estimation	LSTM_layer_2	128
Epoch	125	Dropout_layer _1	128
Batch size	138	Dense_layer _1	128
Loss function	Mean square error	Dropout_layer _2	128
Activation_1	Tanh	Dense_layer _2	4
Activation_2	Linear	Kernel_initializer	Orthogonal
Train set	938	Validation set	235
Test set	293	-	-

The loss function is shown in Figure 9. We used 938 pieces of data to train the LSTM model, and finally, 235 pieces of data were used to verify the prediction effect. At an equal time (00:59:40), the predicted trajectory point of the LSTM model differed from the actual trajectory point by 0.593 nm, which was in line with expectations.

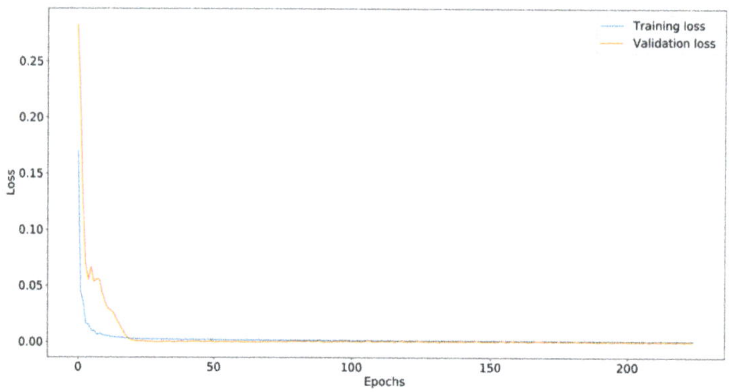

Figure 9. Training and validation loss of LSTM model.

The prediction results are shown in Figure 10, and we further plot these in 2D (Figure 11) and 3D (Figure 12) graphs for visualization.

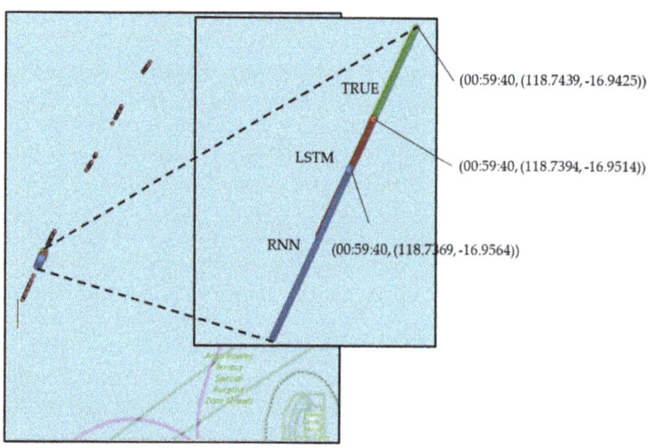

Figure 10. Geographical distribution of predicted results.

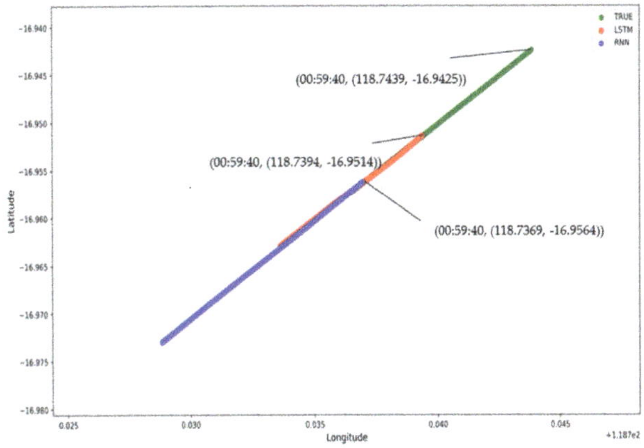

Figure 11. Coordinate distribution of true trajectory and predicted trajectory results (2D).

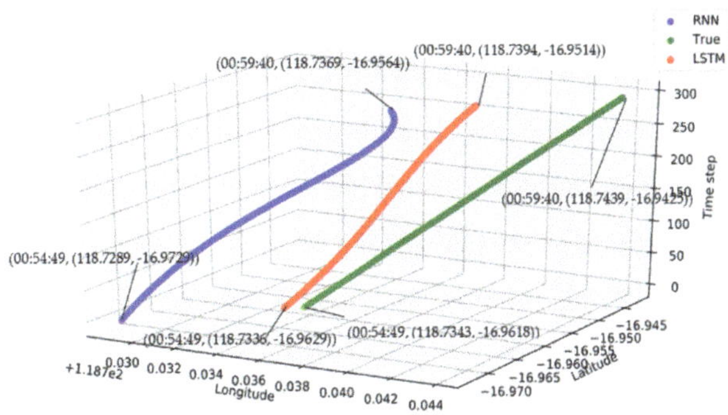

Figure 12. Spatial–temporal distribution of true trajectory and predicted trajectory results (3D).

As Figure 10 shows, the trajectories predicted by the two models roughly coincided with the actual trajectories. In Figures 11 and 12, the green, red and blue lines indicate the trajectory points of the vessel during the period of study, where the green line indicates the true trajectory point, the red point indicates the trajectory point predicted by the LSTM model and the blue point represents trajectory point predicted by the RNN model. The trajectory point predicted by the LSTM (118.7394, −16.9514) is closer to the true value (118.7439, −16.9425) than the trajectory point predicted by the RNN model (118.7369, −16.9564).

In Figure 11, we intercepted the prediction result from original trajectory to visualize the differences. Three trajectories had the same start point (the first orange point at the lower left in Figure 10) but the predicted endpoints were different. The LSTM model has a smaller error (0.593 nautical miles or 1.097 km) than that of the RNN model (0.928 nautical miles or 1.718 km). This can be observed in Figure 12 where the trajectory predicted by the RNN was more concentrated at the end point, causing the trajectories to overlap, while the LSTM model did not have this defect. The time step was the time difference from 00:54:49 (in seconds).

3.4. Carbon Dioxide Estimation

We first calculated the carbon dioxide emissions of the vessel without interpolation by the ITTC; using Equation (24), we could determine the emissions of the vessel at 00:35:10–00:59:40 to be 47,169 kg for the period of study. The carbon dioxide emissions after interpolation totaled 74,926 kg for the same period. This was 27,757 kg more carbon dioxide emissions than determined by traditional estimation methods. Our proposed model was able to interpolate more time intervals, and hence, improve the accuracy of the vessel carbon dioxide emissions. Observing the vessel carbon dioxide emissions from a spatial–temporal perspective, they go down first and then up (Figures 13 and 14).

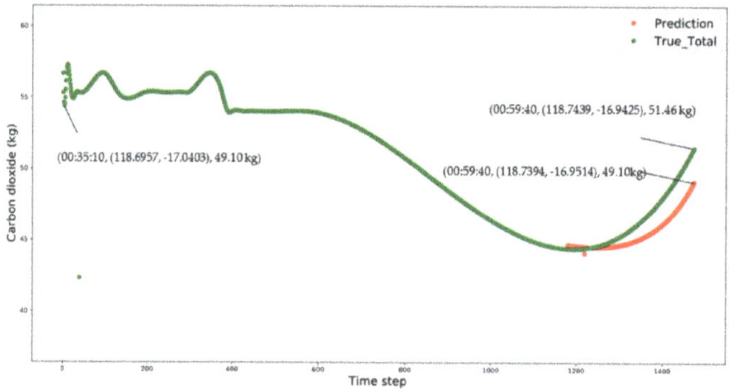

Figure 13. Temporal distribution of total true and predicted results in comparison.

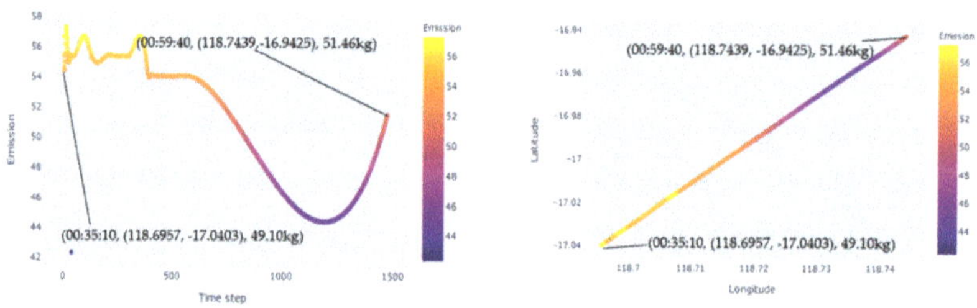

(a) Temporal distribution (b) Spatial distribution

Figure 14. Spatial–temporal distribution of total true results.

Next, we used the trajectories predicted by the LSTM model to estimate the carbon dioxide emissions, to obtain the spatial–temporal distribution of future carbon dioxide emissions from vessels. With the vessel emissions per second obtained, we calculated that the cumulative carbon dioxide emissions from vessels during 00:54:49–00:59:41 were 13,315 kg. We predicted that the highest value would be reached in this area (118.7439, −16.9425).

The predicted carbon dioxide emissions (13,315 kg) and the true emissions (13,616 kg) gave an error of 301 kg. A heat map is shown of the spatial–temporal distribution of carbon dioxide emissions from vessels every second in specific section from 00:54:49 to 00:59:40 in Figure 15. Observing the vessel carbon dioxide emissions from a time perspective, the actual and predicted values increased during this period. We found that the growth rate of the predicted value was slower than the growth rate of the true value. At 00:59:41, the error was 2.36 kg (51.46–49.10 kg).

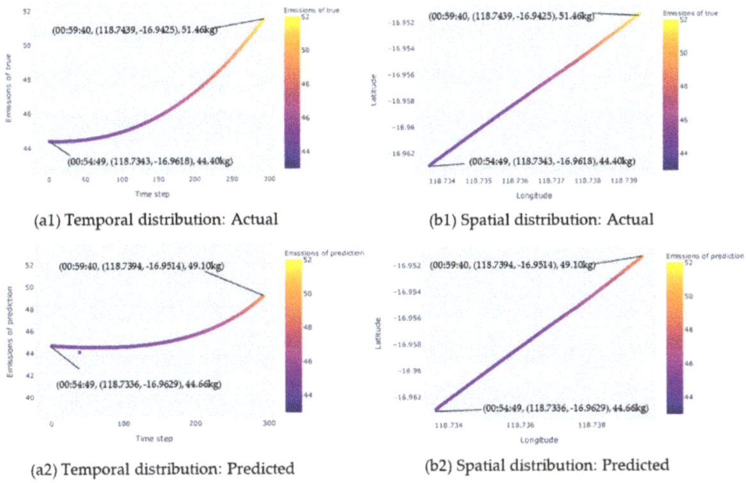

Figure 15. Spatial-temporal distribution of true and predicted results for specific section.

Moreover, it was found that the vessel emissions in the future would continue to increase. The deceleration trend indicated that the vessel would accelerate to the northeast in the future. On the other hand, because we had the data of the carbon dioxide emissions of the vessel in seconds, we could re-establish any time interval to reduce the number of vessel trajectory points if needed. That practice is useful for calculating a larger number of vessel carbon emission trajectories, for example, to calculate the carbon dioxide emissions of the vessel in 3 s, 6 s or 10 s intervals (Figure 16). We can see from Figure 16 that the larger the number of time steps, the faster increase of carbon dioxide emissions, which also indicates that the SOG of the vessel was accelerating. At the same time, the number of trajectory points decreased as the time interval decreased, and the vessel carbon dioxide emissions represented by each track point gradually increased, but the overall trajectory trend remained the same.

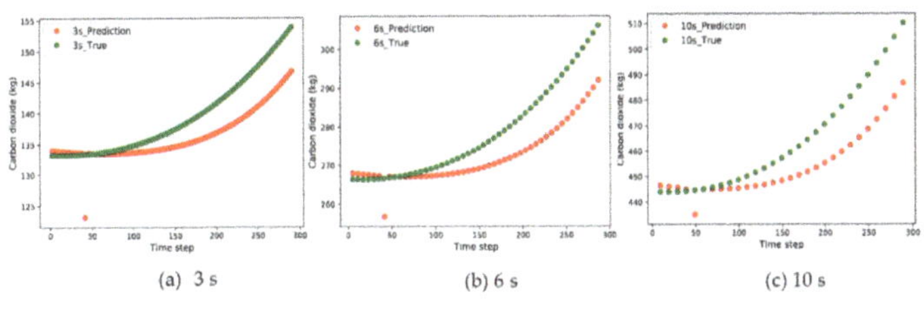

Figure 16. Spatial distribution of true and predicted carbon dioxide emissions with time intervals.

4. Conclusions

In this study, we proposed a method to interpolate AIS data and estimate vessel emissions based on predicted vessel trajectories using deep learning models. The results suggest that the carbon dioxide emissions are underestimated. The method can be applied to monitor a vessel's carbon dioxide emissions and trajectory in real time, and to provide vessels with early warning services relating to carbon dioxide emissions monitoring.

The contributions of the study are fourfold. Firstly, it has novelty in interpolating AIS data points with a non-linear model. AIS data sometimes have time intervals that are too long due to equipment issues and/or human reasons, which limits the accuracy of the vessels' carbon dioxide emissions estimation based on AIS data. To overcome the problem, this study proposes a cubic spline interpolation model to resample missing AIS data points. The repaired trajectories of the vessel (Figure 8) maintain the same trend as the original trajectory in smooth curves, which represent the actual situation well when compared to the traditional linear models in previous studies. Through the interpolation, we could successfully obtain AIS data with a time interval of one second.

Secondly, it reveals that vessel emissions are currently underestimated. In the study, by applying the repaired trajectory data of the vessel (1472 pieces of data), we estimated the carbon dioxide emissions of the vessel to be 74,926 kg. Compared to the emissions calculated from the original trajectory data (23 pieces of data), a 27,757 kg increase of carbon dioxide emissions was identified. This clearly indicates that the actual carbon dioxide emissions are higher than those reported. The finding offers insight into how to accurately measure the emissions of vessels, and hence, better execute a GHG reduction strategy.

Thirdly, it validated the performance of two deep-learning algorithms in predicting vessel trajectories. We conducted prediction with two deep-learning algorithms, LSTM and RNN. The results confirm that the LSTM model performs better. We used the predicted SOG to predict the future carbon dioxide emissions of the vessel. The error was as small as 301 kg, which shows that the LSTM model was suitable for spatial–temporal data prediction with excellent performance.

Fourthly, it offers an alternative solution to predict vessel trajectories when AIS data is not retrievable. The model can also be used to estimate vessel emissions when AIS data are not available on occasions such as the shutdown of AIS for artificial reasons (including active or passive) or when the collection of AIS data fails in certain time periods.

At present, the method has limitations since it has only been tested in cases where the vessel trajectory is not comprehensive. It is unclear about the interpolation effect and trajectory prediction effect on curved trajectories, which need further research with several cases considering the computational time in the future. Future studies may also verify different artificial neural network (ANN) models such as the feedforward neural network (FFNN) or 1D convolutional neural networks (1D-CNN), which may work well on time-series data [14].

Author Contributions: Conceptualization, Y.W. and D.W.; methodology, Y.W. and E.H.; software, Y.W.; validation, Y.W., D.W., E.H. and S.T.; data curation, S.T.; writing—original draft preparation, Y.W.; writing—review and editing, Y.W., D.W., E.H. and S.T.; supervision, D.W. All authors have read and agreed to the published version of the manuscript.

Funding: This research was funded by the Japan Society for the Promotion of Science (JSPS KAKENHI), grant numbers 21H01564, 18K04601 and 17H02042.

Institutional Review Board Statement: Not applicable.

Informed Consent Statement: Not applicable.

Data Availability Statement: Not applicable.

Conflicts of Interest: The authors declare no conflict of interest.

References

1. Mabunda, A.; Astito, A.; Hamdoune, S. Estimating Carbon Dioxide and Particulate Matter Emissions from Ships using Automatic Identification System Data. *Int. J. Comput. Appl.* **2014**, *88*, 27–31.
2. Winther, M.; Christensen, J.H.; Plejdrup, M.S.; Ravn, E.S.; Eriksson, O.F.; Kristensen, H.O. Emission inventories for ships in the arctic based on satellite sampled AIS data. *Atmos. Environ.* **2014**, *91*, 1–14. [CrossRef]
3. International Maritime Organization (IMO). *MEPC 75/7/15: Reduction of GHG Emissions from Ships: Fourth IMO GHG Study 2020—Final Report*; International Maritime Organization (IMO): London, UK, 2020.
4. Coello, J.; Williams, I.; Hudson, D.A.; Kemp, S. An AIS-based approach to calculate atmospheric emissions from the UK fishing fleet. *Atmos. Environ.* **2015**, *114*, 1–7. [CrossRef]
5. Yao, X.; Mou, J.; Chen, P.; Zhang, X. Ship emission inventories in estuary of the Yangtze River using terrestrial AIS data. *TransNav* **2016**, *10*, 633–640. [CrossRef]
6. Wang, Z.H.; Wang, W.X.; Shi, X. AIS data-based ship emission estimation model and real ship verification. *J. Shanghai Marit. Univ.* **2019**, *40*, 12–16.

7. Kim, H.; Watanabe, D.; Toriumi, S.; Hirata, E. Spatial Analysis of an Emission Inventory from Liquefied Natural Gas Fleet Based on Automatic Identification System Database. *Sustainability* **2021**, *13*, 1250. [CrossRef]
8. Laxhammar, R.; Falkman, G.; Sviestins, E. Anomaly detection in sea traffic—A comparison of the Gaussian Mixture Model and the Kernel Density Estimator. In Proceedings of the 12th International Conference on Information Fusion, Seattle, WA, USA, 6–9 July 2009; pp. 756–763.
9. Zhao, S.B.; Tang, C.; Liang, S.; Wang, D. Track prediction of vessel in controlled waterway based on improved Kalman filter. *J. Comput. Appl.* **2012**, *32*, 3247–3250. [CrossRef]
10. Quan, B.; Yang, B.; Guo, C.; Li, Q. Prediction Model of Ship Trajectory Based on LSTM. *Comput. Sci.* **2018**, *45*, 126–131.
11. Nguyen, D.D.; Le Van, C.; Ali, M.I. Vessel trajectory prediction using sequence-to-sequence models over spatial grid. In Proceedings of the 12th ACM International Conference on Distributed and Event-based Systems, Hamilton, New Zealand, 25–29 June 2018; pp. 258–261.
12. Ljunggren, H. Using Deep Learning for Classifying Ship Trajectories. In Proceedings of the 21st International Conference on Information Fusion (FUSION), Cambridge, UK, 10–13 July 2018; pp. 2158–2164.
13. Wang, L.L.; Liu, J. Ship behavior recognition method based on multi-scale convolution. *J. Comput. Appl.* **2019**, *39*, 3691–3696.
14. Theodoropoulos, P.; Spandonidis, C.C.; Themelis, N.; Giordamlis, C.; Fassois, S. Evaluation of Different Deep-Learning Models for the Prediction of a Ship's Propulsion Power. *J. Mar. Sci. Eng.* **2021**, *9*, 116. [CrossRef]
15. Theodoropoulos, P.; Spandonidis, C.C.; Giordamlis, C.; Fassois, S. Stability and Safety of Ships and Ocean Vehicles Enhancing vessel 's operational safety using Convolutional Neural Networks and Big Data techniques. In Proceedings of the 1st International Conference on the Stability and Safety of Ships and Ocean Vehicles (STAB&S), Online, Glasgow, UK, 6–11 June 2021.
16. Bartels, R.; Beatty, J.; Barsky, B. Hermite and Cubic Spline Interpolation: Chapter 3. In *An Introduction to Splines for Use in Computer Graphics and Geometric Modelling*; Morgan Kaufmann: Burlington, MA, USA, 1995; pp. 9–17.
17. Hochreiter, S.; Schmidhuber, J. Long Short-Term Memory. *Neural Comput.* **1997**, *9*, 1735–1780. [CrossRef] [PubMed]
18. Olah, C. Understanding LSTM Networks. 2015. Available online: https://colah.github.io/posts/2015-08-Understanding-LSTMs/ (accessed on 7 August 2021).
19. International Towing Tank Conference. *ITTC—Recommended Procedures and Guidelines*; Procedure 7.5-02-02-01, Revision 04; ITTC Association: Zürich, Switzerland, 2017.
20. Molland, A.F.; Turnock, S.R.; Hudson, D.A. *Ship Resistance and Propulsion: Practical Estimation of Ship Propulsive Power*, 2nd ed.; Cambridge University Press: Cambridge, UK, 2017.
21. Smith, T.W.P.; Jalkanen, J.P.; Anderson, B.A.; Corbett, J.J.; Faber, J.; Hanayama, S.; O'Keeffe, E.; Parker, S.; Johansson, L.; Aldous, L.; et al. *Third IMO GHG Study 2014*; International Maritime Organization: London, UK, 2015.

Journal of
Marine Science and Engineering

MDPI

Article

GMDSS Equipment Usage: Seafarers' Experience

Sanjin Valčić *, Antonio Škrobonja, Lovro Maglić and Boris Sviličić

Faculty of Maritime Studies, University of Rijeka, 51000 Rijeka, Croatia; skrobo@pfri.hr (A.Š.); maglic@pfri.hr (L.M.); svilicic@pfri.hr (B.S.)
* Correspondence: svalcic@pfri.hr

Abstract: Maritime non-commercial communications are performed using the equipment of Global Maritime Distress and Safety System (GMDSS), under the International Convention for the Safety of Life at Sea (SOLAS). This equipment includes numerous devices of different technologies, features and user interfaces, which every officer of the watch (OOW) must know how to use. Besides, the equipment, i.e., its technology has not changed significantly in the last 30 years. Therefore, the purpose of this research was to collect and analyze information on how often GMDSS devices are used and whether any modernization is needed. The main research method was a questionnaire survey intended for experienced officers. Each question, as well as the answers of the respondents, were analyzed in detail and presented graphically. The results of the survey analysis indicate that the VHF (Very High Frequency) radiotelephone, Navigational Telex (NAVTEX) receiver, and Enhanced Group Call (EGC) receiver are still very used, while Digital Selective Calling (DSC) devices and MF/HF (Medium Frequency/High Frequency) radios (telephone and telex) are almost unused onboard. Additionally, more than 80% of respondents believe that GMDSS needs modernization.

Keywords: GMDSS; frequency of use; electronic radio log; VDES; e-Navigation

1. Introduction

The GMDSS, which was fully implemented on 1 February 1999, was expected to represent, at the time, a revolution in maritime communications. Its main purpose is to alert and warn coast centers and surrounding ships in cases of distress, urgency, and safety [1]. It can also be used for routine (commercial) communications. According to the GMDSS requirements, the amount of communication equipment on ships, depends on the area of navigation (A1–A4) and includes terrestrial and satellite devices. These devices are shown in Figure 1.

However, from the day of implementation until today, GMDSS technology has not changed significantly and some parts of it have become, at the very least, obsolete. Thus, the system as a whole has not reached its full potential. Therefore, the International Maritime Organization's (IMO) Maritime Safety Committee (MSC) in 2009 at its 86th session decided to initiate the modernization of the GMDSS, which is still in development phase [2,3]. As for the scientific literature review, there is a scarce research in general on the subject of the GMDSS.

Even before the IMO's decision on GMDSS modernization and shortly after its full implementation, a survey conducted on 100 respondents concluded that the design of the GMDSS and the increasing workload on the ship's bridge, constitute the main drawbacks to its operability [4]. In addition, based on another survey with the same number of respondents, the GMDSS was established to be operationally complex and its design was not satisfactory [5]. However, the same GMDSS devices are still used onboard SOLAS ships since then.

As for the use of GMDSS equipment, the authors found only two researches on that subject. The IMO's Sub-committee on Navigation, Communications and Search and Rescue (NCSR) presented analysis results of user survey on GMDSS modernization submitted by the Republic of Korea [6]. On a sample of 129 respondents, it was found that MF/HF installation is the most unhelpful and underused GMDSS equipment, followed by NAVTEX and portable VHF radiotelephone. However, a questionnaire of 16 not very well-structured open-ended questions was used as a data collection method, on which only a qualitative analysis (which was not even described) could have been conducted. In addition, this study examined how often a respondent uses a DSC device in general. In other words, DSC devices were not separated by the frequency band (VHF, MF/HF). The same applies to the INMARSAT-C and the EGC receiver (which was not even mentioned in the paper). Furthermore, research on the use of GMDSS equipment and the Electronic Chart Display and Information System (ECDIS) on a sample of 260 seafarers on oceangoing vessels was also conducted [7]. It was concluded, based on the opinion of the respondents, that INMARSAT (telex) and MF/HF (telephone and telex) are the most unused devices in the GMDSS. However, the methodology of data collection was not described. It was only stated that a survey was conducted among seafarers on oceangoing vessels. In addition, this study did not examine the frequency of use of any DSC device. Thus, both of these studies have some drawbacks.

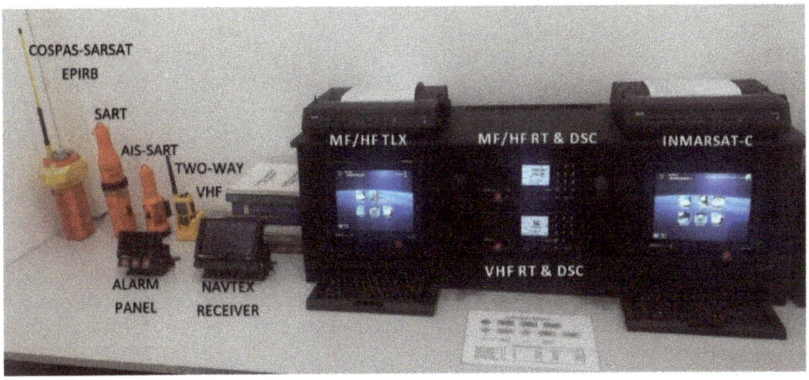

Figure 1. Typical Global Maritime Distress and Safety System (GMDSS) radio station and equipment for sea area A3.

Among other researches related to the modernization of maritime communications (including GMDSS), the authors highlight papers that have analyzed technological advancements and challenges in maritime communications and the integration of some GMDSS equipment with the Integrated Navigation System (INS) [8–11]. However, these researches did not investigate the frequency of use of GMDSS devices and the problem of GMDSS modernization among seafarers.

Prompted by the modernization of the GMDSS and drawbacks of the previous researches, the authors of this paper researched to investigate the extent to which every single existing GMDSS device is used (or not) on ships in international voyages (SOLAS ships), excluding locating devices (SART, AIS-SART, and EPIRB). Additionally, the authors wanted to examine the views of experienced seafarers on GMDSS modernization in the form of spatial equipment reduction, bridge integration, and radio log digitization and automation. Therefore, the main hypothesis of the conducted research, which sought to be confirmed, is that some GMDSS equipment is not practical and efficient for use in modern shipping and the modernization of the GMDSS is required. Furthermore, the main goal of the research is to get first-hand confirmation whether seafarers believe that GMDSS system or a part of it should be modernized or not. An additional goal is to prove that some GMDSS devices are still necessary and frequently used onboard ships.

2. Methodology

As a part of the GMDSS modernization evaluation and proposal, the authors developed an exploratory study to investigate the frequency of use of GMDSS equipment. The main method for data collection was a questionnaire survey conducted in an online environment (FreeOnlineSurveys.com, accessed on 15 October 2020). The questionnaire was anonymous and participation in it was on a voluntary basis. However, each respondent was asked to indicate age, rank, years of experience as an officer, and their nationality for the research purposes (this was indicated in the cover letter). The questionnaire consisted of only 10 questions (items), from which 8 were closed type and 2 were combined, offering closed answers (indicators) and providing an open section for comments. Such a simple questionnaire is not time consuming, and yet very valuable information about seafarers' personal experience can be obtained from it. Therefore, in order to prove the need for GMDSS modernization, the authors in the first 8 questions offered answers based on the 5-point Likert scale to indicate the average frequency of use of each GMDSS device (VHF radiotelephone, VHF DSC, MF/HF radiotelephone, MF/HF radio telex, MF/HF DSC, INMARSAT-C, NAVTEX receiver, and EGC receiver) during one watch. The following numerical answers were offered to the first six questions: "0 times", "1–2 times", "3–4 times", "5–10 times", and ">10 times", with the additional option "Equipment not installed", offered on questions 3., 4., 5., and 6., since not every ship needs to be equipped with all GMDSS devices. The following text-based answers were offered to the seventh and eighth question: "Upon message receipt", "Periodically during watch", "On daily basis", "On weekly basis", and "On monthly basis", with the option "Not reading". All of these answers were used for quantitative data analysis, which involves simple descriptive statistics in Microsoft Excel software. On the other hand, in the last 2 questions, the authors offered answers based on the 3-point Likert scale to examine whether respondents agree with the authors' suggestions regarding spatial reduction and integration of GMDSS equipment and radio log digitization. Those offered text-based answers were: "Completely agree", "Partially agree", and "Disagree at all". Additionally, in these questions, respondents were able to give their own opinions, which were used for qualitative data analysis shown in the Survey Analysis and Results section. The authors used the data triangulation method to better present the opinions of seafarers on the integration and reduction in GMDSS equipment and the digitization of the radio log, since the triangulation facilitates validation of quantitative data through cross-verification using qualitative data.

The authors used the so-called snowball sampling: the link to the online questionnaire was sent via e-mail to the Faculties of Maritime Studies of the Republic of Croatia, selected crewing agencies and seafarers' companies in order to be distributed among experienced OOW's. Therefore, the target group, i.e., the sample frame of this research was the OOW's on SOLAS ships holding a Certificate of Competence for ships of 3000 GT or more (this was also indicated in the cover letter). The questionnaire was available from 4 April 2020, to 4 October 2020. In that period, the answers were collected, exported to the Microsoft Excel software, and analyzed on a sample of 112 respondents. Data on the respondents were derived using descriptive statistics in Microsoft Excel software (Figure 2).

Regarding the limitations of the survey, it is important to note that the use of GMDSS equipment also depends on the navigation area, which was not considered in this survey, i.e., not every ship needs to be equipped with all GMDSS devices. However, it was indicated in the questionnaire that the participants should answer according to their personal experience. Furthermore, it should be emphasized that the survey was conducted anonymously and is based on the subjective responses of the respondents. Therefore, the results should be used with caution. However, the results coincide to a good extent with the already mentioned previous researches (regardless of different sample of the respondents), as will be shown hereinafter.

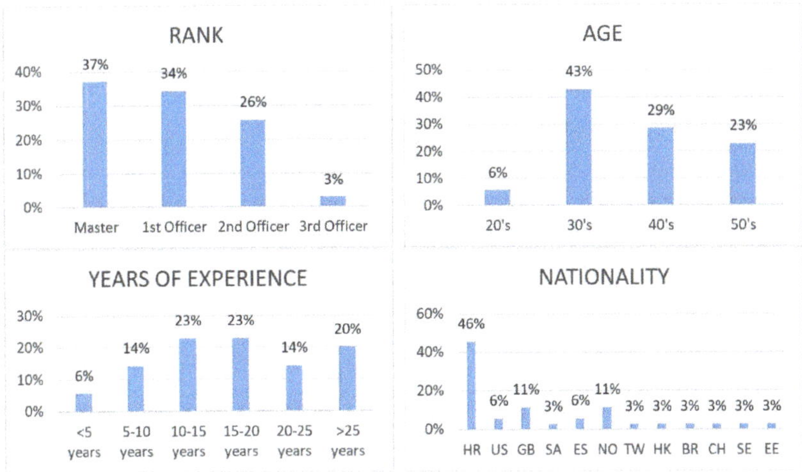

Figure 2. Rank, age, years of experience, and nationality of respondents.

3. Survey Analysis and Results

In this section, each question of the survey is presented, followed by the results and analysis. In the last subsection, the reliability and correlation analyses of the survey are presented.

3.1. *On Average, How Often Do You Use a VHF Radiotelephone for any Type of Communication, during One Watch?*

The authors are convinced that the VHF radiotelephone is still one of the most used methods of terrestrial radiocommunications in the maritime world, which is confirmed by the respondents' answers (Figure 3).

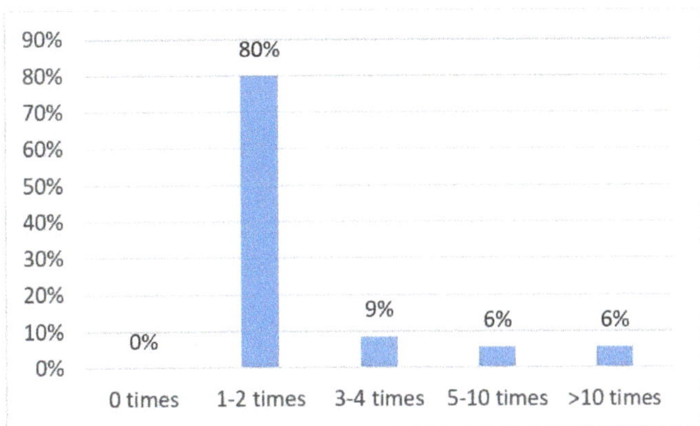

Figure 3. Frequency of use of the VHF radiotelephone.

It can be seen that 80% of the respondents answered that they use a VHF radiotelephone at least once during the watch. However, such result is expected, as it is generally known that the radiotelephone in question is, due to the principle of operation, the most reliable and fastest way to establish a contact with surrounding ships or coast facilities. Additionally, ships subject to the SOLAS Convention must maintain a continuous watch on the VHF channel 16 [12]. The obtained results also correlate with the results of researches conducted in [6] (96% of the respondents answered that they often use the equipment and 91% of the respondents answered that the equipment is absolutely necessary) and [7] (the first most used GMDSS device).

3.2. *On Average, How Often Do You Use a VHF DSC for Calling (Excluding Test Calls), during One Watch?*

VHF DSC, as a special feature of VHF radio station, is extremely rarely used for sending ship-to-shore and/or ship-to-ship calls, excluding safety test calls.

Figure 4 shows that 82% of respondents mainly do not use the DSC functionality of a VHF radio station to send a call, while the remaining 18% of respondents answered that on average, they use VHF DSC only 1–2 times during a single watch. The survey did not ask to specify the reasons for the minimal use of this functionality, but it can be assumed that some of the main reasons are: very low data rate (1200 bps), compared to modern technologies such as Telenor Maritime Radio (133 kbps) or WiMAX in sub-GHz frequencies (over 20 Mbps) [8,9]; very limited range (line of sight), compared to satellite coverage; not very user-friendly interface—in terms of a small display with complex menu trees [13]; and in general, sending a DSC call is a time demanding task compared to a voice call. It should also be noted that the survey conducted in [6] examined how often respondents use DSC in general (either VHF or MF/HF), while the authors of this paper wanted to separate the frequency bands. On the other hand, the survey conducted in [7] did not consider DSC devices at all.

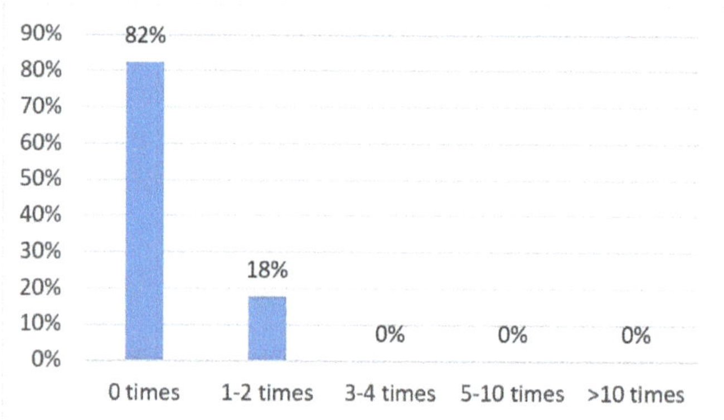

Figure 4. Frequency of use of the VHF Digital Selective Calling (DSC).

3.3. *On Average, How Often Do You Use MF/HF Radiotelephone for any Type of Communication, during One Watch?*

The MF/HF radiotelephone is certainly a communication system that is no longer recognized among seafarers as useful, either because of its complexity of use (ionospheric effects on ground and sky waves) or because of outdated analogue technology (SSB modulation).

Figure 5 shows that almost 90% of respondents never use an MF/HF radiotelephone during one watch, while others possibly use it once or twice. Furthermore, none of the respondents answered that they did not have this device installed onboard, although it is not mandatory in the GMDSS sea area A1. The obtained results on the frequency of use of this device do not correlate with the results of researches conducted in [6] (53% of the respondents answered that they often use the equipment) and [7] (the third most frequently used device). However, respondents in these researches largely agree that the MF/HF installation is unhelpful, difficult to operate, and requires functional improvement.

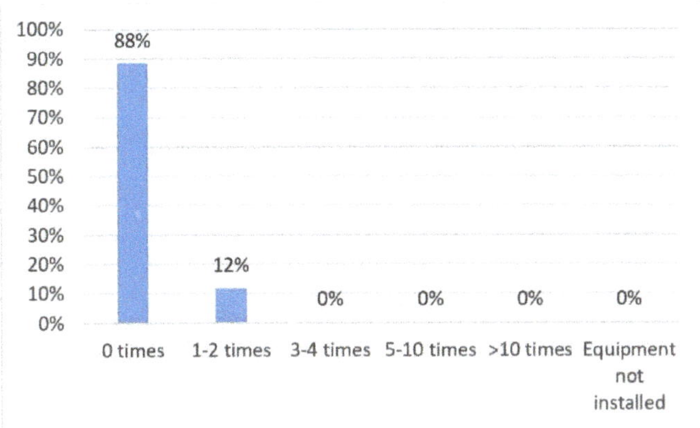

Figure 5. Frequency of use of the MF/HF radiotelephone.

3.4. On Average, How Often Do You Use MF/HF Radio Telex for Calling, during One Watch?

The radio telex as an optional device for SOLAS ships navigating in the GMDSS sea area A3 and mandatory for SOLAS ships navigating in the GMDSS sea area A4 is also rarely used or not used at all, especially among the younger generations of seafarers. The reason is that radio telex requires a complex procedure for today's seafarers, providing slow data transmission (100 bps) and it is generally known that more and more ships use the Internet protocol, i.e., e-mail to send information. In terms of Internet/e-mail availability in the polar (A4) areas, more and more ships are using different terrestrial and/or satellite telecommunications systems that provide these features, among others. This further pushes radio telex to the margins of maritime long-range communications systems, although it is still mandatory for ships in the A3 (if there is no INMARSAT equipment) or A4 sea area.

Figure 6 shows the frequency of use of radio telex, where it can be seen that 88% of respondents never use this device, 9% of respondents use it once or twice during one watch, while only 3% of respondents stated that they did not have this device installed on a ship. The obtained results also correlate with the results of researches conducted in [6] (75.2% of the respondents answered that they do not use the equipment) and [7] (the first most useless device).

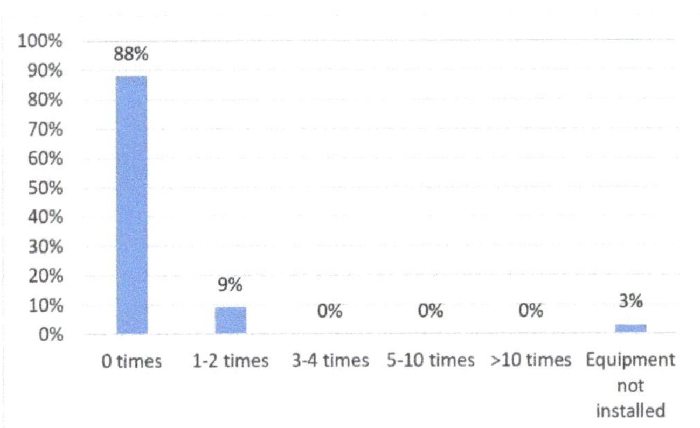

Figure 6. Frequency of use of the MF/HF radio telex.

3.5. On Average, How Often Do You Use MF/HF DSC for Calling (Excluding Test Calls), during One Watch?

The MF/HF DSC as a special possibility of transmitting a digital selective call on MF/HF frequencies also confirms the research hypothesis that MF/HF radio installation as a whole is almost not used in today's maritime radiocommunications, which is confirmed by the survey results shown on Figure 7.

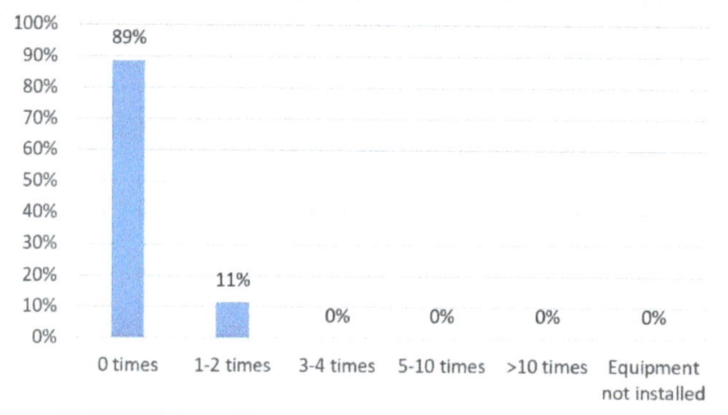

Figure 7. Frequency of use of the MF/HF DSC.

As can be seen, as many as 89% of respondents do not use the MF/HF DSC for calling at all, although no one answered that the device was not installed on the ships on which they navigated. The reasons for this are the same as already stated for VHF DSC with the emphasis that the data rate of MF/HF DSC is 12 times less than the data rate of VHF DSC.

3.6. On Average, How Often Do You Use INMARSAT-C Ship Earth Station (Excluding EGC Receiver) for Any Type of Communication, during One Watch?

The INMARSAT-C ship Earth station (SES) has remained on ships to these days as the only satellite communication device to provide GMDSS communication functions (excluding the COSPAS-SARSAT EPIRB, which is intended for ship-to-shore alerting and locating). Until recently (December 2020), INMARSAT Fleet 77 also existed, but like its predecessors, INMARSAT-A and INMARSAT-B, it was discontinued due to new satellite communications systems and services within GMDSS, which provide more features compared to it (Iridium, Inmarsat Fleet Safety) [14]. Given the poor capabilities of the INMARSAT-C device in terms of data rate (600 bps) and store and forward messaging (when high-speed Internet on board is often available), the device is still used on ships, but to a lesser extent. Its biggest advantage over modern satellite communication devices is its small omnidirectional antenna and economic cost-effectiveness. Besides, INMARSAT-C can be effectively used for Long Range Identification and Tracking (LRIT) and Ship Security Alert System (SSAS) [15].

As can be seen from Figure 8, 55% of respondents stated that they did not use INMARSAT-C, 36% of them answered that they used the device up to 2 times during one watch, 6% of them answered that they used it up to 4 times, while only 3% of respondents stated that they did not have it installed onboard. If these results are compared with the results of researches conducted in [6] (71% of the respondents answered that they often use the equipment) and [7] (the third most frequently used device), a correlation can be seen regarding its use. Namely, this device is still used onboard ships, but the authors in those researches concluded that it requires functional improvement.

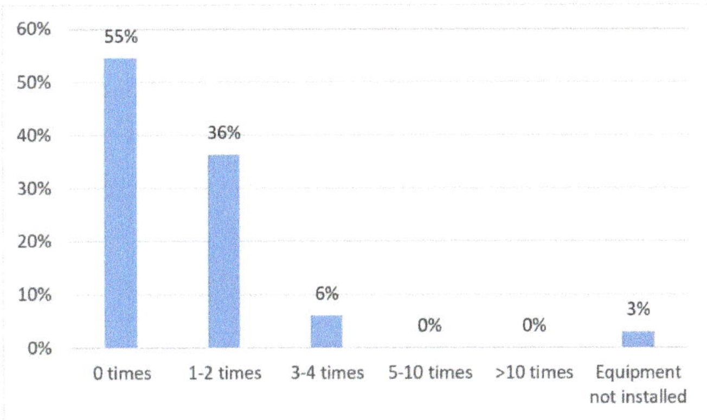

Figure 8. Frequency of use of the INMARSAT-C ship Earth station (SES).

3.7. How Often Do You Read Received NAVTEX Messages?

The NAVTEX system and its receiver on ships, as part of GMDSS terrestrial communications for receiving Maritime Safety Information (MSI), are still very used onboard ships, as shown by the results of the survey (Figure 9).

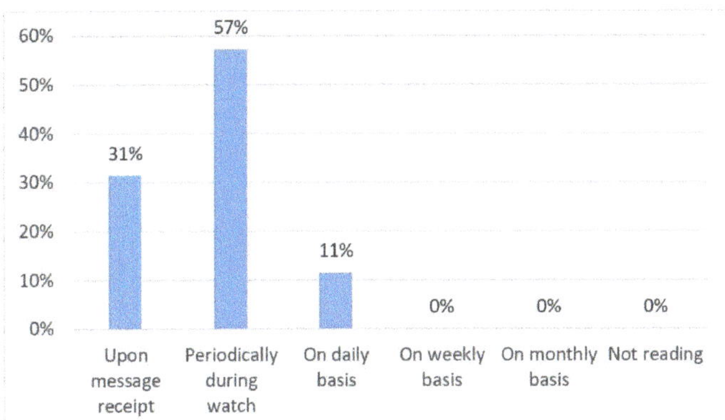

Figure 9. Frequency of use of the Navigational Telex (NAVTEX) receiver.

The authors wanted to examine and show how often respondents read NAVTEX messages received on a NAVTEX receiver, and it can be seen that 31% of respondents read messages immediately upon receipt, 57% read them periodically during one watch, while the remaining 11% read messages daily. As it is known, the device as such has not undergone some serious modernization from the time of the official introduction, but it is also a known fact that a certain degree of integration with other systems, such as ECDIS, has been achieved [16]. In addition, that is the main reason why this GMDSS device is used more when compared to other GMDSS devices. Thus, it can be said that the survey results confirm the justification of the existence and future use of this system.

3.8. How Often Do You Read Received EGC Messages?

When it comes to the INMARSAT, i.e., EGC system, it has to be emphasized that it has the universality in informing ships in the form of weather forecasts, notifications for sailors, fleet or single ship, etc. (SafetyNET and FleetNET). Typically, an EGC receiver is integrated into an INMARSAT-C SES, but it has different functionality, and therefore, the authors wanted to specifically examine the use of both devices.

If the survey responses shown in Figure 10 are compared with the survey responses to the previous question, a correlation can be observed in the use of these two systems, emphasizing that the EGC system is not integrated into any type of navigation system, such as ECDIS. It can be concluded that the system is still used onboard ships because 97% of the respondents answered that they read received messages, while only 3% of respondents do not read them at all.

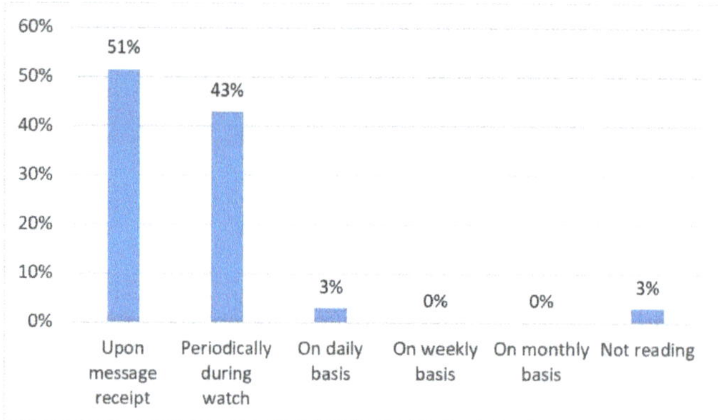

Figure 10. Frequency of use of the Enhanced Group Call (EGC) receiver.

3.9. *According to You, Should GMDSS Equipment Be Dimensionally Reduced and Integrated on the Ship's Bridge, Which Would Then Have a Unique Communication Interface and Thus Serve to Store and Automate the Communication Process?*

After examining the frequency of use of individual GMDSS equipment, the authors wanted to examine the opinions of respondents on the spatial reduction in GMDSS equipment and its integration on a ship's bridge. The answers to this question indicate that modernization in general is necessary, in order to increase the efficiency and ease the information collection and logging process (Figure 11).

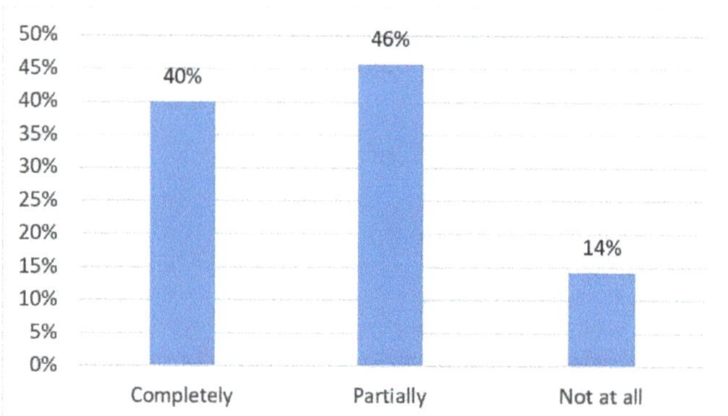

Figure 11. Respondents' compliance with space reduction and integration of GMDSS equipment.

The proposed GMDSS integration aims to reduce the administrative effort of OOW's. Figure 11 shows that 46% of respondents partially agree that integration and reduction in GMDSS equipment dimensions would be good to some extent, while 40% of respondents fully agree with the authors' proposal. However, it is important to note that there is a certain amount of caution (14% of respondents) among seafarers when it comes to further integration. Not only were quantitative answers offered to this question but respondents were also able to write their own opinion. Quantitative and qualitative analysis found that only 40% of the respondents wrote their additional opinion. Qualitative analysis implied content analysis. After all additional answers, i.e., opinions of the respondents were collected, their coding was performed in order to create meaningful categories for easier analysis. Respondents' opinions and comments were coded using 8 codes: NAVTEX integration, VHF DSC integration, EGC integration, Overall simplification, Overall space reduction, Overall integration, Overall modernization, and Changeless. After coding the qualitative data, a triangulation with quantitative data was performed, in order to verify the additional comments of the respondents. Namely, the answers offered to this question quantified the extent to which the respondents agree with the authors' proposal on the reduction and integration of GMDSS equipment: "Completely", "Partially", and "Not at all" (Figure 11). Using the triangulation method, the authors concluded that additional comments (qualitative data) can be compared with the offered answers (quantitative data). The codes NAVTEX integration, VHF DSC integration and EGC integration support the quantitative answer "Partially", since some of the respondents stated that they would only like to integrate some of devices on the ship's bridge. The codes Overall simplification, Overall space reduction, Overall integration, and Overall modernization support the quantitative answer "Completely", since some of the respondents stated that they would like to completely simplify, modernize, reduce, and/or integrate all of the GMDSS devices. Finally, the code Changeless supports the quantitative answer "Not at all", since some of the respondents stated that they would not like to change anything in the GMDSS equipment. The triangulation method proved that some GMDSS equipment is not practical and efficient for use in modern shipping and that partial or complete modernization of GMDSS is required, which was the basic hypothesis of this research.

It is interesting to note that among the respondents who provided their additional opinion, 21% would not change anything regarding GMDSS equipment. What is important to emphasize is that such an opinion cannot be correlated with the age of the respondents, because within that 21% of respondents who believe that nothing should be changed, there are seafarers of all ages (20–50 years). The same applies to respondents who provided comments in the form of changing, integrating, or reducing GMDSS equipment.

Some of the additional comments, which the authors found significant, are "Definitely GMDSS equipment needs to be reduced and integrated into an information system, like ECDIS with a unique communication interface"—56 years; "DSC is unnecessary equipment that distracts the watch with a large number of messages, rather than contributes"—53 years; "It is time for this system to be modernized and adapted to the times in which we live"—35 years; and "It would be good to integrate VHF DSC to send reports automatically"—31 years.

3.10. Do You Agree That the Radio Log Should Be Automated and Exclusively Stored in Electronic form, i.e., in a Form Filled on ECDIS?

This question aimed to prove that a certain type of information centralization is needed, which could be obtained by automated and electronic GMDSS radio log.

As can be seen in Figure 12, 88% of respondents partially or completely agree with the proposed change in this particular example, while as in the previous question, there is a certain caution towards changes (12% of respondents). As with the previous question, the respondents were able to give their own opinion regarding the proposed automated electronic radio log. After all the additional comments of the respondents were collected, the authors performed their coding with the following codes: Digitalization, Automation, Data oversight, and Changeless. Again, a triangulation with quantitative data was performed, in order to verify the additional comments of the respondents. Namely, the answers offered to this question quantified the extent to which the respondents agree with the authors' proposal on electronic and automated radio log: "Fully agree", "Partially agree", and "Disagree at all" (Figure 12). The code Automation matches the quantitative answer "Fully agree", since some of the respondents wrote that they would like complete automation of the radio logging process. The codes Digitalization and Data oversight match the quantitative answer "Partially agree", since some of the respondents wrote that they would like an electronic form of the radio log, but believe that its full automation would lead to data oversight. Again, the code Changeless matches the quantitative answer "Disagree at all", since some of the respondents stated that they would not like to change anything regarding the radio log.

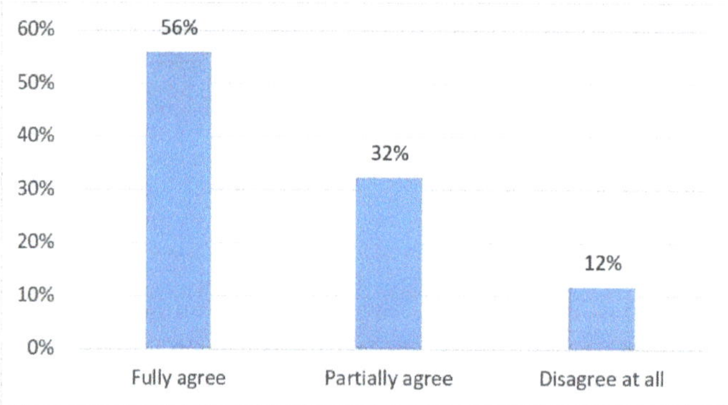

Figure 12. Respondents' compliance with automated electronic radio log.

Of the respondents who gave their additional opinion (29% of respondents), 40% would not want an automated electronic radio log. Again, such an opinion cannot be correlated with the age of the respondents, because within this 40% of the respondents there are seafarers of all ages (20 years–50 years). The same is true for respondents who gave positive comments in terms of paperwork reduction. Some of the additional comments, which the authors found significant are "All ISM should be paperless and electronically filled"—56 years; "The Deck Logbook already exists in electronic form and is used, so in that case, it only needs to be upgraded to contain the radio log in it"—53 years; "I completely agree, for the reason of reducing paperwork and unnecessarily filling in a large number of logs, and devoting more time to navigation"—35 years; and "Given that today more and more navigation is going in the digital direction, communication and 'record-keeping' should be integrated and developed in that direction"—32 years.

3.11. Reliability and Correlation Analyses of the Survey Questions and Answers

Regarding the reliability analysis of the survey, the authors used XLSTAT statistical software for Microsoft Excel. As the questions have different answers, reliability analysis could not be done on the whole dataset. Therefore, first the reliability analysis was done on the first six questions, where the offered answers "0 times", "1–2 times", "3–4 times", "5–10 times", and ">10 times" were numbered from 1 to 5, respectively. The obtained value of Cronbach's alpha was 0.494, which was not acceptable. However, after analyzing the results of the survey, it can be concluded that unlike other devices, the VHF radiotelephone is often used and therefore, the answers to the first question are not consistent with the answers to the other questions. After removing the answer to the first question (for VHF radiotelephone) the value of Cronbach's alpha was 0.751, which indicates a very good reliability of the survey questions.

After that, the reliability analysis was performed on the next two questions, where the answers "Upon message receipt", "Periodically during watch", "On daily basis", "On weekly basis", and "On monthly basis" were also numbered from 1 to 5, respectively. The obtained value of Cronbach's alpha was 0.628, which also indicates a very good reliability of these survey questions.

Finally, the reliability analysis was performed on the last two questions, where the offered answers "Completely agree", "Partially agree", and "Disagree at all" were numbered from 1 to 3, respectively. The obtained value of Cronbach's alpha was 0.758, which also indicates a very good reliability of the survey questions.

Figure 13 represents the Pearson correlation matrix between respondents' answers to every question, including their age. The variable representing the age of the respondents was categorized and numbered (Figure 2), in order to be able to calculate the Pearson correlation coefficients. The values in bold in Figure 13 indicate the significance of the correlation coefficients at a 0.05 significance level.

Correlation matrix (Pearson):

Variables	AGE	VHF RT	VHF DSC	MF/HF RT	MF/HF TLX	MF/HF DSC	INMARSAT-C	NAVTEX	EGC	REDUCTION AND INTEGRATION	ELECTRONIC RADIO LOG
AGE	**1**	0.018	−0.066	−0.101	−0.025	−0.101	**0.428**	−0.101	−0.031	0.179	0.065
VHF RT	0.018	**1**	0.107	−0.057	−0.018	−0.057	−0.144	0.314	**0.409**	−0.278	−0.185
VHF DSC	−0.066	0.107	**1**	**0.622**	**0.456**	**0.622**	0.205	−0.128	−0.075	0.276	**0.488**
MF/HF RT	−0.101	−0.057	**0.622**	**1**	**0.852**	**0.717**	0.301	−0.026	0.164	0.138	0.218
MF/HF TLX	−0.025	−0.018	**0.456**	**0.852**	**1**	**0.530**	0.257	0.101	0.280	0.266	0.185
MF/HF DSC	−0.101	−0.057	**0.622**	**0.717**	**0.530**	**1**	**0.452**	−0.170	−0.164	0.138	**0.350**
INMARSAT-C	**0.428**	−0.144	0.205	0.301	0.257	**0.452**	**1**	−0.115	−0.044	**0.451**	**0.421**
NAVTEX	−0.101	0.314	−0.128	−0.026	0.101	−0.170	−0.115	**1**	**0.461**	0.143	−0.059
EGC	−0.031	**0.409**	−0.075	0.164	0.280	−0.164	−0.044	**0.461**	**1**	−0.114	−0.077
REDUCTION AND INTEGRATION	0.179	−0.278	0.276	0.138	0.266	0.138	**0.451**	0.143	−0.114	**1**	**0.627**
ELECTRONIC RADIO LOG	0.065	−0.185	**0.488**	0.218	0.185	**0.350**	**0.421**	−0.059	−0.077	**0.627**	**1**

Values in bold are different from 0 with a significance level alpha = 0.05

Figure 13. Pearson correlation matrix of the respondents' answers.

To begin with, the low correlation coefficients between the age of the respondents and reduction and integration of GMDSS equipment as well as the electronic radio log confirm the conclusions of the analysis conducted in Sections 3.9 and 3.10—there is no correlation between the age of the respondents and their desire for GMDSS modernization. In addition, according to the coefficients in bold, it is evident that there is a positive correlation in the answers to the last two questions on the reduction and integration of the equipment and electronic radio log, i.e., on the modernization of the GMDSS.

Furthermore, according to the coefficients in bold, a positive correlation can be seen in the responses on the use of VHF DSC, MF/HF radiotelephone, MF/HF radio telex, and MF/HF DSC, which supports the analysis results on the use of these devices in Sections 3.2–3.6, i.e., they are very rarely used during a single watch.

Finally, it can be seen that there is a positive correlation in the answers to the questions on the use of NAVTEX and EGC receivers, i.e., on reading the received safety messages. Again, this correlation supports the results of the analysis in Sections 3.7 and 3.8.

4. Discussion

The main findings of the conducted research are:

1. The VHF radiotelephone, NAVTEX receiver and EGC receiver are often used and therefore still necessary on SOLAS ships;
2. The DSC, in general, as well as MF/HF installation (telephone and telex), are almost unused on SOLAS ships;
3. More than 80% of respondents expect modernization of GMDSS in terms of integration with existing bridge systems and changes to the existing GMDSS technical platform.

Due to its simplicity and efficiency, the VHF radiotelephone is the most widely and frequently used GMDSS communication equipment. Although it uses analogue open VHF channels, it can very easily be used for almost all GMDSS communication functions: alerting (distress call), SAR coordinating communications, on-scene communications, transmission (and reception) of MSI, general communications, and bridge-to-bridge communications. This was confirmed by the analysis of the survey results. Namely, none of the respondents answered that they do not use a VHF radiotelephone, while 6% of them stated that they use it more than 10 times during a single watch. Furthermore, the analysis of the survey results showed that MSI receiving devices (NAVTEX and EGC receiver) are used daily for reading messages related to the safety of navigation either immediately upon receipt or periodically during a single watch.

The DSC (VHF and MF/HF) proved to be a missed modernization concept of the communications at the time of its introduction, because, in addition to not being used for calling, as shown by the survey results, it may act as a distraction during demanding navigational situations [17]. As its name suggests, the device is intended only for calling, not communication. However, it is easier for seafarers to establish a contact by telephone (either satellite or terrestrial) rather than browse and type the necessary data to make a DSC call (MMSI, working channel, etc.). Besides, the analysis of the survey results showed that MF/HF installation (telephone and telex) are seldom used during a single watch on SOLAS ships (almost 90% of respondents are not using that equipment). Although still mandatory in some GMDSS sea areas, outdated technology and the cessation of mandatory watchkeeping have caused their seldom or no use by seafarers.

In addition to examining the frequency of use of GMDSS devices, the authors wanted to examine respondents' opinions on the possible reduction and integration of GMDSS equipment within information systems. The authors believe that the modernization of GMDSS should include the spatial organization of the system (deployment of GMDSS units) to facilitate radiocommunications, both emergency and general. As can be seen from Figure 1, the compact GMDSS console consists of multiple devices, which have different interfaces taking a lot of space. Thus, the authors suggest that the ship's bridge should have a unique interface of GMDSS equipment which could be on the conning display, ECDIS, or some new integrated GMDSS management system. Among all the respondents, only 14% disagree with this authors' proposal. Besides, the authors also suggested the use of an automated electronic radio log, and only 12% of the respondents disagree with that proposal.

It should be emphasized that the modernization of GMDSS has already begun. For example, the NAVTEX receiver can optionally be integrated within the ECDIS for displaying MSI. In terms of its modernization, its successor Navigational Data (NAVDAT), is in the development and testing phase, which will provide additional possibilities for displaying MSI [18]. Besides, the satellite system for receiving MSI, i.e., INMARSAT's EGC SafetyNET, also got its successor EGC SafetyNET II as part of INMARSAT's modernization of GMDSS services [14].

One of the systems that can serve to modernize GMDSS, if included in its infrastructure, is the VHF Data Exchange System (VDES), which was developed as a direct response to congestion of Automatic Identification System (AIS) [19–22]. Namely, a study conducted in 2013 by the International Telecommunication Union (ITU) has proved that in some parts of the world with dense maritime traffic, a limit of 50% of the data link capacity, which represents the critical level of the ship's AIS station, has been reached [23]. The causes of reaching 50% of AIS channel capacity are dense maritime traffic, use of Application Specific Messages (ASM), additional AIS stations such as AIS Aids to Navigation (AtoN), too much coastal AIS station information, as well as atmospheric effects [24]. It can be concluded that the AIS in some areas is at the edge of its capacity, while some GMDSS devices are not used, but must be installed on ships concerning the area of navigation. Therefore, the VDES can be seen as a successor to AIS (AIS represents only one of the VDES possibilities for data exchange) and its capabilities could improve certain services of today's GMDSS. As an example, the survey has shown that the INMARSAT-C SES is used almost exclusively when receiving EGC messages (Figures 10 and 12). According to its features, the VDES could support the reception of EGC messages with the additional possibility of global coverage (VDES will also have a LEO satellite component), which INMARSAT-C cannot satisfy (due to GEO satellites). If the modernization of GMDSS is considered in the function of improving the safety of navigation and information exchange with land services related to port operations and safety of navigation, VDES could serve as a communication platform that ensures the automatic exchange of information between ship and coastal stations [25]. In addition to contributing to the acceleration of the communication process with the ship's environment, it can also serve as a single data platform in achieving e-Navigation, as a given development goal [26,27].

5. Conclusions

We live in the 21st century where technology is advancing day by day making life easier. However, the GMDSS has not changed significantly in technological terms since its introduction (back in 1999). The authors believe that the time has come for GMDSS modernization, not only on a technological level but also on a conceptual level. Therefore, the items in the questionnaire survey were very precisely structured, indicating in each question for what purpose and how often each device was used during one watch. The research revealed the extent to which experienced seafarers use each individual GMDSS device during one of their watches. Thus, unlike other researches on the topic of GMDSS modernization, quantitative answers were obtained on the frequency of use of each individual device during a single watch. The survey results showed that some GMDSS devices are hardly used during the watch on the ship's bridge (MF/HF radio station and DSC in general). In addition, a qualitative analysis of respondents' opinions revealed that experienced seafarers want the GMDSS modernization in terms of spatial equipment reduction and integration on a ship's bridge, as well as in terms of paperwork reduction and process automation, which has not been presented in previous researches.

However, it should be noted that GMDSS is primarily designed for emergencies, which occur extremely rarely. So, the fact that the survey results showed that it is not used a lot does not mean that it is not necessary and useful in distress situations. Thus, the aim of the survey was not to prove that GMDSS equipment is not needed, but that its modernization is needed. The spatial reduction, functional improvement, and integration of GMDSS equipment on the ship's bridge would certainly modernize communications of all priorities (distress, urgency, safety, and routine) and also facilitate OOW's daily navigation. Of course, each change (improvement) of the equipment also requires a certain type of familiarization and training that takes time. However, it should be emphasized that according to the current IMO's Model Course 1.25, the training for the General Operator's Certificate (GOC) for the GMDSS suggests a total of 108 h for lectures, practice, and examination.

One of the systems mentioned in the maritime community in the context of GMDSS modernization is certainly the VDES, which represents the future of maritime communications, both distress/safety and general/commercial. In addition, new maritime terrestrial (NAVDAT) and satellite communication systems (Iridium) and services (INMARSAT Fleet Safety, SafetyNET II) have already started to be used (tested) in some areas.

Therefore, more similar research, in terms of precisely structured survey items, is needed to gather important information from practice (from seafarers) in order to develop and standardize guidelines for future change and upgrade of the GMDSS system. Additional research that will continue in this direction could include a correlation analysis between accidents at sea and the use of GMDSS equipment to report them, to show whether this equipment is really necessary onboard ships.

Author Contributions: Conceptualization, S.V. and A.Š.; methodology, S.V., A.Š., L.M. and B.S.; software, S.V. and A.Š.; validation, S.V., A.Š. and L.M.; formal analysis, S.V., A.Š., L.M. and B.S.; investigation, S.V., A.Š. and L.M.; resources, S.V., A.Š. and L.M.; data curation, B.S.; writing—original draft preparation, S.V. and A.Š.; writing—review and editing, S.V., A.Š., L.M. and B.S.; visualization, A.Š.; supervision, S.V.; project administration, A.Š.; funding acquisition, S.V., A.Š., L.M. and B.S. All authors have read and agreed to the published version of the manuscript.

Funding: This research received no external funding.

Institutional Review Board Statement: Not applicable.

Informed Consent Statement: Not applicable.

Data Availability Statement: Data is contained within the article.

Conflicts of Interest: The authors declare no conflict of interest.

References

1. Kent, P. The Global Maritime Distress and Safety System: Basic Concept and Functions. *J. Navig.* **1990**, *43*, 164–175. [CrossRef]
2. Korcz, K. Concepts of the GMDSS Modernization. In *Information, Communication, and Environment: Marine Navigation and Safety of Sea Transportation*; Weintrit, A., Neumann, T., Eds.; CRC Press-Taylor & Francis Group: Boca Raton, FL, USA, 2015; pp. 75–82.
3. Korcz, K. Some Aspects of the Modernization Plan for the GMDSS. *TransNav* **2017**, *11*, 164–174. [CrossRef]
4. Tzannatos, E. GMDSS operability: The operator equipment interface. *J. Navig.* **2002**, *55*, 75–82. [CrossRef]
5. Tzannatos, E. GMDSS False Alerts: A Persistent Problem for the Safety of Navigation at Sea. *J. Navig.* **2004**, *57*, 75–82. [CrossRef]
6. International Maritime Organization. *Analysis Result of User Survey on GMDSS Modernization. NCSR 1/inf.14*; International Maritime Organization: London, UK, 2014.
7. Lee, B.K.; Kong, G.Y.; Kim, D.H.; Cho, I.S. Survey and Analysis of User Opinion for the Review and Modernization of GMDSS and Implementation of e-Navigation. *J. Korean Soc. Mar. Environ. Saf.* **2015**, *21*, 381–388. [CrossRef]
8. Bekkadal, F. Novel maritime communications technologies. *TransNav* **2010**, *4*, 129–135.
9. Sarkodie, P.A.; Zhang, Z.K.; Benuwa, B.B.; Ghansah, B.; Ansah, E. A Survey of Advanced Marine Communication and Navigation Technologies: Developments and Strategies. *Int. J. Eng. Res. Afr.* **2018**, *34*, 102–115. [CrossRef]
10. Miyusov, M.V.; Koshevoy, V.M.; Shishkin, A.V. Increasing Maritime Safety: Integration of the Digital Selective Calling VHF Marine Radiocommunication System and ECDIS. *TransNav* **2011**, *5*, 159–161.
11. Koshevyy, V.; Shyshkin, O. Standardization of Interface for VHF, MF/HF Communication Using DSC within Its Integration with INS in the Framework of e-Navigation Concept. *TransNav* **2019**, *13*, 593–596. [CrossRef]
12. International Telecommunication Union. *Final Acts of World Radiocommunication Conference—WRC-07*; International Telecommunication Union: Geneva, Switzerland, 2008.
13. Patterson, A.; McCarter, P. Digital Selective Calling: The Weak Link of the GMDSS. *J. Navig.* **1999**, *52*, 28–41. [CrossRef]
14. Valčić, S.; Žuškin, S.; Brčić, D.; Šakan, D. An Overview of Recent Changes in the Global Maritime Distress and Safety System Regarding Maritime Mobile Satellite Service. *Naše More* **2019**, *66*, 135–140. [CrossRef]
15. Cairns, W. AIS and Long Range Identification & Tracking. *J. Navig.* **2005**, *58*, 181–189.
16. Vu, V.; Lützhöft, M.; Emad, G. Frequency of use—The First Step Toward Human-Centred Interfaces for Marine Navigation Systems. *J. Navig.* **2019**, *72*, 1089–1107. [CrossRef]
17. Maglić, L.; Zec, D. The Impact of Bridge Alerts on Navigating Officers. *J. Navig.* **2020**, *73*, 421–432. [CrossRef]
18. International Telecommunication Union. *NAVDAT Guidelines. Report ITU-R. M.2443*; International Telecommunication Union: Geneva, Switzerland, 2018.
19. Valčić, S.; Žuškin, S.; Brčić, D. VHF Maritime Mobile Band: A New System to Declutter AIS Channels. *Sea Technol.* **2018**, *59*, 24–27.
20. International Association of Marine Aids to Navigation and Lighthouse Authorities. *The Technical Specification of VDES, Edition 3. IALA Guideline G1139*; International Association of Marine Aids to Navigation and Lighthouse Authorities: Saint Germain en Laye, France, 2019.
21. Lázaro, F.; Raulefs, R.; Wang, W.; Clazzer, F.; Plass, S. VHF Data Exchange System (VDES): An enabling technology for maritime communications. *CEAS Space J.* **2019**, *11*, 55–63. [CrossRef]
22. Jiang, Y.; Zheng, Y.; Wang, J. A Novel Random Access Algorithm for Very High Frequency Data Exchange (VDE). *J. Mar. Sci. Eng.* **2020**, *8*, 83. [CrossRef]
23. International Telecommunication Union. *Automatic Identification System VHF Data Link Loading. Report ITU-R. M.2287*; International Telecommunication Union: Geneva, Switzerland, 2013.
24. International Association of Marine Aids to Navigation and Lighthouse Authorities. *R0124 (A-124) Appendix 18: VDL Load Management*; International Association of Marine Aids to Navigation and Lighthouse Authorities: Saint Germain en Laye, France, 2011.
25. Škrobonja, A.; Valčić, S.; Žuškin, S.; Brčić, D. On VDES/ECDIS Integration. *Pomorstvo* **2020**, *34*, 195–200. [CrossRef]
26. Hahn, A.; Bolles, A.; Franzle, M.; Froschle, S.; Park, J.H. Requirements for e-Navigation Architectures. *Int. J. e-Navig. Marit. Econ.* **2016**, *5*, 1–20. [CrossRef]
27. Korcz, K. GMDSS as a Data Communication Network for E-Navigation. *TransNav* **2008**, *2*, 263–268.

Journal of
Marine Science and Engineering

MDPI

Article

Capacity of the Weakly Absorbent Turbulent Ocean Channel with the Coaxial Double-Position Power Gaussian Vortex

Qingze Yan, Yun Zhu and Yixin Zhang *

School of Science, Jiangnan University, Wuxi 214122, China; Yanqz0620@163.com (Q.Y.); zhuyun1210@163.com (Y.Z.)
* Correspondence: zyx@jiangnan.edu.cn

Abstract: Turbulence and absorption of seawater are two important factors affecting the signal transmission quality of underwater optical communication link. Here, we study the channel capacity and bit error rate of an underwater extinction communication link with a coaxial double-position power Gaussian vortex carrier based on Rytov approximation theory. The study finds that channel capacity and bit error rate are the nonlinear functions of the dimensionless structural parameter and reach maximum and minimum values at $|\alpha| = 1$, respectively. The seawater absorption has a great influence on the channel capacity but not bit error rate. The communication link with large receiving aperture, small transmitting beam diameter, long wavelength of light source in a seawater window, and more OAM channels has high channel capacity.

Keywords: oceanic turbulence; seawater absorption; position power Gaussian vortex; channel capacity

1. Introduction

Orbital angular momentum (OAM) is a new dimension that can be encoded, which can be adopted by optical [1–4] and acoustic [5–7] communication systems to improve the information capacity of links. Although acoustic OAM has the advantages of low loss of underwater communication and perfect theoretical [5–7] and experimental model compared with underwater optical OAM transmission, its fatal weakness is the low channel capacity and signal transmission rate of underwater acoustic communication link. Therefore, as underwater communication and detection technology requires higher communication link channel capacity and signal transmission rate, OAM optical communication has become one of the new focuses in the field of underwater communication.

However, due to the presence of various constituents in the ocean and oceanic turbulence, seawater scattering and absorption inevitably limit the transmission of information carried by OAM modes over optical communication links [8–18]. Therefore, in order to reduce the above negative effects of seawater and consider the relationship between signal transmission probability and channel capacity, a lot of studies have been conducted from the perspectives of OAM detection probability and link channel capacity. The disturbance degree of ocean turbulence to the detection probability of OAM mode of partially coherent Laguerre–Gaussian beams increases with the increase in beam radial mode order, OAM mode topological charge and ocean turbulence strength, and fully coherent vortex beams provided better performance than partially coherent ones [8]. The partially coherent elegant Laguerre–Gaussian beams are also more affected by turbulence as compared to the fully coherent elegant Laguerre–Gaussian beams, and one can choose optimum beam source to mitigate the influence of oceanic turbulence [9]. Although the detection probability of fractional OAM mode is lower than that of adjacent integer OAM mode, the channel capacity of the link can be increased by using fractional OAM mode when the available OAM topological charges are limited [10]. Compared with Laguerre–Gaussian beams, the Bessel Gaussian beam transmitted in turbulent seawater has stronger anti-turbulence and better OAM transmission performance [11]. For partially coherent modified Bessel correlated vortex beam, the effect of high-anisotropic ocean turbulence on OAM signal is weaker than that of low-anisotropic ocean turbulence. In addition, the mode with low OAM quantum number can be selected to improve the information transmission of the link, and the OAM mode with large energy level difference between neighboring OAM can be used to reduce the impact of turbulence, to improve the information capacity of the optical communication link [12]. The channel capacity of the link increases as the pulse width of the half-modulated input pulse increases and the Bessel cone angle decreases. Channel capacity is more sensitive to salinity fluctuations than temperature fluctuations. In anisotropic and isotropic turbulence, the internal and external scales of turbulence have different effects on channel capacity [13]. An airy-OAM beam with a smaller topological charge, larger main ring radius, and longer wavelength has larger detection probability [14]. The Lommel Gaussian beam, which has a small asymmetric coefficient, low OAM quantum number, optimal waist width, and long transmission window wavelength, has a low probability of crosstalk introduced by turbulence. [15]. The self-focusing characteristic of a perfect optical vortex beam is beneficial to the propagation of the OAM mode, and the perfect optical vortex beam with smaller topological charge, smaller initial radius, and optimized half-ring width can reduce the degradation effect of the PV beam caused by turbulence [16]. The normalized received probability of signal OAM modes carried by finite energy frozen wave is independent of the quantum number of OAM modes in weak seawater turbulence, and the higher normalized received probability achieved in the smaller received diameter of the receiver, the larger the transverse size and longer the signal wavelength and the signal OAM modes [17]. An HyGG vortex beam with smaller topological charge and longer wavelength has higher transmission quality [18]. Compared with the Laguerre–Gaussian beam and the Bessel–Gaussian beam, a diffraction- and attenuation-resistant random vortex beam is a more resistant to seawater turbulence and seawater absorption and the optical communication link, which consists of a long-wave length anti-diffraction and anti-attenuation random vortex carrier and a small receiving diameter, has high channel capacity [19]. More importantly, by designing a longitudinal profile of finite energy frozen beams, we can compensate the attenuation loss of the transmittance [20]. Moreover, the localized wave of Bessel–Gaussian amplitude envelope (or Bessel–Gaussian localized vortex) is a more suitable beam for the vortex mode communication than conventional vortex waves [21,22]. The disturbance of turbulent seawater to the OAM modes of anti-diffraction and anti-attenuation random beam with a large truncated Gaussian aperture is less than the corresponding OAM modes carried by the Laguerre–Gaussian beam [23].

The position-power Gaussian vortex is a new kind of vortex beam, which is different from all the previous research. The disturbance of seawater turbulence when the OAM mode carried by the beam is transmitted in the underwater channel is also a problem worth discussing. Furthermore, most of these studies only focus on the transport of a single OAM vortex or multiple OAM vortices with the same rotation direction in turbulent seawater. When the vortex carrier carries two OAM modes with different amplitudes and rotation directions in the absorption of turbulent seawater, whether the transmission characteristics are different from the above research conclusions, that is, whether there are more factors that can optimize the communication system. However, as far as we know, there are no reports on the channel capacity of underwater communication links with a coaxial dual-position power vortex Gaussian carrier.

In this paper, the influence of coaxial double-position power vortex Gaussian carrier parameters and seawater turbulence parameters on the channel capacity and bit error rate (BER) of the communication link is studied. The remainder of this paper is organized as follows. In Section 2, we derive the random coaxial double-position power Gaussian vortex (RCDPPGV) laser based on the Huygens–Fresnel diffraction integral formula and the Rytov approximation theory. We derive the BER of OAM mode and the channel capacity of the optical communication link with a RCDPPGV carrier in Section 3. In Section 4, we discuss the effects of the parameters of turbulent absorption seawater and beam structure on the channel capacity as well as BET. Section 5 provides the main conclusions of the paper.

2. Kolmogorov Ocean Spectrum and Random Coaxial Double-Position Power Gaussian Vortex

This section mainly discusses two parts of your content. First, we briefly discuss the seawater spectrum. Secondly, the random coaxial double-position power Gaussian vortex is established by Rytov theory.

2.1. Kolmogorov Ocean Spectrum

For a liner, isotropic, homogeneous, and clean absorbing seawater, the refractive index of seawater is [19–21]

$$n_{oc} = n_r + in_i \tag{1}$$

where n_r is the mean value of real part of refractive index of seawater, n_i is the mean value of imaginary part of refractive index of seawater that depends on the type of seawater, and $i = \sqrt{-1}$ is the imaginary symbol.

The turbulent characteristics of seawater are described by the following turbulent spectrum equation [24]

$$\begin{aligned}\Phi_n(\kappa) &= 0.267 \times 10^{-8} \frac{\varepsilon^{-1/3}\chi_T\kappa^{-11/3}}{\varpi^2}\left[1 + 4.6(\kappa l_0)^{2/3}\right]\left[1 - \exp\left(-\frac{\kappa^2}{\kappa_0^2}\right)\right] \\ &\times \left[\varpi^2 \exp\left(-\frac{\kappa^2}{\kappa_T^2}\right) + 1\exp\left(-\frac{\kappa^2}{\kappa_S^2}\right) - 2\varpi \exp\left(-\frac{\kappa^2}{\kappa_{TS}^2}\right)\right] \quad 0 < \kappa < \infty,\end{aligned} \tag{2}$$

where κ is the spatial frequency of turbulent fluctuations, ε is the rate of dissipation of kinetic energy per unit mass of fluid ranging from 10^{-10} m^2/s^3 to 10^{-1} m^2/s^3, χ_T is the dissipation rate of the mean-squared temperature and has the range from 10^{-10} K^2/s to 10^{-2} K^2/s, l_0 is the inner scale, $\kappa_0 = 2\pi/L_0$, L_0 is the outer scale of turbulence, $\kappa_T = R_T/l$, $\kappa_S = R_S/l_0$, and $\kappa_{TS} = R_{TS}/l_0$. $R_j = \sqrt{3}[W_j - 1/3 + 1/(9W_j)]^{3/2}Q^{-3/2}$, $(j = T, S, TS)$, $W_j = \left\{\left[\frac{\mathrm{Pr}_j^2}{(6\beta Q^{-2})^2} - \frac{\mathrm{Pr}_j}{81\beta Q^{-2}}\right]^{1/2} - \left[\frac{1}{27} - \frac{\mathrm{Pr}_j}{6\beta Q^{-2}}\right]\right\}^{1/3}$, $Q = 2.35$, Pr_T and Pr_S, respectively, represent the Prandtl number of the temperature and salinity, $\mathrm{Pr}_{TS} = 2\mathrm{Pr}_T\mathrm{Pr}_S/(\mathrm{Pr}_T + \mathrm{Pr}_S)$, ϖ defines the ratio of temperature and salinity contributions to the refractive index spectrum, which can vary in the interval [−5,0] in the seawaters, $\varpi = -5$ temperature is dominant in the induction of optical turbulence, while $\varpi = 0$ salinity is dominant in the induction of optical turbulence.

Compared with the atmospheric turbulence spectrum [24], Equation (2) can be rewritten into the following ocean spectrum

$$\Phi_n(\kappa) = \varphi_n(\kappa) f(\varpi, \kappa, \eta), \quad 0 < \kappa < \infty, \tag{3}$$

with "Kolmogorov ocean spectrum"

$$\varphi_n(\kappa) = 0.033 C_{\varepsilon\chi_T}^2 \kappa^{-11/3}, \quad 1/\eta < \kappa < 1/L_0, \tag{4}$$

Additionally,

$$\begin{aligned}f(\varpi, \kappa, \eta) &= \left[1 + 4.6(\kappa l_0)^{2/3}\right](1 - \varpi)^{-2}\left[1 - \exp\left(-\frac{\kappa^2}{\kappa_0^2}\right)\right] \\ &\times \left[\exp\left(-\frac{\kappa^2}{\kappa_T^2}\right) + \frac{1}{\varpi^2}\exp\left(-\frac{\kappa^2}{\kappa_S^2}\right) - \frac{2}{\varpi}\exp\left(-\frac{\kappa^2}{\kappa_{TS}^2}\right)\right].\end{aligned} \tag{5}$$

Here, $C_{\varepsilon\chi_T}^2 = 0.809 \times 10^{-7}\varepsilon^{-1/3}\chi_T\varpi^{-2}(1 - \varpi)^2$ defines the structural constants of the salinity–temperature fluctuation of oceanic turbulence.

Consider that the physical mechanism behind the oceanic turbulence and atmospheric turbulence are same and in conjunction with the above discussion, using Equation (5), we can just write Rytov variance for a plane wave in a homogeneous isotropic weakly turbulent ocean as

$$\sigma_{1\varepsilon\chi T}^2 = 1.23 C_{\varepsilon\chi T}^2 k^{7/6} z^{11/6} \quad (6)$$

Formally, Equation (6) is the same as the Rytov variance of plane waves in the turbulent atmosphere, except that $C_{\varepsilon\chi T}^2$ is replaced by the structure constant of atmospheric turbulence C_n^2. Of course, from the turbulence effect theory of light transmission in the atmosphere [25], we can directly obtain the conditions that seawater is in a weak fluctuation state, that is, $\sigma_{1\varepsilon\chi T}^2 < 1$.

2.2. Random Coaxial Double-Position Power Gaussian Vortex

In Figure 1, we give a schematic diagram of the rectangular coordinates (x', y', z') and the communication link for signal transmission from the light source plane (x', y', z') to the receiving plane (x, y, z).

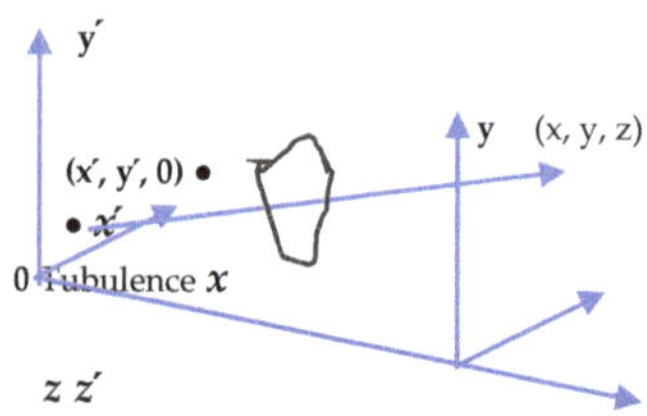

Figure 1. The schematic diagram of a signal transmission from the light source plane (x', y', z') to the receiving plane (x, y, z).

In rectangular coordinates (x', y', z') and at light source plane, we consider two position power Gaussian vortexes that have the same amplitude at their waist [26]

$$E(x', y') = (x' \pm iy')^{l_0} \exp\left(-\frac{x'^2 + y'^2}{w_0^2}\right), \; x', y' > 0 \quad (7)$$

where "+" represents right- spin vortex, "−" represents the left—spin vortex, l_0 is the topological charge of vortex, and w_0 is the waist of these vortexes. Essentially, $(x' \pm iy')^{l_0}$ is a transmission factor and is a position power vortex. The position power vortex can be produced from "fork" hologram [27].

In the cylindrical coordinate system (r', θ', z') these two position power Gaussian vortexes can be rewritten as follows

$$E(r', \theta', 0) = r'^{l_0} \exp\left(-\frac{r'^2}{w_0^2} \pm i l_0 \theta'\right) \; \theta' > 0 \quad (8)$$

where $r' = \sqrt{x'^2 + y'^2}$, and $\theta' = \arctan(y'/x')$.

Now, according to Equation (8), we form a coaxial double-position power Gaussian vortex with a normalized field and opposite helices in the source plane $z' = 0$, as shown in the following equation (see Appendix A)

$$E(r', \theta', 0) = N\left[\frac{1-|\alpha|}{2} r'^{l_0} \exp\left(-\frac{r'^2}{w_0^2} - i l_0 \theta'\right) + \frac{1+|\alpha|}{2} r'^{l_0} \exp\left(-\frac{r'^2}{w_0^2} + i l_0 \theta'\right)\right] \quad (9)$$

where $N = \left\{\pi\left[(1+|\alpha|)^2+(1-|\alpha|)^2\right]2^{-l_0-3}w_0^{2l_0+2}\gamma\left(l_0+1,\frac{D^2}{2w_0^2}\right)\right\}^{-1/2}$ is normalization coefficient, D is the diameter of the receiver, $\gamma\left(l_0+1,\frac{D^2}{2w_0^2}\right)$ is the incomplete gamma function, and α denotes a dimensionless structural parameter. When $|\alpha| = 1$, the beam degenerates into left spin or right spin single-position power Gaussian vortex, respectively; when $|\alpha| \neq 1$, the beam is a coaxial double-position power Gaussian vortex with reverse helix and different amplitude, and when $|\alpha| = 0$, the beam is a coaxial double-position power Gaussian vortex with inverse spiral and equal amplitude.

The field of the coaxial double-position power Gaussian vortex in free space at any receiving lane $x' = x, y' = y, z' = z$ is given by (see Appendix B)

$$\begin{aligned} E(r,\theta,z) &= -\frac{ik}{2Nz}\exp\left[i\left(kz+\frac{kr^2}{2z}\right)\right]\left[(1-|\alpha|)(-i)^{l_0}e^{-il_0\theta}+(1+|\alpha|)(i)^{l_0}e^{il_0\theta}\right] \\ &\times\frac{zk^{l_0}w_0^{2(l_0+1)}r^{l_0}}{\left(2z-ikw_0^2\right)^{l_0+1}}{}_1F_1\left(l_0+1;l_0+1;-\frac{k^2r^2w_0^2}{2z\left(2z-ikw_0^2\right)}\right) \end{aligned} \tag{10}$$

where $\Gamma\left(\frac{v+\mu+1}{2}\right)$ is gamma function, and ${}_1F_1\left(\frac{v+\mu+1}{2};v+1;-\frac{b^2}{4a}\right)$ is confluent hypergeometric function. k is the wave number, and λ is the wavelength of light in the vacuum.

In weak turbulence and absorbent seawater and by Rytov approximation [25], the field of the RCDPPGV is expressed as (see Appendix C)

$$\begin{aligned} \widetilde{E}(r,\theta,z) &= \frac{\pi(n_i-in_r)}{N\lambda}i^{l_0}e^{i\frac{2\pi}{\lambda}(n_r+in_i)\left(z+\frac{r^2}{2z}\right)+\psi(r,\theta,z)}\left[(1-|\alpha|)(-1)^{l_0}e^{-il_0\theta}+(1+|\alpha|)e^{il_0\theta}\right] \\ &\times\frac{\left[\frac{2\pi}{\lambda}(n_r+in_i)\right]^{l_0}w_0^{2(l_0+1)}r^{l_0}}{\left[2z+\frac{2\pi w_0^2}{\lambda}(n_i-in_r)\right]^{l_0+1}}{}_1F_1\left(l_0+1;l_0+1;-\frac{\left[\frac{2\pi}{\lambda}(n_r+in_i)\right]^2r^2w_0^2}{4z^2+\frac{4\pi}{\lambda}z(n_i-in_r)w_0^2}\right) \end{aligned} \tag{11}$$

where $\psi(r,\theta,z)$ represents the complex phase disturbance caused by seawater turbulence.

3. Bit Error Rate of the OAM Link and Channel Capacity

In this section, the series expansion method of RCDPPGV random spiral mode is used to obtain the mathematical model of bit error rate and channel capacity.

3.1. Bit Error Rate of OAM Links

To understand the OAM mode redistribution generated by turbulent seawater disturbance, we expanded the disturbed field into a series of the spiral harmonic $\exp(il\theta)$ [18,19]

$$\widetilde{E}(r,\theta,z) = \frac{1}{\sqrt{2\pi}}\sum_{l=-\infty}^{\infty}a_l(r,z)e^{il\theta} \tag{12}$$

with

$$a_l(r,z) = \frac{1}{\sqrt{2\pi}}\int_0^{2\pi}E(r,\theta,z)e^{-il\theta}d\theta \tag{13}$$

Note that $l \neq l_0$ is the topological charge of an OAM crosstalk mode.

The ensemble average of $|a_l|^2$ represents the condition probability distribution of OAM modes in turbulent as well as absorbent seawater and at receiving plane, which has the form of

$$\begin{aligned} p_{l/l_0}(r) &= \left\langle|a_l(r,z)|^2\right\rangle \\ &= \frac{1}{2\pi}\left|\int_0^{2\pi}\int_0^{2\pi}E(r,\theta,z)E^*(r,\theta,z)\exp[-il(\theta-\theta')]\right. \\ &\left.\times\exp\left[-2r^2[1-\cos(\theta-\theta')]/\rho_{0c}^2\right]d\theta d\theta'\right| \end{aligned} \tag{14}$$

where * denotes the complex conjugate, and ρ_{oc} is the spatial coherence radius of waves propagating in isotropic and turbulent as well as absorbent seawater and is given by [13]

$$\begin{aligned}\rho_{oc}^{-2} =& \frac{2.511\times10^{-7}\varepsilon^{-1/3}\chi_T(n_r^2+n_i^2)z}{\lambda^2}\left[\kappa_0^{1/3}U\left(2;\frac{7}{6};\frac{\kappa_0^2\eta^2}{R_T^2}\right)\right.\\ &+4.6\eta^{2/3}\kappa_0\Gamma\left(\frac{7}{3}\right)U\left(\frac{7}{3};\frac{3}{2};\frac{\kappa_0^2\eta^2}{R_T^2}\right)+\frac{1}{\varpi^2}\kappa_0^{1/3}U\left(2;\frac{7}{6};\frac{\kappa_0^2\eta^2}{R_S^2}\right)\\ &+\frac{4.6}{\varpi^2}\eta^{2/3}\kappa_0\Gamma\left(\frac{7}{3}\right)U\left(\frac{7}{3};\frac{3}{2};\frac{\kappa_0^2\eta^2}{R_S^2}\right)-\frac{2}{\varpi}\kappa_0^{1/3}U\left(2;\frac{7}{6};\frac{\kappa_0^2\eta^2}{R_{TS}^2}\right)\\ &\left.-\frac{9.2}{\varpi}\eta^{2/3}\kappa_0\Gamma\left(\frac{7}{3}\right)U\left(\frac{7}{3};\frac{3}{2};\frac{\kappa_0^2\eta^2}{R_{TS}^2}\right)\right].\end{aligned} \tag{15}$$

Substituting Equation (9) into Equation (12), through integration operations, we obtain the probability distribution of OAM modes (see Appendix D)

$$\begin{aligned}p_{l/l_0}(r) =& \frac{2\pi^3(n_r^2+n_i^2)\left[\left(\frac{2\pi}{\lambda}\right)^2(n_r^2+n_i^2)\right]^{l_0}w_0^{4(l_0+1)}r^{2l_0}}{N_2\lambda^2\left[\left(2z+\frac{2\pi}{\lambda}n_iw_0^2\right)^2+\left(n_r\frac{2\pi}{\lambda}w_0^2\right)^2\right]^{l_0+1}}\\ &\times\left|{}_1F_1\left(l_0+1;l_0+1;-\frac{\left[\frac{2\pi}{\lambda}(n_r+in_i)\right]^2r^2w_0^2}{4z^2+\frac{4\pi}{\lambda}z(n_i-in_r)w_0^2}\right)\right|^2\exp\left[-\frac{2\pi}{\lambda}n_i\left(z+\frac{r^2}{2z}\right)-\frac{2r^2}{\rho_{oc}^2}\right]\\ &\times\left|\left[(|\alpha|-1)^2I_{l+l_0}\left(\frac{2r^2}{\rho_{oc}^2}\right)+(|\alpha|+1)^2I_{l-l_0}\left(\frac{2r^2}{\rho_{oc}^2}\right)\right]\right|\end{aligned} \tag{16}$$

The receiving probability of OAM mode carried by RCDPPGV and measured by the detector with detection diameter D can be expressed as

$$P_{l/l_0} = 2\pi\int_0^{D/2} p_{l/l_0}(r)rdr \tag{17}$$

In addition, for sufficiently large transmit power, the signal-to-noise ratio (SNR) of OAM channel is approximately [28]

$$SNR(l,l_0) = \frac{P_{l/l_0,l=l_0}}{\sum\limits_{l=-\infty}^{\infty} P_{l/l_0,l\neq l_0}} \tag{18}$$

For signal modulation of on–off keying (OOK), based on (18), the bit error rate (BER) of OAM channels is derived as [29]

$$BER(l,l_0) = \frac{1}{2}\mathrm{erfc}\sqrt{\frac{SNR(l,l_0)}{2}} \tag{19}$$

where erfc(·) denotes the complementary error function.

3.2. Channel Capacity

Consider a line-of-sight communication link using OAM eigenstates in the range of topological charge number $l_0 = 0,1,\ldots,M-1$ that means M dimension Hilbert space for different signal values. Further, by classical Shannon information theory of a stationary discrete memory-less system [3] and Equation (17), the channel capacity of a communication link with coaxial double-position power Gaussian vortex in weakly absorbent seawater can be given by [3]

$$C = \log_2 M + \frac{1}{M}\sum_{l=-\infty}^{\infty}\sum_{l_0=0}^{M-1} P_{l|l_0}\left[\log_2 P_{l|l_0} - \log_2\sum_{l_0} P_{l|l_0}\right]. \tag{20}$$

with the entropy of the signal vortex mode l_0

$$\max_{\{P_{l_0}\}} H_{l_0} = \log_2 M, \tag{21}$$

and conditional entropy of the l mode for given signal mode l_0

$$\max_{\{P_{l_0}\}} H_{l_0|l} = -\frac{1}{M}\sum_{l=-\infty}^{\infty}\sum_{l_0=0}^{M-1} P_{l|l_0}\left[\log_2 P_{l|l_0} - \log_2 \sum_{l_0} P_{l|l_0}\right]. \tag{22}$$

4. Numeric Analysis

To explore the influence of beam structure parameters and seawater parameters on BER and the channel capacity of OAM links in isotropic turbulent as well as absorbent seawater, in this section, we numerically study the BER and channel capacity of the link with a RCDPPGV carrier. In the following analysis, we set the calculation parameters as: $|\alpha| = 1$, $w_0 = 0.05$ m, $D = 0.1$ m, $z = 100$ m, $\lambda = 561$ nm, $\chi_t = 10^{-8}$ K^2/s, $\varepsilon = 10^{-4}$ m^2/s^3, $\eta = 1$ mm, $l_0 = 1$, $L_0 = 10$ m, $\varpi = -4$, $M = 5$, $n_r = 1.34$, and $n_i = 0.6741 \times 10^{-10}$, unless other variable parameters are specified in calculation. Moreover, in the following numerical analysis, we first analyze the influence of system parameters and turbulent seawater on the BER of OAM channels and, then, analyze the influence of system parameters and turbulent seawater on channel capacity.

4.1. Effects of Seawater Turbulence and Beam Parameters on the BER

Figure 2 shows variation curve of the BER of the random double–Gaussian optical vortex carrier communication link with the dissipation rate of the mean–squared temperature χ_T and topological charge of the RCDPPGV vortex. The variation curve shows that with the increase in transmission distance z, the BER of this link increases. When the transmission distance is less than 50 m, BER increases rapidly and, then, tends to be flat. Additionally, a larger l_0 means a higher BER. The physical reason for this result is that when the topological charge (orbital angular momentum quantum number) of OAM signal mode is larger, the OAM energy level of the signal is also higher. Therefore, the probability of transition form to another OAM energy level is also higher when it is disturbed.

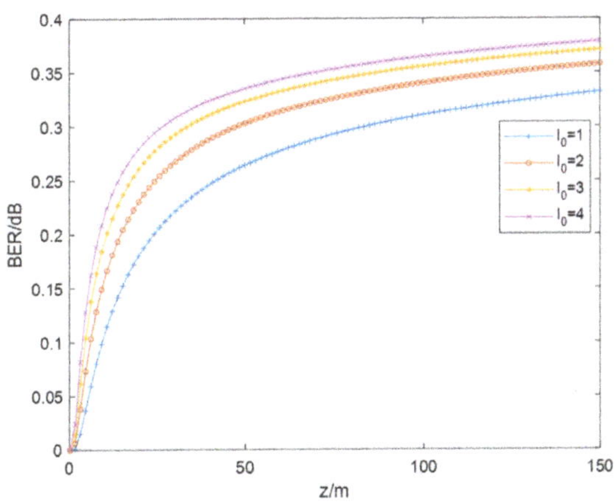

Figure 2. BER of the coaxial double-position power Gaussian vortex carrier communication link versus dissipation rate of the mean-squared temperature χ_T for different topological charge of vortex.

Under certain topological charge of vortex laser, we use Figure 3 to show the variation curve of BER of random double–Gaussian optical vortex carrier communication link with dimensionless parameter α. From Figure 2, we can see that BER has two minimum values at $|\alpha| = 1$. This result indicates that the link with a single-position power Gaussian vortex has minimum BER.

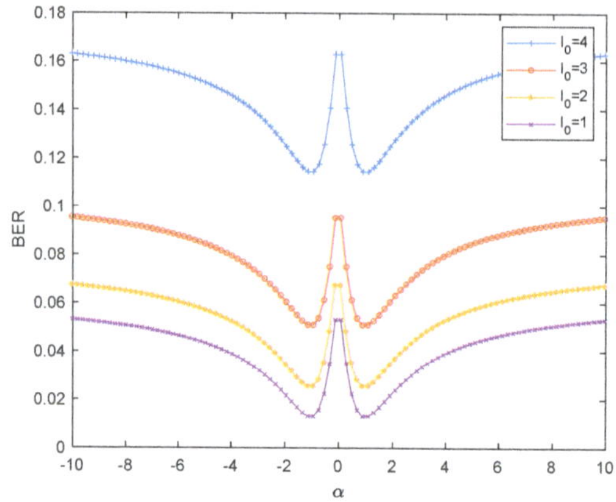

Figure 3. BER of the coaxial double-position power Gaussian vortex carrier communication link versus dimensionless parameter $|\alpha|$ for different topological charge of vortex.

In Figure 4, we investigate the effects of the ε and l_0 on the BER of RCDPPGV carrier link for given topological charge of vortex. From Figure 3, we can see that the value of the BER decreases with an increase in the ε. The reason for this phenomenon is the increase in ε, the turbulence strength of seawater decreases, and the influence of turbulence on the transmission vortex decreases.

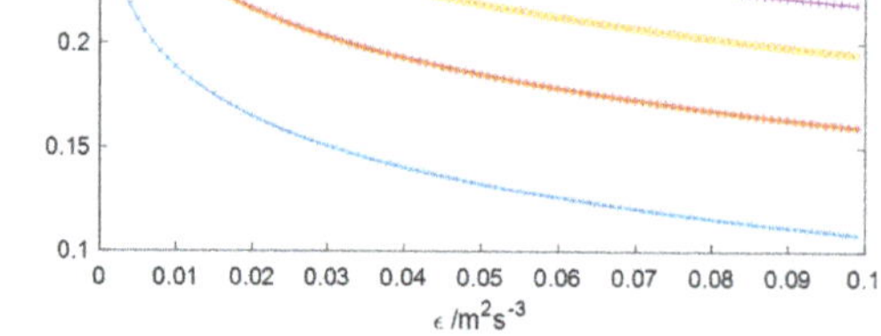

Figure 4. BER of the coaxial double-position power Gaussian vortex carrier communication link versus dissipation rate of kinetic energy per unit mass of fluid ε for different topological charge of vortex.

In Figure 5, we investigate the effects of dissipation rate of the mean-squared temperature χ_T on the channel capacity of the RCDPPGV carrier link for the given topological charge of vortex. It is clear from Figure 5 that the BER of the channel capacity increases with increasing χ_T. The reason for this result comes from that with the increase in χ_T, the turbulence strength of seawater increases, and the influence of turbulence on the transmission vortex increases.

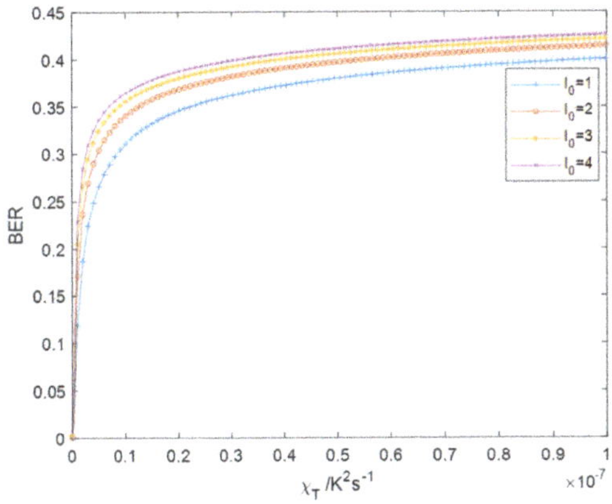

Figure 5. BER of the coaxial double-position power Gaussian vortex carrier communication link versus dissipation rate of the mean-squared temperature χ_T for different topological charge of vortex.

4.2. Effects of Seawater Turbulence and Beam Parameters on Channel Capacity

Figure 6 illustrates the impact of the dimensionless structural parameters α on the channel capacity of the RCDPPGV carrier with different values of the channel numbers M. Figure 6 shows that the channel capacity in the $[-10, 10]$ interval of α is a nonlinear function of α and reaches two maximum values at $|\alpha| = 1$. The difference value between the maximum channel capacity of each link increases with the number of OAM channels M. However, with the decrease in $|\alpha|$ value, channel capacity tends to be constant, this constant is channel capacity at $|\alpha| = 0$. This result indicates that the link with single-position power Gaussian vortex has maximum channel capacity. Similar to any other type of beam [9–22], Figure 6 shows that the channel capacity also increases with the increase in channel number.

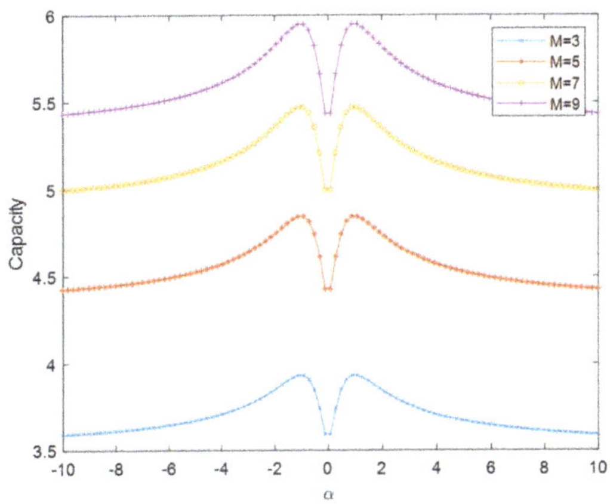

Figure 6. Channel capacity of the link with the coaxial double-position power Gaussian vortex carrier versus dimensionless parameter $|\alpha|= 1$ for different channel numbers M.

Figure 7 shows that channel capacity decays rapidly as absorption increases. Seawater absorption has a great influence on the channel capacity, so in the detector design of the underwater communication system, it should take into account the inconsistency between the absorption of the offshore area and the open sea area.

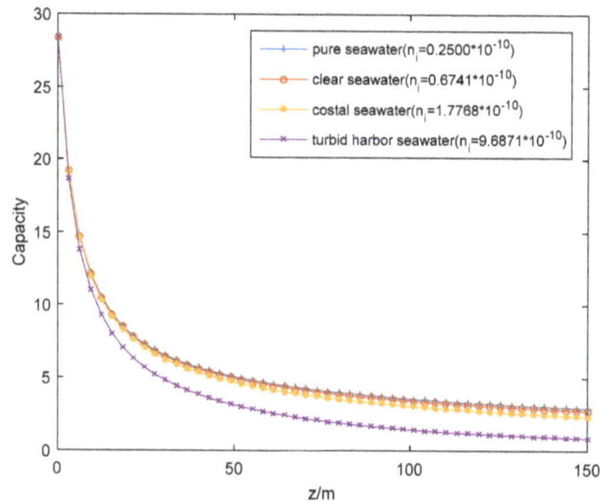

Figure 7. Channel capacity of the link with the random double-position power Gaussian vortex carrier and wavelength $\lambda = 561$ nm versus pure seawater, clear seawater, costal seawater, and turbid harbor seawater.

Figure 8 shows the channel capacity of the RCDPPGV carrier versus the initial beam width w_0 for different values of received diameter D. We can find that the channel capacity of the link with RCDGV carrier decreases with increase in the w_0. This phenomenon can be explained as follows: when the initial beam size w_0 is larger, the difference of turbulent seawater within the beam diameter is also larger. Of course, the wave front distortion of the transmitting beam is also large, so that the difference between signal probability and crosstalk probability tends to zero [30]. Figure 8 also shows that the channel capacity of the link with RCDPPGV carrier increases with an increase in D. This result can be explained by the following reasons: because the channel of OAM mode with higher topological charge is farther away from the axis of communication link [1,26], the receiver with the larger receiving aperture can receive the information of the OAM mode channel with higher topological charge.

In Figure 8, we investigate the effects of the temperature–salinity contribution ratios and different wavelength λ on the channel capacity of the RCDPPGV carrier in the "seawater light transmission window". Similar to any other type of beam [9–22], Figure 8 shows that the channel capacity of the RCDPPGV carrier increases with increasing λ. This is because the long-wavelength beam has lower scintillations [25]. As ϖ increases, channel capacity decreases. This study is similar to studies for the single vortex information carrier [9–22], and salinity fluctuations have a greater impact on the channel capacity with a larger number of channels.

In Figure 9, we investigate the effects of the outer scale of turbulence L_0 and the inner scale of turbulence η on the channel capacity of RCDPPGV carrier. From Figure 5, we can see that the value of the channel capacity increases with an increase in η and a decrease in L_0. This shows that the channel capacity increases with the increase in the inner scale of turbulence. According to the turbulence theory, with the increase in the inner scale of turbulence, the uniform area in the seawater increases when the beam passes through the channel, that is, the influence of the channel on the transmitted beam decreases. We can also find that the channel capacity decreases with the increase in the outer scale, but the range of change is small. The smaller the outer scale of turbulence is, the less wave front distortion will be generated, so the crosstalk probability will be smaller. The reasons for this phenomenon are as follows: When the turbulent outer scale increases, the deflection of light ray in the beam is growing, and the frequency of light rays going through different paths also increases. As a result, the optical path difference between the light rays increases, and the wave front distortion of the beam increases accordingly. However, because the refractive index difference is small in the outer scale of turbulence, the wave front distortion of the beam increases but not much.

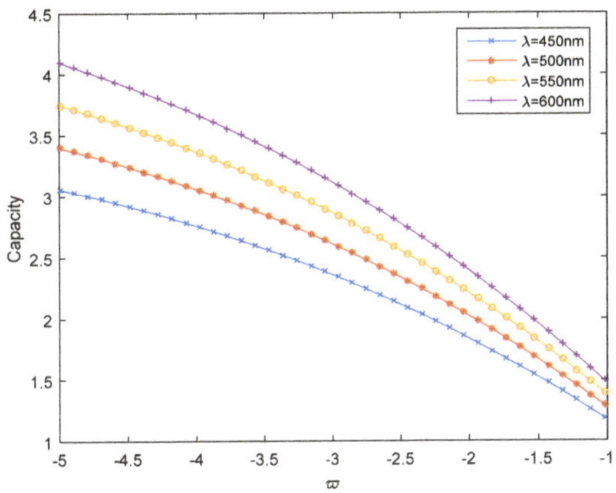

Figure 8. Channel capacity of the link with the random double-position power Gaussian vortex carrier communication system versus temperature–salinity contribution ratios and different values of wavelength λ.

In Figure 10, we investigate the effects of dissipation rate of kinetic energy per unit mass of fluid and dissipation rate of the mean-squared temperature χ_t on the channel capacity of the RCDPPGV carrier. It is clear from Figure 6 that the magnitude of the channel capacity decreases with increasing χ_t and decreasing ε. The reason for this result comes from that with the increase in χ_t and the decrease in ε, the turbulence of seawater increases, and the influence of turbulence on the beam increases [31].

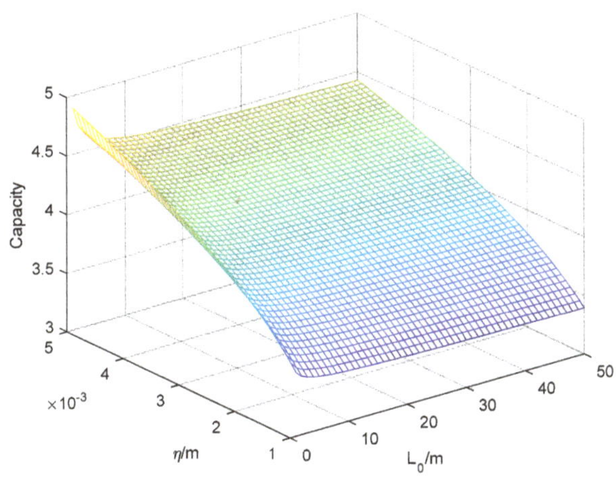

Figure 9. Channel capacity of the link with the random double-position power Gaussian vortex carrier optics communication system versus the inner scale of turbulence η and the outer scale of turbulence L_0.

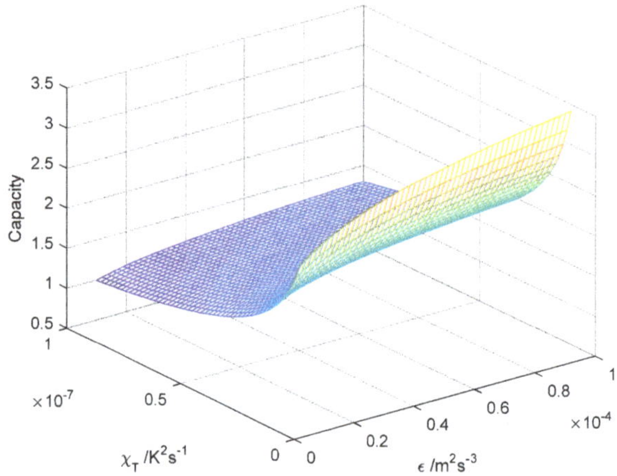

Figure 10. Channel capacity of the link with the random double-position power Gaussian vortex carrier optics communication system versus dissipation rate of kinetic energy per unit mass of fluid ε and dissipation rate of the mean-squared temperature χ_T.

5. Conclusions

We focused on the propagation properties of OAM modes carried by the RCDPPGV carrier in absorptive, isotropic, and weak turbulent ocean link and derived the channel capacity of this link. The results show that seawater absorption has a great influence on the channel capacity of the optics communication link, and the larger the seawater absorption, the faster the absorption of channel capacity with the transmission distance. With the optics communication link with single-position power Gaussian vortex carrier, the channel capacity of the link reached an extreme value. The channel capacity increases with the increase in the ε and η, and the change in L_0 has little effect on the channel capacity. The channel capacity decreases with increasing ϖ and χ_t. The large number of OAM channels M leads to higher channel capacity. The channel capacity decreases with the increase in initial beam width and the decrease in D. Our results also show that with the increase in transmission distance z, the topological charge of vortex, and ε, the BER of this link increases; however, the BET of the link increases with the increase in χ_T and is minimum at $\alpha = 1$ or -1. Our paper also points out that we need to further study the influence of the helical direction of non-coaxial multi-helical combination vortices on the channel capacity of the communication link.

Author Contributions: Conceptualization, Y.Z. (Yixin Zhang) and Q.Y.; methodology, Y.Z. (Yixin Zhang); software, Q.Y.; validation, Y.Z. (Yixin Zhang), Q.Y. and Y.Z. (Yun Zhu); formal analysis, Q.Y.; investigation, Q.Y.; resources, Y.Z. (Yixin Zhang); data curation, Q.Y.; writing—original draft preparation, Y.Z. (Yun Zhu); writing—review and editing, Y.Z. (Yixin Zhang); visualization, Q.Y.; supervision, Y.Z. (Yixin Zhang); project administration, Y.Z. (Yixin Zhang); funding acquisition, Y.Z. (Yixin Zhang). All authors have read and agreed to the published version of the manuscript.

Funding: This work was supported by the National Natural Science Foundation of China (61871202).

Institutional Review Board Statement: Not applicable.

Informed Consent Statement: Not applicable.

Data Availability Statement: Not applicable.

Conflicts of Interest: The authors declare no conflict of interest.

Appendix A

According to Equation (8), we form coaxial double-position power Gaussian vortex with the normalization field and opposite spirals at the source plane $z = 0$ as follows

$$E(r',\theta',0) = \left[\frac{1-|\alpha|}{2} r'^{l_0} \exp\left(-\frac{r'^2}{w_0^2} - il_0\theta'\right) + \frac{1+|\alpha|}{2} r'^{l_0} \exp\left(-\frac{r'^2}{w_0^2} + il_0\theta'\right)\right] \quad \text{(A1)}$$

Consider the following normalization relations

$$\int_0^{D/2} \int_0^{2\pi} E(r',\theta',0) E^*(r',\theta',0) d\theta r dr = 1 \quad \text{(A2)}$$

Applying the integral relation [32] $\int_0^u x^m e^{-bx^2} dx = \frac{\gamma\left(\frac{m+1}{2}, bu^2\right)}{2b^{\frac{m+1}{2}}}$ in the above equation, the normalization field and opposite spirals at the source plane $z = 0$

$$E(r',\theta',0) = N\left[\frac{1-|\alpha|}{2} r'^{l_0} \exp\left(-\frac{r'^2}{w_0^2} - il_0\theta'\right) + \frac{1+|\alpha|}{2} r'^{l_0} \exp\left(-\frac{r'^2}{w_0^2} + il_0\theta'\right)\right] \quad \text{(A3)}$$

with normalization coefficient

$$N = \left\{2^{-l_0-3}\pi w_0^{2l_0+2}\left[(1+|\alpha|)^2 + (1-|\alpha|)^2\right]\gamma\left(l_0+1, \frac{D^2}{2w_0^2}\right)\right\}^{-1/2} \quad \text{(A4)}$$

Appendix B

By Huygens–Fresnel diffraction integral Equation (9), the field of the coaxial double-position power Gaussian vortex in free space at the receiving plane is given by [31,33]

$$\begin{aligned} E(r,\theta,z) &= -\tfrac{ik}{2\pi z} \exp\left[i\left(kz + \tfrac{kr^2}{2z}\right)\right] \int_0^\infty \int_0^{2\pi} E(r',\theta',0) \\ &\quad \times \exp\left\{i\tfrac{k}{2z}\left[r'^2 - 2rr'\cos(\theta-\theta')\right]\right\} r' dr' d\theta' \\ &= -\tfrac{ik}{2\pi N z} e^{i\left(kz+\frac{kr^2}{2z}\right)} \int_0^\infty r'^{l_0+1} e^{-\left(\frac{1}{w_0^2} - \frac{ik}{2z}\right) r'^2} \\ &\quad \left\{\tfrac{|\alpha|-1}{2} \int_0^{2\pi} e^{-i\left[l_0\theta' + \frac{krr'}{z}\cos(\theta-\theta')\right]} d\theta' + \tfrac{|\alpha|+1}{2} \int_0^{2\pi} e^{-i\left[-l_0\theta' + \frac{krr'}{z}\cos(\theta-\theta')\right]} d\theta'\right\} dr' \end{aligned} \quad \text{(A5)}$$

Consider integrals [32]

$$\int_0^{2\pi} e^{-i\left[l_0\theta' + \frac{krr'}{z}\cos(\theta-\theta')\right]} d\theta' = (-i)^{l_0} 2\pi J_{l_0}\left(\frac{krr'}{z}\right) e^{-il_0\theta} \quad \text{(A6)}$$

$$\int_0^{2\pi} e^{-i\left[-l_0\theta' + \frac{krr'}{z}\cos(\theta-\theta')\right]} d\theta' = (i)^{l_0} 2\pi J_{l_0}\left(\frac{krr'}{z}\right) e^{-il_0\theta} \quad \text{(A7)}$$

Equation (A1) becomes

$$\begin{aligned} E(r,\theta,z) &= -\tfrac{ik}{2Nz} \exp\left[i\left(kz + \tfrac{kr^2}{2z}\right)\right] \\ &\quad \times \left\{\left[(1-|\alpha|)(-i)^{l_0} e^{-il_0\theta} + (1+|\alpha|)(i)^{l_0} e^{il_0\theta}\right]\right. \\ &\quad \left. + \int_0^\infty r'^{l_0+1} \exp\left[-\left(\tfrac{1}{w_0^2} - \tfrac{ik}{2z}\right) r'^2\right] J_{l_0}\left(\tfrac{krr'}{z}\right) dr'\right\} \end{aligned} \quad \text{(A8)}$$

Further, consider the following integrals further [32]

$$\int_0^\infty r'^\mu e^{-ar'^2} J_\nu(br') dr' = \frac{b^\nu \Gamma\left(\frac{\nu+\mu+1}{2}\right)}{2^{\nu+1} a^{\frac{1}{2}(\nu+\mu+1)} \Gamma(\nu+1)} {}_1F_1\left(\frac{\nu+\mu+1}{2}; \nu+1; -\frac{b^2}{4a}\right), \ \mathrm{Re}\, a > 0 \quad \text{(A9)}$$

Equation (A4) becomes

$$\begin{aligned} E(r,\theta,z) &= -\frac{ik}{2Nz}\exp\left[i\left(kz+\frac{kr^2}{2z}\right)\right]\left[(1-|\alpha|)(-i)^{l_0}e^{-il_0\theta}+(1+|\alpha|)(i)^{l_0}e^{il_0\theta}\right] \\ &\times\frac{zk^{l_0}w_0^{2(l_0+1)}r^{l_0}}{\left(2z-ikw_0^2\right)^{l_0+1}}{}_1F_1\left(l_0+1;l_0+1;-\frac{k^2r^2w_0^2}{2z\left(2z-ikw_0^2\right)}\right) \end{aligned} \tag{A10}$$

where $\Gamma\left(\frac{v+\mu+1}{2}\right)$ is gamma function, and ${}_1F_1\left(\frac{v+\mu+1}{2};v+1;-\frac{b^2}{4a}\right)$ is confluent hypergeometric function. k is the wave number, and λ is the wavelength of light in the vacuum.

Appendix C

In weak turbulence and extinction seawater, by using the Rytov approximation [25] and Equation (10), the field of random coaxial double-position power Gaussian vortex (RCDPPGV) laser is expressed as

$$\begin{aligned} E(r,\theta,z) &= -\frac{ik}{2Nz}e^{i\left(kz+\frac{kr^2}{2z}\right)+\psi(r,\theta,z)}\left[(1-|\alpha|)(-i)^{l_0}e^{-il_0\theta}+(1+|\alpha|)(i)^{l_0}e^{il_0\theta}\right] \\ &\times\frac{zk^{l_0}w_0^{2(l_0+1)}r^{l_0}}{\left(2z-ikw_0^2\right)^{l_0+1}}{}_1F_1\left(l_0+1;l_0+1;-\frac{k^2r^2w_0^2}{2z\left(2z-ikw_0^2\right)}\right) \end{aligned} \tag{A11}$$

where $k=\frac{2\pi}{\lambda}(n_r+in_i)$, and $\psi(r,\theta,z)$ represents the complex phase disturbance caused by seawater turbulence.

In Equation (A11), the wave number of seawater k is substituted, and Equation (A11) is rewritten as

$$\begin{aligned} \widetilde{E}(r,\theta,z) &= \frac{\pi(n_i-in_r)}{N\lambda z}i^{l_0}e^{\frac{2\pi}{\lambda}(in_r-n_i)\left(z+\frac{r^2}{2z}\right)+\psi(r,\theta,z)}\left[(1-|\alpha|)(-1)^{l_0}e^{-il_0\theta}+(1+|\alpha|)e^{il_0\theta}\right] \\ &\times\frac{z\left[\frac{2\pi}{\lambda}(n_r+in_i)\right]^{l_0}w_0^{2(l_0+1)}r^{l_0}}{\left[2z+\frac{2\pi}{\lambda}(n_i-in_r)w_0^2\right]^{l_0+1}}{}_1F_1\left(l_0+1;l_0+1;-\frac{\left[\frac{2\pi}{\lambda}(n_r+in_i)\right]^2r^2w_0^2}{4z^2+\frac{4\pi}{\lambda}z(n_i-in_r)w_0^2}\right) \end{aligned} \tag{A12}$$

Appendix D

Rewrite the condition probability distribution of OAM modes Equation (14) in turbulent as well as absorbent seawater and at the receiving plane

$$\begin{aligned} p(l/l_0) &= \left\langle|a_l(r,z)|^2\right\rangle \\ &= \frac{1}{2\pi}\left|\int_0^{2\pi}\int_0^{2\pi}\widetilde{E}(r,\theta,z)\widetilde{E}^*(r,\theta,z)\exp[-il(\theta-\theta')]\,d\theta d\theta'\right| \end{aligned} \tag{A13}$$

Substitute Equation (11) into Equation (A13) and make a simple arrangement to obtain the following equation

$$\begin{aligned} p(l/l_0) &= \frac{\pi\left(n_r^2+n_i^2\right)\left[\left(\frac{2\pi}{\lambda}\right)^2\left(n_r^2+n_i^2\right)\right]^{l_0}w_0^{4(l_0+1)}r^{2l_0}}{2N^2\lambda^2\left[\left(2z+\frac{2\pi}{\lambda}n_iw_0^2\right)^2+\left(n_r\frac{2\pi}{\lambda}w_0^2\right)^2\right]^{l_0+1}}e^{-\frac{2\pi}{\lambda}n_i\left(z+\frac{r^2}{2z}\right)-\frac{2r^2}{\rho_{0c}^2}} \\ &\times\left|{}_1F_1\left(l_0+1;l_0+1;-\frac{\left[\frac{2\pi}{\lambda}(n_r+in_i)\right]^2r^2w_0^2}{4z^2+\frac{4\pi}{\lambda}z(n_i-in_r)w_0^2}\right)\right|^2 \\ &\times\left|\int_0^{2\pi}\int_0^{2\pi}e^{2r^2\frac{\cos(\theta-\theta')}{\rho_{0c}^2}}\left[\left(1-|\alpha|^2\right)e^{-i(l+l_0)(\theta-\theta')}+(-1)^{l_0}\left(1-|\alpha|^2\right)e^{-i2l_0\theta'}e^{i(l-l_0)(\theta-\theta')}\right.\right. \\ &\left.\left.+\left(1-|\alpha|^2\right)(-1)^{l_0}e^{2il_0\theta'}e^{-i(l+l_0)(\theta-\theta')}+\left(1+|\alpha|^2\right)e^{-i(l-l_0)(\theta-\theta')}\right]d\theta d\theta'\right| \end{aligned} \tag{A14}$$

By integrals $\int_0^{2\pi} e^{-im\theta+x\cos(\theta-\phi)}\,d\theta = 2\pi e^{-im\phi} I_m(x)$ and $\int_0^{2\pi} e^{-i(m-n)\theta}\,d\theta = 2\pi\delta_{mn}$ [32], where δ_{mn} is the Kronick notation, we have the conditional probability distribution $p(l/l_0)$ of the OAM modes of the RCDPPGV laser

$$\begin{aligned} p(l/l_0) &= \frac{2\pi^3\left(n_r^2+n_i^2\right)\left[\left(\frac{2\pi}{\lambda}\right)^2\left(n_r^2+n_i^2\right)\right]^{l_0} w_0^{4(l_0+1)} r^{2l_0}}{N^2\lambda^2\left[\left(2z+\frac{2\pi}{\lambda}n_i w_0^2\right)^2+\left(n_r\frac{2\pi}{\lambda}w_0^2\right)^2\right]^{l_0+1}} e^{-\frac{2\pi}{\lambda}n_i\left(z+\frac{r^2}{2z}\right)-\frac{2r^2}{\rho_{0c}^2}} \\ &\times\left|{}_1F_1\left(l_0+1;l_0+1;-\frac{\left[\frac{2\pi}{\lambda}(n_r+in_i)\right]^2 r^2 w_0^2}{4z^2+\frac{4\pi}{\lambda}z(n_i-in_r)w_0^2}\right)\right|^2 \\ &\times\left|\left[(1-|\alpha|)^2 I_{l+l_0}\left(\frac{2r^2}{\rho_{0c}^2}\right)+(1+|\alpha|)^2 I_{l-l_0}\left(\frac{2r^2}{\rho_{0c}^2}\right)\right]\right|. \end{aligned} \tag{A15}$$

where $\Gamma\left(\frac{\nu+\mu+1}{2}\right)$ is the gamma function, and ${}_1F_1\left(\frac{\nu+\mu+1}{2};\nu+1;-\frac{b^2}{4a}\right)$ is the confluent hypergeometric function.

References

1. Allen, L.; Beijersbergen, M.W.; Spreeuw, R.J.C.; Woerdman, J.P.; Barnett, S.M.; Padgett, M.J. Orbital angular momentum of light and the transformation of Laguerre-Gaussian laser modes. *Optical. Angular. Momentum.* **2004**, *45*, 8185–8189. [CrossRef]
2. Krenn, M.; Fickler, R.; Fink, M.; Handsteiner, J.; Malik, M.; Scheidl, T.; Ursin, R.; Zeilinger, A. Communication with spatially modulated light through turbulent air across Vienna. *New J. Phys.* **2014**, *16*. [CrossRef]
3. Paterson, C. Atmospheric Turbulence and Orbital Angular Momentum of Single Photons for Optical Communication. *Phys. Rev. Lett.* **2005**, *94*, 153901. [CrossRef]
4. Krenn, M.; Handsteinera, J.; Fink, M.; Fickler, F.; Ursin, R.; Malik, M.; Zeilinge, A. Twisted light transmission over 143 km. *PNAS* **2016**, *113*, 13648–13653. [CrossRef]
5. Qasem, Z.A.H.; Leftah, H.A.; Sun, H.; Qi, J.; Esmaiel, H. X-Transform Time-Domain Synchronous IM-OFDM-SS for Underwater Acoustic Communication. *IEEE Syst. J.* **2021**, *99*, 1–12. [CrossRef]
6. Qasem, A.H.; JWang, J.; Kuai, X.; Sun, H.; Esmaiel, H. *Enabling Unique Word OFDM for Underwater Acoustic Communication*; IEEE Wireless Communications Letters: Xiamen, China, 2021. [CrossRef]
7. Qasem, A.H.; Leftah, H.A.; Sun, H.; Qi, J.; Wang, J.; Esmaiel, H. Deep learning-based code indexed modulation for autonomous underwater vehicles systems. *Veh. Commun.* **2021**, *28*, 100314.
8. Cheng, M.; Guo, L.; Li, J.; Huang, Q.; Cheng, Q.; Zhang, D. Propagation of an optical vortex carried by a partially coherent Laguerre–Gaussian beam in turbulent ocean. *Appl. Opt.* **2016**, *55*, 4642–4648. [CrossRef]
9. Li, Y.; Cui, Z.; Han, Y.; Hui, Y. Channel capacity of orbital-angular-momentum-based wireless communication systems with partially coherent elegant Laguerre–Gaussian beams in oceanic turbulence. *J. Opt. Soc. Am. A* **2019**, *36*, 471–477. [CrossRef]
10. Zhang, W.; Wang, L.; Wang, W.; Zhao, S. Propagation property of Laguerre-Gaussian beams carrying fractional orbital angular momentum in an underwater channel. *OSA Contin.* **2019**, *2*, 3281–3287. [CrossRef]
11. Cheng, M.; Guo, L.; Li, J.; Zhang, Y. Channel Capacity of the OAM-Based Free-Space Optical Communication Links with Bessel–Gauss Beams in Turbulent Ocean. *IEEE Photon. J.* **2016**, *8*, 1–11. [CrossRef]
12. Chen, M.; Zhang, Y. Effects of anisotropic oceanic turbulence on the propagation of the OAM mode of a partially coherent modified Bessel correlated vortex beam. *Waves Random. Complex. Media.* **2019**, *29*, 694–705. [CrossRef]
13. Deng, S.; Yang, D.Y.; Zhang, Y. Capacity of communication link with carrier of vortex localized wave in absorptive turbulent seawater. *Waves Random. Complex. Media.* **2020**, 1–14. [CrossRef]
14. Wang, X.; Yang, Z.; Zhao, S. Influence of oceanic turbulence on propagation of Airy vortex beam carrying orbital angular momentum. *Optik* **2019**, *176*, 49–55. [CrossRef]
15. Yu, L.; Zhang, Y. Analysis of modal crosstalk for communication in turbulent ocean using Lommel-Gaussian beam. *Opt. Express* **2017**, *25*, 22565–22574. [CrossRef]
16. Hu, Z.; Liu, H.; Xia, J.; He, A.; Li, H.; Du, Z.; Chen, T.; Li, Z.; Lü, Y. Propagation characteristics of the perfect vortex beam in anisotropic oceanic turbulence. *Appl. Opt.* **2020**, *59*, 9956–9962. [CrossRef]
17. Liang, Q.; Yang, D.; Zhang, Y.; Zheng, Y.; Hu, L. Probability of orbital angular momentum modes carried by a finite energy frozen wave in turbulent seawater. *OSA Contin.* **2020**, *3*, 2429. [CrossRef]
18. Wang, X.; Wang, L.; Zhao, S. Research on Hypergeometric-Gaussian Vortex Beam Propagating under Oceanic Turbulence by Theoretical Derivation and Numerical Simulation. *J. Mar. Sci. Eng.* **2021**, *9*, 442. [CrossRef]
19. Zhang, Y.; Zheng, Y.; Yan, Q.; Yu, L. Average capacity of an absorptive turbulent ocean channel with diffraction-and attenuation-resistant beam carriers. *Opt. Commun.* **2021**, *499*, 127291. [CrossRef]
20. Deng, S.; Yang, D.; Hu, L.; Zhang, Y. *Transmittance of Diffraction- and Attenuation-Resistant Vortex Frozen Beams in Oceanic Turbulence*; PIERS Fall: Rome, Italy, 2019; pp. 2528–2621.

21. Deng, S.; Zhu, Y.; Zhang, Y. Received Probability of Vortex Modes Carried by Localized Wave of Bessel–Gaussian Amplitude Envelope in Turbulent Seawater. *J. Mar. Sci. Eng.* **2019**, *7*, 203. [CrossRef]
22. Yang, D.; Zhang, Y.; Shi, H. Capacity of turbulent ocean links with carrier Bessel–Gaussian localized vortex waves. *Appl. Opt.* **2019**, *58*, 9484–9490. [CrossRef]
23. Zheng, Y.; Yang, D.; Qin, S.; Zhang, Y. Received probability of orbital-angular-momentum modes carried by diffraction- and attenuation resistant beams in weak turbulent oceans. *J. Mar. Sci. Eng.* **2020**, *8*, 701. [CrossRef]
24. Li, Y.; Zhang, Y.; Zhu, Y. Oceanic spectrum of unstable stratification turbulence with outer scale and scintillation index of Gaussian-beam wave. *Opt. Express* **2019**, *27*, 7656–7672. [CrossRef]
25. Andrews, L.C.; Phillips, R.L. *Laser Beam Propagation through Random Media*; SPIE Optical Engineering Press: Bellingham, WA, USA, 1998; pp. 47–50.
26. Kotlyar, V.V.; Kovalev, A.A.; Porfirev, A.P. *Vortex Laser Beams*; Taylor & Francis Group; CRC Press: Boca Raton, FL, USA, 2019.
27. Bazhenov, V.; Vasnetsov, M.V.; Soskin, M.S. Laser-beam with screw dislocations in the wavefrom. *JETP Lett.* **1990**, *52*, 429–431.
28. Yousif, B.B.; Elsayed, E.E. Performance enhancement of an orbital-angular-momentum-multiplexed free-space optical link under atmospheric turbulence effects using apatial-mode multiplexing and hybrid diversity based on adaptive MIMO equalization. *IEEE Acess* **2020**, *7*, 84401–84412. [CrossRef]
29. Anguita, J.A.; Neifeld, M.A.; Vasic, B.V. Turbulence-induced channel crosstalk in an orbital angular momentum-multiplexed free-space optical link. *Appl. Opt.* **2008**, *47*, 2414–2429. [CrossRef]
30. Tyler, A.G.; Boyd, R.W. Influence of atmospheric turbulence on the propagation of quantum states of light carrying orbital angular momentum. *Opt. Lett.* **2009**, *34*, 142–144. [CrossRef]
31. Keskin, A.; Baykal, Y. Effects of underwater turbulence on average transmittance of cos-Gaussian and cosh-Gaussian optical beams. *Waves Random. Complex. Media.* **2020**, 1–12. [CrossRef]
32. Gradshteyn, I.S.; Ryzhik, I.M. *Table of Integrals, Series, and Products*; Academic Press: Cambridge, MA, USA, 2007.
33. Korotkova, O. *Random Light Beams Theory and Applications*; CRC Press: Boca Raton, FL, USA, 2014.

Journal of
Marine Science
and Engineering

Article

Absorptive Turbulent Seawater and Parameter Optimization of Perfect Optical Vortex for Optical Communication

Qingze Yan †, Yixin Zhang †, Lin Yu and Yun Zhu *

School of Science, Jiangnan University, Wuxi 214122, China
* Correspondence:zhuyun1210@163.com
† They are the co-first authors.

Abstract: In this paper, the optimization of perfect optical vortex (POV) parameter for underwater wireless optical communication link under M-QAM by average bit-error rate (ABER) and the effect of seawtaer turbulence on link information capacity are investigated. The link is absorbent, weakly turbulent, and bandwidth-limited. In investigating, we use the spectral absorption coefficient to describe the wavelength effect of seawater absorption. Specifically, under the paraxial approximation and Rytov approximation conditions, we define the average signal-to-noise-crosstalk ratio including the system bandwidth factor and derive the bandwidth-limited ABER of the OAM carrier link. Capitalizing on the defined average signal-to-noise crosstalk ratio and the derived bandwidth-limited ABER of link, the novel closed-form expression for the average information capacity of the perfect optical vortex link under M-QAM modulation is proposed. Through the numerical analysis of the ABER and the average information capacity, the POV optimization parameters in specific communication links are obtained and new conclusions are drawn that the average information capacity is restricted by both signal wavelength and the seawater absorption coefficient.

Keywords: perfect optical vortex; absorptive turbulent seawater; parameter optimization; average information capacity; modulation

1. Introduction

Vortex beams with infinite N qubit orbital angular momentum (OAM) base [1–8] have attracted the attention of underwater wireless optical communication (WOC) researchers because of their strong resistance to environmental interference, high security, wide band width and low cost [8–12]. However, because the modes of OAM are determined by the wave-front spatial structure of the carrier, the wave-front distortion of vortex beams caused by ocean turbulence will inevitably lead to crosstalk among OAM modes [8–14]. To this end, the attempt to use vortex beams such as Gauss-Schell model [8], Bessel-Gauss [9], Hermite-Gauss [10], Lommel [13] and local wave [14] as signal sources to reduce the turbulent sea effect of OAM mode transmission and improve the transmission quality of underwater OAM mode has become one of the most popular research topics in this field.

In 2013, Ostrovsky et al. put forward perfect optical vortex (POV) new beam with dark hollow radius independent of topological charge, and approximately realized this beam in Fourier transform optical experiments using computer-controlled liquid crystal spatial light modulation [15]. After, Pravin Vaity et al. deduced the mathematical model of POVs by Fourier transform of the Bessel beam, and verified the mathematical model of POVs by Fourier transform of the Bessel beam experimentally. At present, this model has become a general model for theoretical research on POVs characteristics [16]. POVs have the characteristic of having the same initial ring width and radius under different topological charges, and in the source plane, there is no sidelobe of intensity distribution. These new characteristics of POV have been studied extensively through the evolution of transmission media such as no turbulence free space [17,18], turbulent atmosphere [19], and underwater [20–23]. These studies show that: The POV carrying OAM can be used as communication link information carrier by OFDM 16-QAM modulation [17]. In 2-channel OAM multiplexed WOC link, POVs carrier links perform better than Laguerr-Gaussian (LG) carrier links [18]. In a turbulent atmosphere experiment, the POV beam retains its original characteristics and its propagation distance depends on the power of the light source, the parameters of the Fourier objective, and the model of spatial light modulator [19]. POV transmission experiments under different underwater conditions show that the bit-error rate (BER) of OAMs is higher than its ideal value at lower SNR, even in stagnant water, and the bit-error rate reaches its maximum value when the beam propagates in bubble underwater links [20]. In an anisotropic oceanic turbulence, the self-focusing property of POVs improves to the transmission quality of OAM modes [21]. Although the wavelength, topological charge and radius-thickness ratio of POVs hardly affect the propagation quality of POV beam in ocean turbulence, the radius of POV beam does [22]. In the short link of absorbent turbulent seawater, POVs have a long wavelength within the "seawater window wavelength" interval, a small topological charge of OAM, a large annular radius, and a small aperture of transmitter and receiver; the probability of receiving POV is high [23]. Recently, the underwater propagation properties of PLG, a new beam similar to POV, have also received attention. In the case of selected beam types, selecting appropriate system parameters is also one of the important methods to effectively control the impact of ocean turbulence [24], but the parameter optimization of existing POVs has not been studied [15–23]. However, as far as we know, the optimization of source parameters of UWOC link with M-QAM and POV signal carrier, and the influence of wavelength dependence of seawater absorption on the average information capacity (AIC) of POV carrier link have not been reported.

The main contributions of this article are mainly summarized as follows: combined with the definition of signal-to-noise ratio(SNR) for bandwidth-constrained systems and the discussion of zero mean additive Gaussian channel, we defined a new average signal-crosstalk-noise ratio (ASCNR) of OAM channel that is connected to the signal bandwidth. The AIC and ABER of bandwidth-constrained link with signal modulation of M-QAM has been derived in a closed form. For a given bandwidth-constrained link, we find that choosing a suitable modulation order M can reduce the bandwidth limited ABER caused by the channel of turbulent absorption seawater.

Seawater scattering, bubbles, physio-chemical properties of seawater, and pointing errors are other important factors affecting optical communication in seawater, but seawater scattering, bubbles and physio-chemical properties of seawater are very complex factors affecting optical communication transmission [25–29], so we only consider a POV in clean or ignore the complicated factors of optical communication problems in offshore underwater channels. Therefore, the optimization of source parameters and AIC of underwater wireless optical communication absorption link with carrier of POV and M-QAM are investigated under Rytov approximation. In Section 2, normalized amplitude of POV in weak absorptive turbulence is presented. In Section 3, the bandwidth limited ABER of POV carrier link using M-QAM is derived. In Section 4, the AIC of the POV carrier and bandwidth-constrained link using M-QAM is proposed. Based on the numerical simulation of bandwidth-limited ABER of M-QAM link and the AIC of the link, the optimized POV source parameters and the influence of seawater turbulence and absorption on the AIC of the communication link are given in Section 5, respectively. Finally, conclusions are given in Section 6.

2. Normalized Amplitude Perfect Optical Vortex in Weak Turbulent and Absorptive Seawater

Figure 1 depicts a schematic representation of an underwater wireless optical communication system model.

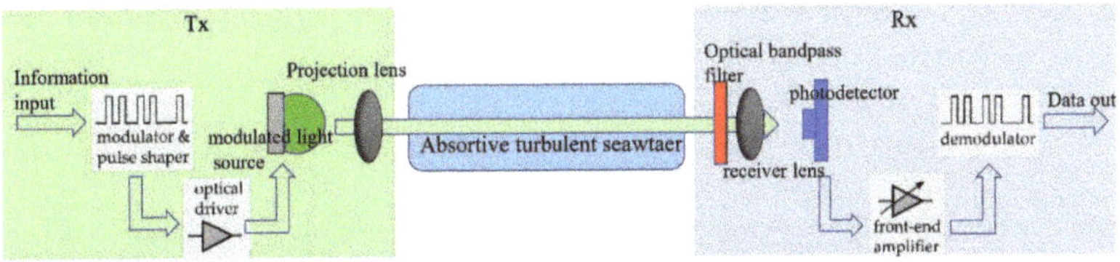

Figure 1. Schematic of underwater wireless optical communication system model.

In this model, the transmitter (Tx) is composed of M-QAM modulator, optical driver, modulated light source and projection lens. The receiver (Rx) is made of an optical bandpass filter, receiver lens, photodetector, front-end amplifier and demodulator. The light POV is studied in weakly scintillated turbulent seawater, $\sigma^2_{1\epsilon\chi T} = 1.23C^2_{\epsilon\chi T}k^{7/6}z^{11/6} < 1$ [30]. Paraxial transmission and Rytov approximations are adopted [31]. In cylindrical coordinates (ρ, θ, z) and turbulent absorbent seawater, the normalized complex amplitude of POV propagation is expressed as [23]

$$
\begin{aligned}
E_{om}(\rho,\theta,z) &= \frac{\mathrm{i}^{m_0}}{w(z)}\sqrt{\frac{2\exp(\frac{r_0^2}{\tilde{w}^2})}{\pi\left|\mathrm{I}_{m_0}(\frac{r_0^2}{\tilde{w}^2})\right|}}\mathrm{J}_{m_0}\left[\frac{2\rho r_0}{\tilde{w}w(z)}\exp(\mathrm{i}\Theta)\right]\exp\left[\mathrm{i}m_0\theta+\mathrm{i}\Theta+\frac{\mathrm{i}kn_R z}{2R(z)}(\rho^2+r_0^2)\right] \quad (1)\\
&\times\exp\left[-k_0 n_I z\left(1+\frac{\rho^2+r_0^2}{2R(z)}\right)+\mathrm{i}k_0 n_R z\right]\exp\left[-\frac{\rho^2-(zr_0/z_{oR})^2}{w(z)^2}+\Psi(\rho,\theta,z)\right],
\end{aligned}
$$

where $w(z) = \sqrt{1+z^2/z^2_{oR}}$, $\tilde{w}$ represents the half beam-width of the POV at the image plane of the Fourier lens, $z_{oR} = k(n_R + \mathrm{i}n_I)\tilde{w}^2/2$ is the Rayleigh range, k_0 is the wave number of light transmitted in a vacuum , n_R and n_I are the real part and the imaginary part of the n_{oc}, respectively; m_0 is the OAM topological charge of initial OAM signal, r_0 is the ring radius of POV at the object plane of the Fourier lens, J_{m_0} is the m_0 th-order first kind Bessel function, $\Theta = arctan(z/z_R)$, $R(z) = z + z^2_{oR}/z$, and Ψ is the complex phase perturbations caused by seawater turbulence. $n_I = \lambda c_{oc}(\lambda)/(4\pi)$ [32], and $c_{oc}(\lambda)$ is the spectrum absorption coefficient of seawater.

3. Average Bit-Error Rate of Link

By Fourier's orthogonal series expansion theory, we can expand the complex amplitudes expressed in Equation (1) into series based on orthogonal plane helical basis $\exp(im\theta)$ [14]

$$E_m(\rho,\theta,z) = \sum_m \alpha_m(\rho,z)\exp(im\theta), \quad (2)$$

where m is a new OAM topological charge of the random POV in turbulent seawater.

The expansion coefficient $\alpha_m(\rho,z)$ in Equation (2) is a random function of position and distance and is given by

$$\alpha_m(\rho,z) = \frac{1}{2\pi}\int_0^{2\pi} E_m(\rho,\theta,z)\exp(-im\theta)d\theta. \quad (3)$$

The probability distribution function of receiving OAM mode is obtained by $p(m/m_0)$ = $\langle\alpha_m(\rho,z)\alpha^*_m(\rho,z)\rangle$, where $\langle\cdot\rangle$ represents the mean of turbulence ensemble, and its mathematical expression is

$$p(m/m_0) = \frac{1}{4\pi^2}\int_0^{2\pi}\int_0^{2\pi} E_m(\rho,\theta,z)E^*_m(\rho,\theta',z)\exp[-\mathrm{i}m(\theta-\theta')d\theta d\theta']. \quad (4)$$

For stable stratification and isotropic turbulence, and taking into account the integral [33]

$$\int_0^{2\pi} \exp[\tau\cos(\theta-\phi) - im\theta]d\theta = 2\pi \mathrm{I}_m(\tau)\exp(-im\phi), \tag{5}$$

and applying Equation (1) in (4), we have the conditional probability distribution of the OAM mode carried by the POV [23]

$$\begin{aligned} p(m/m_0) &= \frac{2\exp\left(\frac{r_0^2}{\varpi^2} - 2k_0 n_I z\left(1+\frac{r_0^2}{2R(z)}\right)\right)}{\pi w(z)^2\left|\mathrm{I}\left(\frac{r_0^2}{\varpi^2}\right)\right|}\exp\left[\frac{8(zr_0)^2 n_R^2}{w(z)^2 k_0^2 \varpi^4 (n_R^2+n_I^2)}\right] \\ &\times\exp\left(-\frac{\rho^2}{w_{eff}^2}\right)\left|\mathrm{J}_{m_0}\left[\frac{2\rho r_0}{\varpi w(z)}\right]\right|^2 \mathrm{I}_{m-m_0}\left[\frac{2\rho^2}{\rho_0{}^2}\right], \end{aligned} \tag{6}$$

where $\mathrm{I}_{m_0}(x)$ is the modified Bessel function of the first kind and m_0-th orders.

In Equation (6) ρ_0 is the transverse spatial coherence radius of the plane wave and is given by

$$\begin{aligned} \rho_0 = &\left\{\frac{9.444C_{\epsilon\chi_T}^2({n_r}^2+n_I^2)z}{\lambda^2(1-\overline{\omega})^2}\left[\overline{\omega}^2\kappa_0^{1/3}\mathrm{U}(2;\frac{7}{6};\frac{\kappa_0^2\eta^2}{R_T^2}) + \overline{\omega}^2 4.6\eta^{2/3}\kappa_0\Gamma(\frac{7}{3})\mathrm{U}(\frac{7}{3};\frac{3}{2};\frac{\kappa_0^2\eta^2}{R_T^2})\right.\right. \\ &\times\kappa_0^{1/3}\mathrm{U}(2;\frac{7}{6};\frac{\kappa_0^2\eta^2}{R_S^2}) + 4.6\eta^{2/3}\kappa_0\Gamma(\frac{7}{3})\mathrm{U}(\frac{7}{3};\frac{3}{2};\frac{\kappa_0^2\eta^2}{R_S^2}) - 2\overline{\omega}\kappa_0^{1/3}\mathrm{U}(2;\frac{7}{6};\frac{\kappa_0^2\eta^2}{R_{TS}^2}) \\ &\left.\left.-\,9.2\eta^{2/3}\kappa_0\Gamma(\frac{7}{3})\mathrm{U}(\frac{7}{3};\frac{3}{2};\frac{\kappa_0^2\eta^2}{R_{TS}^2})\right]\right\}, \end{aligned} \tag{7}$$

where $C_{\varepsilon\chi_T}^2 = 0.809\times10^{-7}\varepsilon^{-1/3}\chi_T\overline{\omega}^{-2}(1-\overline{\omega})^2\left(\mathrm{K}^2/\mathrm{m}^{2/3}\right)$ is the structural constant of the salinity-temperature fluctuation of oceanic turbulence that is a function of the rate of dissipation of kinetic energy per unit mass of fluid ϵ in the range $[10^{-10}, 10^{-1}]\mathrm{m}^2/\mathrm{s}^3$ the dissipation rate of the mean-squared temperature χ_T in the range $[10^{-10}, 10^{-2}]\mathrm{K}^2/\mathrm{s}$ and $\overline{\omega}$ is the ratio of temperature and salinity contributions to the refractive index fluctuations, which can vary in the interval $(-5, 0)$. Here $\overline{\omega} = -5$ temperature fluctuation is dominant in the induction of optical turbulence, while $\overline{\omega} = 0$ salinity fluctuation is dominant in the induction of optical turbulence. κ is the scalar spatial wave number of turbulent fluctuation, $\kappa_0 = 2\pi/L_0$, $\kappa_T = R_T/\eta$, $\kappa_S = R_S/\eta$, $\kappa_{TS} = R_{TS}/\eta$, η is turbulent inner scale, L_0 is the outer scale of turbulence, $R_j = \sqrt{3}[W_j - 1/3 + 1/(9W_j)]^{3/2}/Q^{-3/2}(j = T, S, TS)$, $W_j = \left\{\left[\frac{\mathrm{Pr}_j^2 Q^4}{(6\beta)^2} - \frac{\mathrm{Pr}_j Q^2}{81\beta}\right]^{1/2} - \left[\frac{1}{27} - \frac{\mathrm{Pr}_j Q^2}{6\beta}\right]\right\}^{1/3}$, Q is the non-dimensional constant; Pr_T and Pr_S respectively represent the Prandtl number of the temperature and salinity, $\mathrm{Pr}_{TS} = 2\,\mathrm{Pr}_T\,\mathrm{Pr}_S\,/(\mathrm{Pr}_T + \mathrm{Pr}_S)$; $w_{\mathrm{eff}} = \left(\frac{2}{w^2(z)} + \frac{2}{\rho_0^2} + \frac{k_0 n_I z}{R(z)}\right)^{-1/2}$ is defined as the effective radius of the POV in the absorbed seawater turbulence.

Note the structural constants $C_{\varepsilon\chi_T}^2$ of the salinity-temperature fluctuation of oceanic turbulence increases with the increase of the salinity fluctuation, but it decreases with the increase of the temperature fluctuation.

The received probability of OAM modes of POV can be obtained

$$P(m \mid m_0) = 2\pi\int_0^{D/2} p(m \mid m_0)\rho d\rho, \tag{8}$$

where D is the receiver diameter. For $m \neq m_0$, $P(m \mid m_0)_{m\neq m_0}$ is the received probability of OAM crosstalk modes that describes the probability of the photon migrating from the

OAM signal mode to the new OAM mode, and $m = m_0$, $P(m \mid m_0)_{m=m_0}$ is the received probability of OAM signal modes.

It has been shown in [34] that OAM crosstalk distribution follows the zero-mean Gaussian distribution. Therefore, the communication link studied in this paper is still a zero mean additive Gaussian channel. Combined with the definition of SNR for bandwidth-constrained systems in [35] and the discussion of zero mean additive Gaussian channel in [30], we define a new average signal-crosstalk-noise ratio (ASCNR) of OAM channel as

$$\gamma_{m,m_0} = \frac{P^2(m/m_0)_{m=m_0}}{[\sum_{m=-\infty}^{\infty} P(m/m_0) + N_0/P_{TX}]2B}, \tag{9}$$

where P_{TX} is the transmit power and N_0 is the transmitter and receiver noise power, $B = 2R_b/\log_2 M$ is the bandwidth occupied by modulated signal, and R_b is the bit rate [36].

The Equation (9) can also be rewritten in terms of the bit rate

$$\gamma_{m,m_0} = \frac{P^2(m/m_0)_{m=m_0}\log_2 M}{[\sum_{m=-\infty}^{\infty} P(m/m_0) + N_0/P_{TX}]4R_b}. \tag{10}$$

Equation (10) indicates that when other system parameters remain unchanged, to maintain the ASCNR of the transmitted OAM signal, when the length of the communication channel or the turbulence strength within the channel increases, we must reduce the communication bandwidth or bit rate or increase the modulation order M.

For signal modulation of M-QAM, based on (9) [37], we can obtain the bandwidth-limited ABER of the OAM mth channel carried by the POV as

$$\begin{aligned} P_B = {} & \frac{2}{\log_2 M}\left(1 - \frac{1}{\sqrt{M}}\right) \operatorname{erfc}\sqrt{\frac{3\log_2 M}{2(M-1)}\gamma_{m,m_0}} \\ & \times \left[1 - \frac{1}{2}\left(1 - \frac{1}{\sqrt{M}}\right) \operatorname{erfc}\sqrt{\frac{3\log_2 M}{2(M-1)}\gamma_{m,m_0}}\right], \end{aligned} \tag{11}$$

where $\operatorname{erfc}(\cdot)$ denotes the complementary error function.

4. Average Information Capacity of Link

Figure 2 shows the discrete memoryless channel among the energy levels of signal OAM and received OAM, which is characterized by channel (transition) probabilities. Let $X = \{m_{00}, m_{01} \cdots, m_{0i-1}\}$ and $Y = \{m_0, m_1 \cdots, m_{J-1}\}$ be the input and the output letters channel, respectively; the channel is completely characterized by the following set of transition probabilities [36]:

$$p(m_j/m_{oi}) = P(Y = m_j/X = m_{0i}), 0 \le p(m_j/m_{0i}) \le 1, \tag{12}$$

where $i \in \{0, 1, \cdots, I-1\}, j \in \{0, 1, \cdots, J-1\}$, while I and J denote the sizes of input and output alphabets, respectively. The transition probability $p(m_j \mid m_{0i})$ represents the conditional probability that $Y = m_j$ for given input $X = m_{0i}$.

The information capacity is obtained by maximization of mutual information over all possible input distributions [36]:

$$C = \max_{\{p(m_{0i})\}} I(X,Y) = \max_{\{p(m_{0i})\}} [H(X) - H(X \mid Y)], \tag{13}$$

where $H(X) - \langle \log_2 P(X) \rangle$ represents the uncertainty about the channel input before observing the channel output, while $H(X \mid Y)$ represents the uncertainty of the channel input after it has been received.

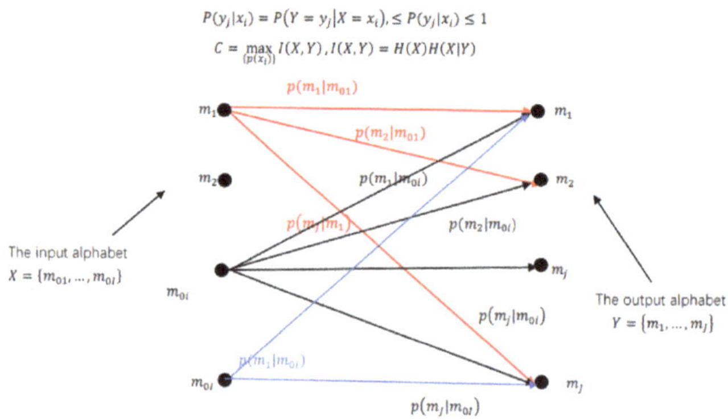

Figure 2. Discrete memoryless channel among the energy levels of signal OAM and received OAM.

In Equation (13), $I(X;Y)$ is defined as mutual information. The mathematical expression for mutual information is expressed as follows:

$$\begin{aligned} I(X;Y) &= H(X) - H(X \mid Y) \\ &= \sum_{i=1}^{M} p(m_{0i}) \log_2\left[\frac{1}{p(m_{0i})}\right] - \sum_{j=1}^{N} p(m_j) \sum_{i=1}^{M} p(m_{0i} \mid m_j) \log_2\left[\frac{1}{p(m_{0i} \mid m_j)}\right]. \end{aligned} \tag{14}$$

Equation (14) represents that the mutual information is obtained as the output information minus information lost in the channel.

Considering that (1) POV constitutes an infinite OAM channel and the channels used for communication are finite: (2) when POV is transmitted in weakly turbulent seawater, the difference between the energy level corresponding to OAM crosstalk formed by seawater turbulence disturbance and the energy level of OAM signal is very small. Therefore, the OAM channel composed of $N = 2m + 1$ input and $N = 2m + 1$ output is a symmetric channel (MSC). Further, let constellation points be marked with gray codes, and the symbol error causes the corruption of only 1 bit out of $\log 2M$ bit comprising the symbol [38]. We have that $p(m_j \mid m_{0i}) = P_B \log_2 M/(M-1)$ and $p(m_j \mid m_{0j}) = 1 - P_B \log_2 M$, where $P_B \log_2 M$ is symbol error probability, P_B is the bit-error rate (ABER).

The AIC of a bandwidth-constrained link is described by the following formula

$$C = \log_2 N + (1 - P_B \cdot \log_2 M) \cdot \log_2(1 - P_B \cdot \log_2 M) + P_B \cdot \log_2 M \cdot \log_2 \frac{P_B \cdot \log_2 M}{N-1}. \tag{15}$$

Taking into account Equations (11) and (15), we can obtain the AIC of the POV carrier link with M-QAM as

$$\begin{aligned} C = \log_2 N &+ \left\{1 - 2\left(1 - \frac{1}{\sqrt{M}}\right) \operatorname{erfc}\sqrt{\frac{3\log_2 M}{2(M-1)}\gamma_{m,m_0}}\left[1 - \frac{1}{2}\left(1 - \frac{1}{\sqrt{M}}\right) \operatorname{erfc}\sqrt{\frac{3\log_2 M}{2(M-1)}\gamma_{m,m_0}}\right]\right\} \\ &\times \log_2\left\{1 - 2\left(1 - \frac{1}{\sqrt{M}}\right) \operatorname{erfc}\sqrt{\frac{3\log_2 M}{2(M-1)}\gamma_{m,m_0}}\left[1 - \frac{1}{2}\left(1 - \frac{1}{\sqrt{M}}\right) \operatorname{erfc}\sqrt{\frac{3\log_2 M}{2(M-1)}\gamma_{m,m_0}}\right]\right\} \\ &+ \left\{1 - 2\left(1 - \frac{1}{\sqrt{M}}\right) \operatorname{erfc}\sqrt{\frac{3\log_2 M}{2(M-1)}\gamma_{m,m_0}}\left[1 - \frac{1}{2}\left(1 - \frac{1}{\sqrt{M}}\right) \operatorname{erfc}\sqrt{\frac{3\log_2 M}{2(M-1)}\gamma_{m,m_0}}\right]\right\} \\ &\times \log_2 \frac{2}{N-1}\left\{1 - 2\left(1 - \frac{1}{\sqrt{M}}\right) \operatorname{erfc}\sqrt{\frac{3\log_2 M}{2(M-1)}\gamma_{m,m_0}}\left[1 - \frac{1}{2}\left(1 - \frac{1}{\sqrt{M}}\right) \operatorname{erfc}\sqrt{\frac{3\log_2 M}{2(M-1)}\gamma_{m,m_0}}\right]\right\}, \end{aligned} \tag{16}$$

where $N = -m, -m+1 \cdots, 0, \cdots m-1, m,$ and AIC represents the maximum data rate that the system can achieve.

5. Numerical Analysis

In this section, we will search for the optimal light source parameters of M-QAM modulated POV communication links in the central Pacific [39] under weak turbulence through numerical simulation of bandwidth-limited ABER, and discuss the influence of seawater turbulence and absorption on the AIC of the POV carrier link with M-QAM. For the sake of argument, unless otherwise specified, we choose the parameters of the absorbable turbulent seawater link as Table 1, $\tilde{w} = \tilde{w}_{opt}(z = z_0)$ and we adopted the current reachable experimental distance of $z = 150$ m [40,41] and modulation order $M = 256$ as the link length and modulation order [40]. Note that the maximum transmission distance achieved in the existing experiments is 200 m [42], but the turbulence effect in seawater is not taken into account. Considering that seawater turbulence will cause crosstalk loss of the OAM signal, and the wider the bandwidth, the greater the crosstalk loss of the OAM signal, we adopt a narrower bandwidth ($B = 0.5$ GHZ) and a shorter distance ($z = 150$ m) in numerical calculation.

Table 1. Major system parameters.

Symbol	Parameters	Value
Pr_T	Prandtl number of the temperature	0.72
Pr_S	Prandtl number of the salinity	700
Q	Non-dimensional constant	2.5
B	Bandwidth	0.05 GHZ
$C^2_{\varepsilon\chi_T}$	Turbulence strength	$10^{-13}\ \mathrm{K}^2/\mathrm{m}^{2/3}$
ε	Rate of dissipation of kinetic energy per unit mass of fluid	$10^{-3}\ \mathrm{m}^2/\mathrm{s}^3$
χ_T	Dissipation rate of the mean-squared temperature	$10^{-7}\ \mathrm{K}^2/\mathrm{s}$
$\overline{\omega}$	Ratio of temperature and salinity	-4.5
k_0	Wave number	$2\pi/\lambda$
w_{eff}	Effective radius	$\left(\frac{2}{w^2(z)} + \frac{2}{\rho_o^2} + \frac{k_0 n_I z}{R(z)}\right)^{-1/2}$
$w(z)$	Half beam-width	$\sqrt{1 + z^2/z_{oR}^2}$
r_0	Ring radius	0.03 m
η	Inner scale	0.01 m
L_0	Outer scale	10 m
m_0	OAM topological charge	1
D	Receiver diameter	0.05 m
λ	Wavelength	470 nm
n_I	Imaginary part of the n_{oc}	0.4487×10^{-9}
z	Propagation distance	150 m

5.1. Probability Distribution and Received Probability

Figure 3 shows the transformation of received probability of the OAM signal and crosstalk modes carried by POVs with wavelength $\lambda = 470$ nm ($n_I = 0.4487 \times 10^{-9}$) at transmission $z = 150$ m in seawater. In Figure 3a the diameter of receiver is 0.02 m, while in Figure 3b the diameter of is 0.005 m. By comparing the bar charts in Figure 3a,b, it can be found that (1) the optical communication link adopting the receiver of small aperture will cause serious loss of OAM signal owning high order topological charge, and it will cause a large loss of all OAM signals; (2) the receiving probability of the OAM signal is independent of topological charge when the communication link adopts a larger aperture to receive the OAM signal. This result is very beneficial to the mode division multiplexing of communication link and is the purpose of designing the POV. The curves in Figure 3a show that even though the channel has reached its experimental length 150 m for $D = 0.02$ m, OAM crosstalk produced by seawater turbulence is still very small. This result indicates that POVs transmitted in turbulent absorbing seawater still have the characteristics of their topological charges transmitted in free space [15,16], that is, intensity is independent of their topological charges.

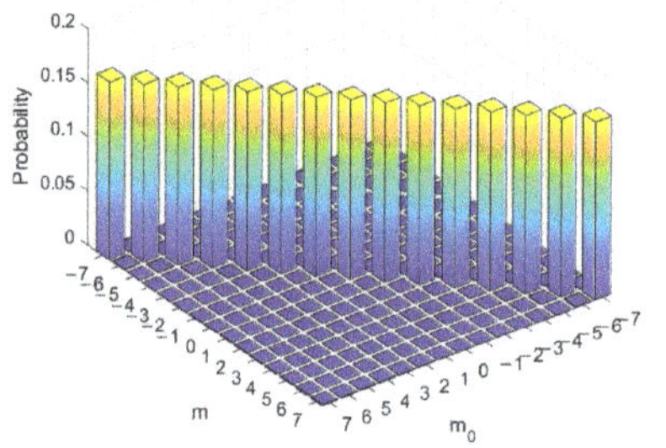

(a)

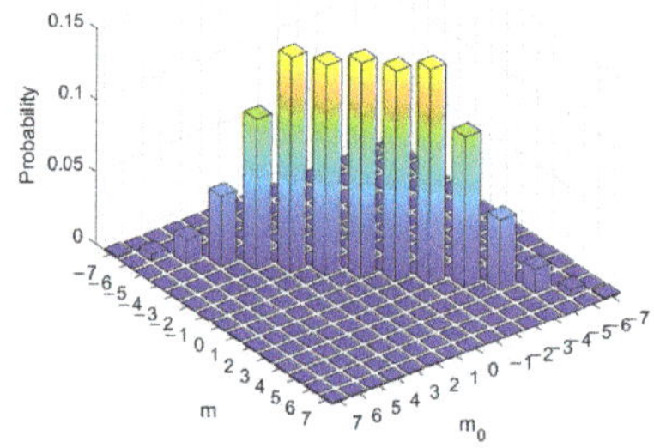

(b)

Figure 3. The received probability of OAM signal modes carried by POV versus different topological charge of initial OAM modes and topological charge of new OAM modes for $r_0 = 0.03$ m, $\overline{\omega} = -4.5$, $n_I = 0.4487 \times 10^{-9}$. (**a**) $z = 150$ m, $D = 0.02$ m; (**b**) $z = 150$ m, $D = 0.005$ m.

5.2. The Bandwidth, Modulation Order, SNR of Optical System and ABER

In this sub-section, we will analyze the influence of modulation order M , the bandwidth and link channel on ABER.

To analyze which of M-QAM is more advantageous for the underwater POV transmission channel, in Figure 4 we give the numerical results for the bandwidth-limited ABER of POV carrier link with M-QAM versus the propagation distance and turbulence strength for $B = 0.05$ GHz. The Figure 4 shows the bandwidth-limited ABER of link increases with the increase of propagation distance z for each modulation order M. Further, in $C^2_{\varepsilon\chi T} = 10^{-13}\ \mathrm{K}^2/\mathrm{m}^{2/3}$, for $z \leq 123$ m, ABER of 4-QAM is the lowest, but for $z \geq 123$ m, ABER of 512-QAM is the lowest. In $C^2_{\varepsilon\chi T} = 10^{-15}\ \mathrm{K}^2/\mathrm{m}^{2/3}$ (see Figure 4a), ABER of 4-QAM is always the lowest. That is, low order modulation has better performance over shorter distances, while high order modulation has advantages over long distances.

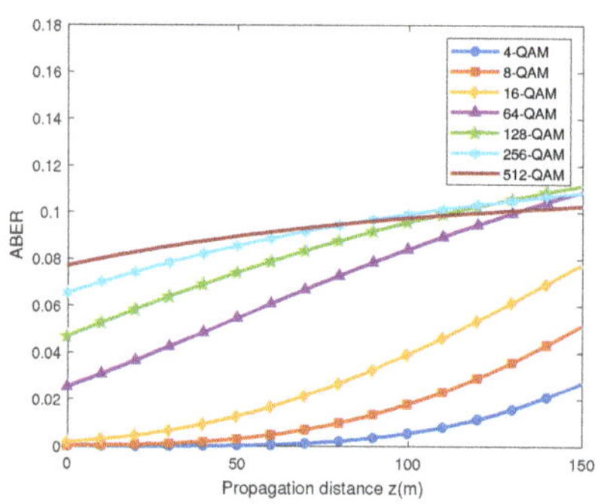

(a)

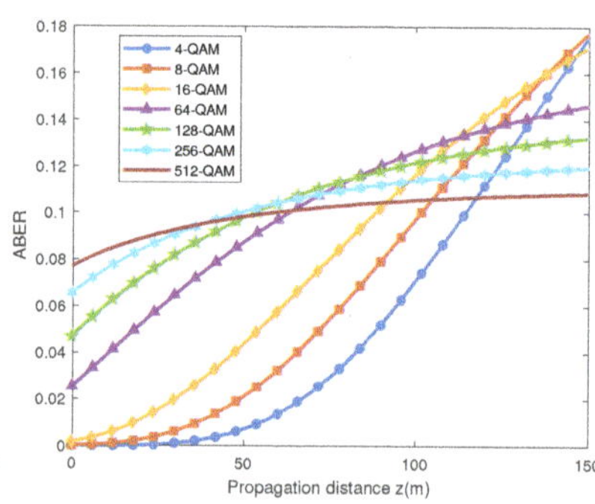

(b)

Figure 4. The bandwidth-limited ABER of the POV carrier link with M-QAM versus the propagation distance in the turbulence strength of (**a**) $C^2_{\varepsilon\chi T} = 10^{-15}\ \mathrm{K}^2/\mathrm{m}^{2/3}$, (**b**) $C^2_{\varepsilon\chi T} = 10^{-13}\ \mathrm{K}^2/\mathrm{m}^{2/3}$.

Figure 5 show the evolutive curves of bandwidth-limited ABER of link with variation of the M-QAM and the turbulence strength. It can be seen from Figure 5 that the bandwidth-limited ABER increases with the increase of turbulence strength for M-QAM. Figure 5a shows that in $z = 100$ m, from $C^2_{\varepsilon\chi T} = 10^{-15}\ \mathrm{K}^2/\mathrm{m}^{2/3}$ to $10^{-13}\ \mathrm{K}^2/\mathrm{m}^{2/3}$, the bandwidth-limited ABER of 4-QAM is always the lowest. While Figure 5b shows that, in

$z = 150$ m, as $C^2_{\varepsilon\chi_T} \geq 10^{-14}$ K^2/m$^{2/3}$, when $C^2_{\varepsilon\chi_T}$ continues to increase, the modulation order M of M-QAM corresponding to the minimum value of bandwidth-limited ABER increases successively.

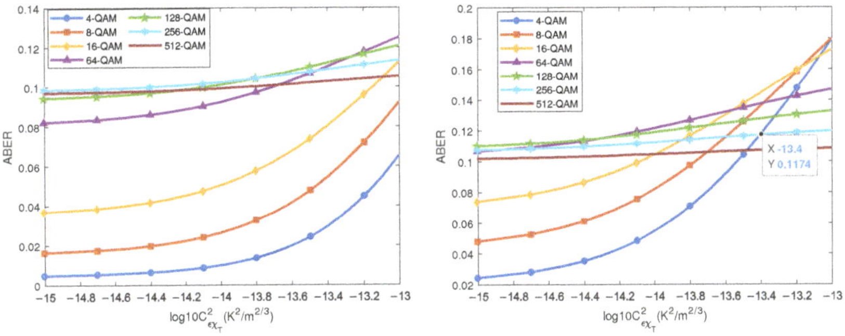

(a) (b)

Figure 5. The bandwidth-limited ABER of POV carrier link with M-QAM versus turbulence strength in the link length of (**a**) $z = 100$ m, (**b**) $z = 150$ m.

More importantly, both Figures 4 and 5 indicate that, in large turbulence fluctuation $\sigma^2_{1\varepsilon\chi_T} = 1.23C^2_{\varepsilon\chi_T}k^{7/6}z^{11/6}$, high-order QAM has better anti-interference ability than low-order QAM.

In Figure 6 we give the numerical results for the bandwidth-limited ABER of POV carrier link with M-QAM versus the bandwidth B for different propagation distance and turbulence strength. Figure 6 displays that the bandwidth limited ABER of link increases with the increase of bandwidth B, this is obviously because the increase of lead to the decrease of SCNR (see Equation (9)). Specifically, comparing the change in propagation distance between Figure 6a,b and the change in turbulence strength between Figure 6b,c, we find that for the short length of communication link with narrow communication bandwidth in weak seawater turbulence, QAM of low order M should be adopted. However, for long communication links with wide communication bandwidth in strong seawater turbulence, QAM of high order M should be adopted.

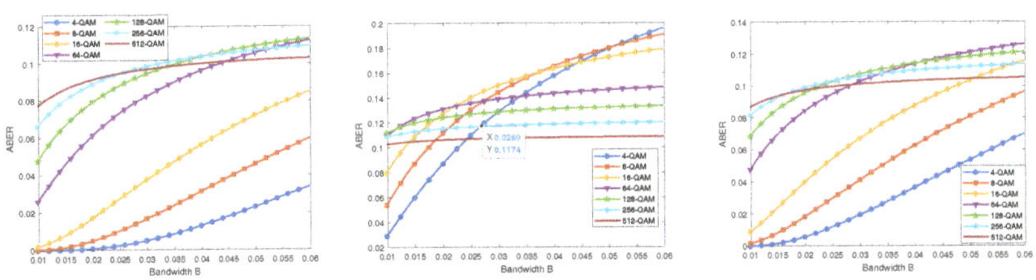

(a) (b) (c)

Figure 6. The bandwidth-limited ABER of POV carrier link with M-QAM versus (**a**) the propagation distance (**b**) turbulence strength. (**a**) $z = 75$ m, $C^2_{\varepsilon\chi_T} = 10^{-13}$ K^2/m$^{2/3}$, (**b**) $z = 150$ m, $C^2_{\varepsilon\chi_T} = 10^{-13}$ K^2/m$^{2/3}$, (**c**) $z = 150$ m, $C^2_{\varepsilon\chi_T} = 10^{-14}$ K^2/m$^{2/3}$.

In Figure 7, we give the numerical results for the bandwidth-limited ABER of the POV carrier link with M-QAM versus the SNR of optical system P_{TX}/N_0. Figure 7 displays that the bandwidth limited ABER of the link decreases with the increase of the P_{TX}/N_0, that is, the increase of SNR P_{TX}/N_0 of optical system contributes to the ability of the communication system to resist turbulent channel disturbance. This is obviously because the decrease of P_{TX}/N_0 leads to the decrease of SCNR (see Equation (9)). Figure 7 also shows that for modulation order $M \leq 16$ and the SNR $P_{TX}/N_0 < 25$ dB, the ABER of the link decreases with the increase of SNR P_{TX}/N_0, but when $P_{TX}/N_0 > 25$ dB, the increase of P_{TX}/N_0 does not improve the link ABER performance. For modulation order $M > 16$, the effect of improving link ABER performance by increasing P_{TX}/N_0 is no longer obvious. The reason for this result is that the total noise in ASCNR is the sum of the noise components of the optical system and the turbulence-induced OAM crosstalk noise components (see Equation (9)). When the P_{TX} is fixed, the noise of the optical system decreases with the increase of P_{TX}/N_0, thus the contribution of the noise component of the optical system to the total noise decreases, the turbulence-induced OAM crosstalk noise is the dominant noise. Further from the relationship between ABER and the ASCNR (Equation (11)), it is not difficult to understand the numerical results of this graph.

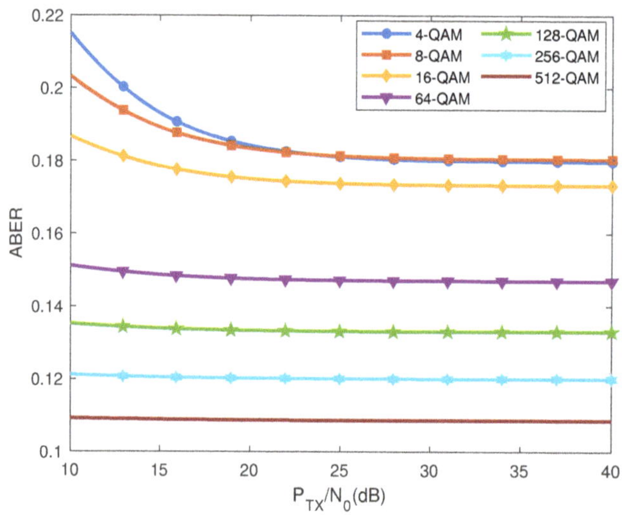

Figure 7. The bandwidth limited ABER of POV carrier link with M-QAM versus the SNR of optical system P_{TX}/N_0.

5.3. System Parameter Optimization

In this sub-section, we will analyze the influence of link channel on ABER to get the optimal signal parameters of communication link.

Figures 8 show the variation of bandwidth-limited ABER of the POV carrier link using 4-QAM/256-QAM with a half beam-width of POV, topological charge m_0 of initial signal modes and channel length z. In Figure 8a we draw curves of bandwidth-limited ABER of POV communication link as a function of the half beam-width of the POV at the ends of four different channel lengths. Figure 8a,c show that there exists optimal with minimum bandwidth-limited ABER as the function of the channel length (see Table 2). Figure 8b,d indicate that $\tilde{w}_{opt}$ is related to the OAM topological charge m_0 of initial signal modes (see Table 3). Combining the conclusions drawn from Figure 8a–d, we conclude that optimization $\tilde{w}$ is a function of channel length z and the OAM topological charge of signal modes, that is $\tilde{w}_{opt}$. In the subsequent discussion of this paper, we will replace half beam-width with the optimization half beam-width.

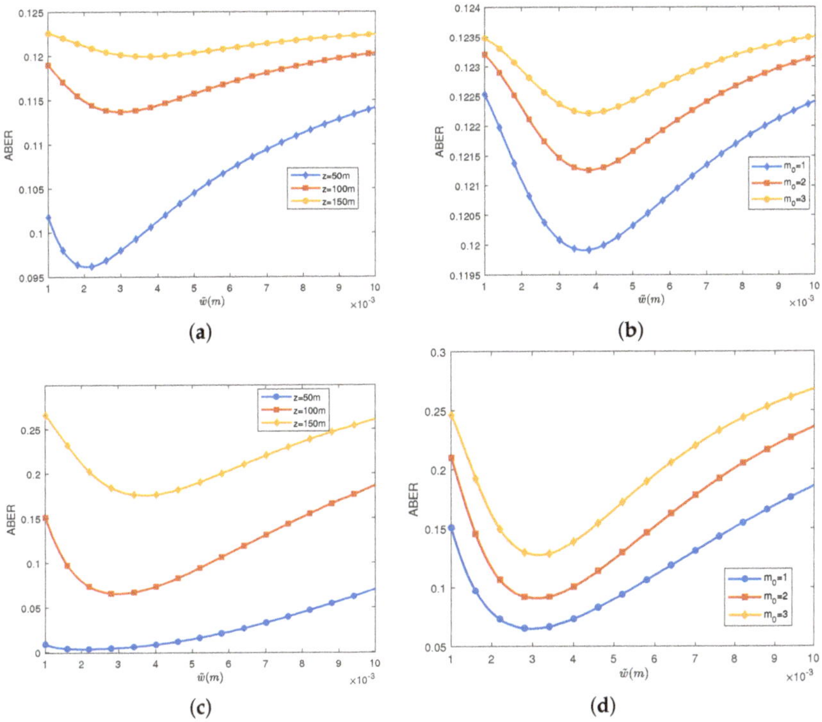

Figure 8. The bandwidth-limited ABER of POV carrier link versus the half beam-width $\tilde{w}$ and topological charge m_0 under modulation of (**a**) 256-QAM, (**b**) 256-QAM, (**c**) 4-QAM and (**d**) 4-QAM.

Table 2. The Relationship between the Optimal Half-width of the POV and the Channel Length.

Modulation System	**Length of the Channel z (m)**	**Optimum Half Width $\tilde{w}_{opt}$ (m)**
4-QAM/256-QAM	50	0.0021
4-QAM/256-QAM	100	0.0032
4-QAM/256-QAM	150	0.0038

Table 3. The Relationship between the Optimal Half Beam-width of POV and the Topological Charge of OAM.

Modulation System	**OAM Topological Charge m_0**	**Optimum Half Width $\tilde{w}_{opt}$ (m)**
4-QAM/256-QAM	1	0.0036
4-QAM/256-QAM	2	0.0037
4-QAM/256-QAM	3	0.0038

5.4. Channel Capacity of the Link

In this sub-section, we apply the conclusions obtained in the previous section, select the optimized source parameter value, and select the 4-QAM/256-QAM modulation schemes to discuss the influence of various factors in weak turbulence absorption seawater on the average information capacity of OAM link.

We know that the turbulence fluctuation of long wavelength OAM signal is lower than that of short wavelength signal [31] for a given absorption coefficient, and the loss of OAM signal transmitted in long absorption seawater channel is higher than that in short channel for a given signal wavelength. However, the absorption coefficient of seawater depends on the wavelength of light [39]. Therefore, seawater absorption as well as signal wavelength are two important factors for the performance research of optical communication system with OAM mode carrier.

In Figure 9, we give the evolution of capacity of underwater communication link using 4-QAM/256-QAM modulation under four absorption seawaters conditions; the four absorption seawater and four signal wavelength are 410 nm ($n_I = 0.3588 \times 10^{-9}$), 470 nm ($n_I = 0.4487 \times 10^{-9}$), 510 nm ($n_I = 1.0549 \times 10^{-9}$) and 570 nm ($n_I = 3.4010 \times 10^{-9}$). Figure 9 reveals that as the absorption coefficients of seawater in the two links are similar, the link with a long signal wavelength has a large capacity, while when the signal wavelengths in the two links are similar, the link with a small absorption coefficient has a large capacity. This result suggest that the carrier wavelength should be adjusted according to the absorptivity of seawater in order to keep the underwater communication link in the optimal state of large capacity.

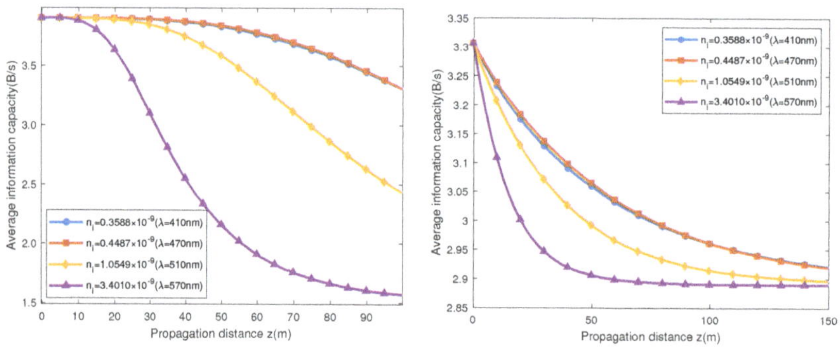

(a) (b)

Figure 9. The capacity of the POV carrier link versus propagation distance for wavelength and image refractive index under (**a**) 4-QAM, $z = 100$ m, (**b**) 256-QAM, $z = 150$ m.

Figure 10 show the AIC curves of the 4-QAM/256-QAM modulation system for different turbulence strength versus the ratio of temperature and salinity fluctuation. As can be found, the AIC decreases with the increase of temperature and salinity fluctuation. That is, the salinity fluctuation has a greater impact on the AIC than the temperature fluctuation, this conclusion is consistent with the research conclusions of other types of optical vortices [11,12].

The evolution of the AIC of OAM signal carried by POV with the turbulent inner scale and turbulent outer scale of seawater is shown in Figure 11. As can be seen from Figure 11, AIC increases with the increase of turbulent inner scale, but it decreases with the increase of turbulent outer scale. According to the theory of turbulence effect, the turbulent inner scale mainly produces the forward scattering of the emitted beam. Therefore, as the turbulent inner scale increases, the uniform area in the seawater increases, and the wavefront distortion of OAM signal mode passing through the channel decreases; that is, the transmittance of the OAM signal mode increases. On the other hand, the turbulent outer scale mainly leads to the random deflection of the beam propagation path, and the value of the random deflection is directly proportional to the size of the turbulent outer scale. As a result, the transmission channel with large outer scale of turbulence will cause a large random deflection of light rays, resulting in large random optical path difference between the sub-beams of OAM mode, resulting in large wavefront distortion of the OAM mode and large OAM crosstalk. Of course, the AIC loss of the link increases as a result.

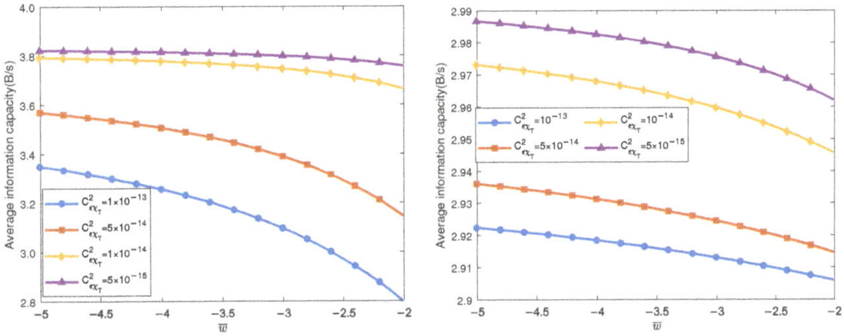

(a) (b)

Figure 10. The capacity of the POV carrier link versus the ratio of temperature and salinity fluctuation contributions to the refractive index spectrum for the structural constant of the salinity-temperature fluctuation $C^2_{\epsilon\chi_T} = 5\times10^{-15}$, $C^2_{\epsilon\chi_T} = 10^{-14}$, $C^2_{\epsilon\chi_T} = 5\times10^{-14}$, $C^2_{\epsilon\chi_T} = 10^{-13}$ under (**a**) 4-QAM, $z = 100$ m, (**b**) 256-QAM, $z = 150$ m.

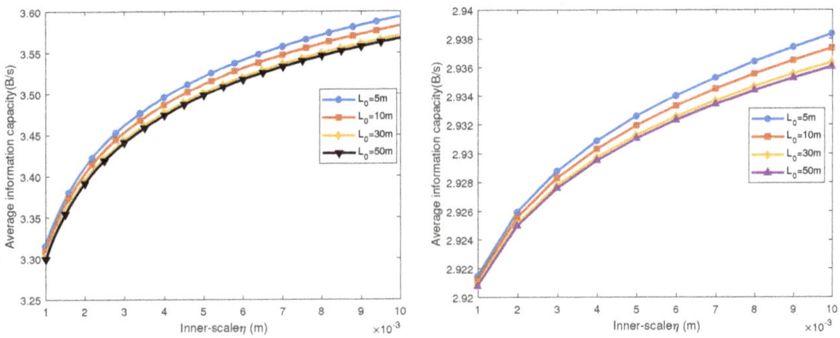

(a) (b)

Figure 11. The capacity of the POV carrier link versus the inner scale for different outer scale under (**a**) 4-QAM, $z = 100$ m, and (**b**) 256-QAM, $z = 150$ m.

It can be seen from Figure 9–11, the influence of the absorption coefficient of seawater, $\tilde{w}$, the inner scale and outer scale of turbulence on the average information capacity of OAM links with the low-order QAM modulation is greater than that of OAM links modulated by high-order QAM.

6. Conclusions

In this paper, we focus on the parameter optimization of POV's source for the underwater transmission system using M-QAM based on bandwidth-limited ABER and study the influence of seawater turbulence and seawater absorption on AIC of bandwidth-constrained communication systems under the optimal parameters of POV's source. The numerical analysis of the ABER and AIC of bandwidth-constrained link is obtained: There exists optimizing half beam-width $\tilde{w}_{opt}(z_0, m_0)$ of POV that is the function of the topological charge m_0 and transmission distance. The ABER of bandwidth-constrained link increases with the increase of bandwidth B and the decrease of P_{TX}/N_0, that is the increase of signal-to-noise ratio (SNR) of the optical system P_{TX}/N_0 enhances the ability of the communication system to resist turbulent channel disturbance. For communication links with short link length, weak seawater turbulence and narrow communication bandwidth, it is more beneficial to adopt QAM of low order M. However, for communication systems with long links, strong seawater turbulence and wide communication bandwidth, QAM of high order M should be adopted. High-order modulation is less affected by the turbulent strength, which indicates that high-order modulation has better anti-interference ability than low-order modulation for long link. Bandwidth-limited ABER increases and AIC decreases with the increase of seawater turbulence strength, and the salinity fluctuation has a greater impact on the AIC than the temperature fluctuation. The decrease of the outer scale and increase of the inner scale as well as channel number have a positive gain for AIC. The results of this paper show that the loss of AIC of underwater transmission system can be reduced by selecting appropriate modulation order of QAM and optimized parameters of POV.

Author Contributions: Conceptualization, Y.Z. (Yixin Zhang); methodology, Y.Z. (Yixin Zhang); software, Q.Y.; validation, Y.Z. (Yixin Zhang), Q.Y. and Y.Z. (Yun Zhu); formal analysis, Q.Y.; investigation, Q.Y.; resources, Y.Z. (Yixin Zhang); data curation, Q.Y.; writing—original draft preparation, Y.Z. (Yun Zhu) and L.Y.; writing—review and editing, Y.Z. (Yixin Zhang); visualization, Q.Y.; supervision, Y.Z. (Yixin Zhang) and Y.Z. (Yun Zhu); project administration, Y.Z. (Yixin Zhang) and Y.Z. (Yun Zhu); funding acquisition, Y.Z. (Yixin Zhang) and L.Y. All authors have read and agreed to the published version of the manuscript.

Funding: This research was funded by National Natural Science Foundation of China (61871202, 11904136).

Institutional Review Board Statement: Not applicable.

Informed Consent Statement: Not applicable.

Data Availability Statement: Not applicable.

Conflicts of Interest: The authors declare no conflict of interest.

References

1. Allen, L.; Beijersbergen, M.; Spreeuw, R.; Woerdman, J. Orbital angular momentum of light and the transformation of La-guerre–Gaussian laser modes. *Phys. Rev. A* **1992**, *45*, 8185–8189. [CrossRef] [PubMed]
2. Paterson, C. Atmospheric turbulence and orbital angular momentum of single photons for optical communication. *Phys. Rev Lett.* **2005**, *45*, 153901-1–153901-14. [CrossRef]
3. Sawant, A.; Lee, I.; Jung, B.C.; Choi, E. Ultimate capacity analysis of orbital angular momentum channels. *IEEE Wirel. Commun* **2021**, *28*, 90–96. [CrossRef]
4. Amhoud, E.M.; Ooi, B.S.; Alouini, M.S. A unified statistical model for atmospheric turbulence-induced fading in orbital angular momentum multiplexed FSO systems. *IEEE Wirel. Commun.* **2020**, *19*, 888–900. [CrossRef]
5. Rusch, L.A.; Rad, M.; Allahverdyan, K.; Fazal, I.; Bernier, E. Carrying data on the orbital angular momentum of light. *IEEE Wirel Commun.* **2018**, *56*, 219–224. [CrossRef]
6. Zhang, Y.; Shan, L.; Li, Y.; Yu, L. Effects of moderate to strong turbulence on the Hankel-Bessel-Gaussian pulse beam with orbital angular momentum in the marine atmosphere. *Opt. Express* **2017**, *25*, 33469–33479. [CrossRef]
7. Yan, B.; Lu, Z.; Hu, J.; Xu, T.; Zhang, H.; Lin, W.; Yue, Y.; Liu, H.; Liu, B. Orbital angular momentum (OAM) carried by asymmetric vortex beams for wireless communications: Theory, experiment and current challenges. *IEEE J. Sel. Top. Quant. Quantum* **2021**, *27* 7600310-1–7600310-10. [CrossRef]
8. Huang, Y.; Zhang, B.; Gao, Z.; Zhao, G.; Duan, Z. Evolution behavior of Gaussian Schell model vortex beams propagating through oceanic turbulence. *Opt. Express* **2014**, *22*, 17723–17734. [CrossRef] [PubMed]
9. Cheng, M.; Guo, L.; Li, J.; Zhang, Y. Channel capacity of the OAM based free-space optical communication links with Bessel–Gauss beams in turbulent ocean. *IEEE Photon. J.* **2016**, *8*, 7901411-1–7901411-12. [CrossRef]
10. Li, Y.; Yu, L.; Zhang, Y. Influence of anisotropic turbulence on the orbital angular momentum modes of Hermite-Gaussian vortex beam in the ocean. *Opt. Express* **2017**, *25*, 12203–12215. [CrossRef]
11. Cui, X.; Yin, X.; Chang, H.; Guo, Y.; Zheng, Z.; Sun, Z.; Liu, G.; Wang, Y. Analysis of an adaptive orbital angular momentum shift keying decoder based on machine learning under oceanic turbulence channels. *Opt. Commun.* **2018**, *429*, 138–143. [CrossRef]
12. Willner, A.E.; Zhao, Z.; Ren, Y.; Li, L.; Xie, G.; Song, H.; Liu, C.; Zhang, R.; Bao, C.; Pang, K. Underwater optical communications using orbital angular momentum-based spatial division multiplexing. *Opt. Commun.* **2018**, *408*, 21–25. [CrossRef]
13. Yu, L.; Zhang, Y. Analysis of modal crosstalk for communication in turbulent ocean using Lommel-Gaussian beam. *Opt. Express* **2017**, *25*, 22565–22574. [CrossRef] [PubMed]
14. Deng, S.; Yang, D.; Zhang, Y. Capacity of communication link with carrier of vortex localized wave in absorptive turbulent seawater. *Waves Random Complex Media* **2020**. [CrossRef]
15. Ostrovsky, A.S.; Ricrenstorff-Parrao, C.; Arrizón, V. Generation of the 'perfect' optical vortex using a liquid-crystal spatial light modulator. *Opt. Lett.* **2013**, *38*, 534–536. [CrossRef] [PubMed]

16. Vaity, P.; Rusch, L. Perfect vortex beam: Fourier transformation of a Bessel beam. *Opt. Lett.* **2015**, *40*, 597–600. [CrossRef] [PubMed]
17. Zhu, F.; Huang, S.; Shao, W.; Zhang, J.; Chen, M.; Zhang, W.; Zeng, J. Free-space optical communication link using perfect vortex beams carrying orbital angular momentum (OAM). *Opt. Commun.* **2017**, *396*, 50–57. [CrossRef]
18. Shao, W.; Huang, J.; Liu, X.; Chen, M. Free-space optical communication with perfect optical vortex beams multiplexing. *Opt. Commun.* **2018**, *427*, 545–550. [CrossRef]
19. Yang, C.; Lan, Y.; Jiang, X.; Long, H.; Hou, J.; Chen, S. Beam-holding property analysis of the perfect optical vortex beam transmitting in atmospheric turbulence. *Opt. Commun.* **2020**, *472*, 125879-1–125879-6. [CrossRef]
20. Karahroudi, M.K.; Moosavi, S.A.; Mobashery, A.; Parmoon, B.; Sghhafifar, H. Performance evaluation of perfect optical vortices transmission in an underwater optical communication system. *Appl. Opt.* **2018**, *57*, 9148–9154. [CrossRef] [PubMed]
21. Hu, Z.; Liu, H.; Xia, J.; He, A.; Li, H.; Du, Z.; Chen, T.; Li, Z.; Lu, Y. Propagation characteristics of the perfect vortex beam in anisotropic oceanic turbulence. *Appl. Opt.* **2020**, *59*, 9956–9962. [CrossRef] [PubMed]
22. Wang, W.; Wang, P.; Pang, W.; Pan, Y.; Nie, Y.; Guo, L. Evolution properties and spatial-mode UWOC performances of the perfect vortex beam subject to oceanic turbulence. *IEEE Trans. Commun.* **2021**, *69*, 7647–7658. [CrossRef]
23. Yang, H.; Yan, Q.; Zhang, Y.; Hu, L. Received probability of perfect optical vortex in absorbent and weak turbulent seawater links. *Appl. Opt.* **2021**, *60*, 10772–10779. [CrossRef]
24. Yang, H.; Yan, Q.; Wang, P.; Hu, L.; Zhang, Y. Bit-error rate and average capacity of an absorbent and turbulent underwater wireless communication link with perfect Laguerre-Gauss beam. *Opt. Express* **2022**, *30*, 9050–9064. [CrossRef] [PubMed]
25. Ali, M.F.; Jayakody, D.N.K.; Chursin, Y.A.; Affes, S.; Dmitry, S. Recent advances and future directions on underwater wireless communications. *Arch. Comput. Methods Eng.* **2020**, *27*, 1379–1412. [CrossRef]
26. Gupta, A.; Sharma, N.; Garg, P.; Jayakody, D.N.K.; Aleksandrovich, C.Y.; Li, J. Asymmetric satellite-underwater visible light communication system for oceanic monitoring. *IEEE Access* **2019**, *7*, 133342–133350. [CrossRef]
27. Ali, M.F.; Ponnimbaduge Perera, T.D.; Jayakody, D.N.K. O2O: An underwater VLC approach in baltic and north sea. *Electronics* **2022**, *7*, 321. [CrossRef]
28. Ali, M.F.; Jayakody, D.N.K.; Li, Y. Recent trends in underwater visible light communication (UVLC) systems. *IEEE Access* **2022**, *10*, 22169–22225. [CrossRef]
29. Kumar, A.; Jayakody, D.N.K. Secure NOMA assisted multi-LED underwater visible light communication. *IEEE Trans. Veh. Technol.* **2022**, *71*, 7769–7779. [CrossRef]
30. Yan, Q.; Zhu, Y.; Zhang, Y. Capacity of the weakly absorbent turbulent ocean channel with the coaxial double-position power Gaussian vortex. *J. Mar. Sci. Eng.* **2021**, *9*, 1117-1–1117-17. [CrossRef]
31. Andrews, L.C.; Phillips, R.L. *Laser Beam Propagation through Random Media*, 2nd ed.; SPIE: Washington, DC, USA, 2005.
32. Fewell, M.P.; Trojan, A.V. Absorption of light by water in the region of high transparency: Recommended values for photon-transport calculations. *Appl. Opt.* **2019**, *58*, 2408–2421. [CrossRef] [PubMed]
33. Jeffrey, A.; Zwillinger, D. *Table of Integrals, Series, and Products*; Academic: Cambridge, MA, USA, 2007.
34. Anguita, J.A.; Neifeld, M.A.; Vasic, B.V. Turbulence-induced channel crosstalk in an orbital angular momentum-multiplexed free-space optical link. *Appl. Opt.* **2008**, *47*, 2414–2429. [CrossRef] [PubMed]
35. Karp, S.; Gagliardi, R.M.; Moran, S.E.; Stotts, L.B. *Optical Channels: Fibers, Clouds, Water, and the Atmosphere*; Springer: New York, NY, USA, 1988.
36. Djordjevic, I.B. *Advanced Optical and Wireless Communications Systems*; Springer: New York, NY, USA, 2018.
37. Krouk, E.; Semenov, S. *Modulation and Coding Techniques in Wireless Communications*; Wiley: New Delhi, India, 2011.
38. Ho, K.P. *Phase-Modulated Optical Communication Systems*; Springer: New York, NY, USA, 2001.
39. Woz'niak, B.; Dera, J. *Light Absorption in Sea Water*; Springer: New York, NY, USA, 2007.
40. Zhu, S.; Chen, X.; Liu, X.; Zhang, G.; Tian, P. Recent progress in and perspectives of underwater wireless optical communication. *Prog. Quantum Electron.* **2020**, *73*, 100274. [CrossRef]
41. Chen, X.; Yang, X.; Tong, Z.; Dai, Y.; Li, X.; Zhao, M.; Zhang, Z.; Zhao, J.; Xu, J. 150 m/500 Mbps underwater wireless optical communication enabled by sensitive detection and the combination of receiver-side partial response shaping and TCM technology. *J. Light. Technol.* **2021**, *39*, 4614–4621. [CrossRef]
42. Dai, Y.; Chen, X.; Yang, X.; Tong, Z.; Du, Z.; Yu, W.; Zhang, C.; Zhang, H.; Zou, H.; Cheng, Y.; et al. 200-m/500-Mbps underwater wireless optical communication system utilizing a sparse nonlinear equalizer with a variable step size generalized orthogonal matching pursuit. *Opt. Express* **2021**, *29*, 32228–32243. [CrossRef]

Journal of
Marine Science and Engineering

MDPI

Article

Capacity Analysis of Incentive Schemes in Opportunistic Networks

Ruoyu Feng, Shengming Jiang * and Zhichao Zheng

College of Information Engineering, Shanghai Maritime University, Shanghai 201306, China
* Correspondence: smjiang@shmtu.edu.cn

Abstract: Many incentive schemes address the selfishness issue in opportunistic networks and show performance improvement by simulations. However, the insights of incentive schemes that affect network performance are not clear. Network capacity analysis can reveal how factors affect performance, which is a guideline for new designs. To analyze incentive schemes, a well-defined mathematical model is necessary, which cannot be achieved by existing analytical models based on empirical formulas or types of incentive schemes. First, this paper proposes such a model to show the incentive degree with the incentive scheme, cooperation degree, energy usage, buffer usage, and security based on a quantum game model. Verification compares the model with delivery ratios that show impacts on selfish nodes in simulations under two typical incentive schemes. Then, network capacity is analyzed with this model and a sparse clustering regime that has similar mobility to opportunistic networks in order to show factors for future designs.

Keywords: opportunistic network; analytical model; incentive scheme; network capacity

1. Introduction

Mobility affects network capacity, which refers to the maximum achievable throughput due to the dynamic cooperation of nodes. Mobile models are discussed often [1]. For instance, sparse clustering regimes show that grouping nodes can increase the cooperation degree, and trading throughput for delay in clusters does not decrease network capacity as a whole [2]. While these studies usually assume nodes are rational, the cooperation degree may decrease due to selfish nodes, especially in opportunistic networks where end-to-end paths between sources and destinations change over time and are occasionally even disrupted [3]. Incentive schemes force nodes to forward messages in order to avoid long waits for the next opportunity, which increases co-operation as a way to prompt delivery ratios and decrease end-to-end contact times. To design more efficient incentive schemes, it is important to discuss factors affecting performance. This paper analyzes network capacity, with incentive schemes as the goal. In addition, our study can be applied to filter candidates for schemes that raise prediction-based, hybrid, or other networks with opportunistic networks such as opportunistic routing based on prediction [4].

To analyze network capacity, a mathematical model of incentive schemes is necessary for sufficient insight. Existing incentive schemes only show improvement by simulations, and existing analyatical models are not detailed enough. For example, a framework for multiple-criteria decision-making problem to rank protocol performance [5] lacks numerical formulas involving parameters. Game theory has been used to state nodes' strategies to analyze network traffic in transmission control protocol (TCP) with selfish nodes and peer-to-peer (P2P) networks with free-riding problem [6,7], and several analytical models for different categories based on classical game theory have been proposed [8–10] which do not find common characters of incentive schemes. A generic model can be built with quantum game theory [11], as quantum game theory extends the strategy space to search for optimal strategies in a wider range and uses the concept of entanglement to depict the complex interaction among nodes accurately in the decision-making process. A model with these features can evaluate most incentive schemes and reveal their common character. However, the above model uses an empirical utility formula and focuses on numerical analysis, which simply places datasets into the model without other comparisons. In other words, the model does not have enough factors to work as a guideline to design new incentive schemes.

A well-defined formula should consider the characteristics of incentive schemes as well as other factors that affect cooperation. A node usually spends payoff, such as virtual money, reputation, exchanging messages, or something else, to stimulate others to forwarding, which can obtain the necessary currency. Cost and reward can represent an incentive scheme in a mathematical model with other properties of nodes. Several incentive schemes take energy usage and buffer usage into account. Therefore, this paper proposes an analytical model based on quantum game theory considering the energy usage and buffer usage, reward, cost, and cooperation degree, which shows the incentive degree and compares incentive degrees with the delivery ratio by simulations.

Network capacity analysis using the proposed model shows the impact of incentive schemes on network performance. During analysis, a mobile model in opportunistic networks is needed as well. The sparse clustering regime is similar to the mobility of opportunistic networks with nodes that work in groups with high cooperation possibility and clusters that move randomly where communications in groups are stable and movement is otherwise needed for opportunities. Furthermore, the scheduling policy is similar to the scenario in [4] for nautical wireless ad hoc networks. Therefore, the analytical model for incentive schemes is applied to a sparse clustering regime to analyze network capacity and discuss various factors.

As a summary, the contributions of this paper are as follows: (1) An analytical model based on quantum game theory considering the energy usage, buffer usage, reward, cost and cooperation degree is proposed. It is verified by a comparison between the incentive degree and delivery ratios according to simulations. (2) Network capacity is analyzed with the above model to show how incentive schemes affect network performance. The factors can be used as a guideline for designing efficient incentive schemes by considering these relations.

The rest of this paper is organized as follows. The mathematical model is detailed in Section 2. Verification with simulations is described in Section 3. Network capacity is analyzed with the model and sparse clustering regime in Section 4. The factors are discussed in Section 5. Finally, we conclude the paper in Section 6.

2. Mathematical Model for Incentive Schemes

2.1. Typical Incentive Schemes

Before detailing the mathematical model, we briefly touch on typical incentive schemes to see what can be taken into account. Most existing incentive schemes are reputation-based or credit-based [10]. They use currency-like metrics to reveal the participation degree for nodes forwarding messages [12].

In credit-based schemes, relay nodes can obtain credits by forwarding messages, then send their own messages with these additional credits. The better design and calculation of reward and price remains a challenge. Payoff according to the time-to-live of a message has been proposed in [12], which is direct to price difference. A Markov chain can be used to calculate price according to the society degree, selfish degree, and cooperation degree with buffer, energy, and interaction to stimulate nodes with high possibility [13]. The clearance credit center rewards relay nodes with extra credits to avoid nodes being unable to obtain enough credits to send messages due to selfish nodes who hold most credits and are not willing to forward messages. Therefore, attackers are not considered in our model or in other similar proposals, such as [8,10].

Other incentive schemes, such as reputation-based ones, usually replace credits with internal bounds for the same purpose. Punishment is used to restrain free-riding nodes in peer-to-peer (P2P) networks via game theory [7]. Grouping nodes by bilinear pairing and cooperation probability prediction has been proposed in [14]. Evolutionary game theory replaces the classical theory for more precise cooperation evaluation [15]. With the exception of node behaviour under different policies, these bounds, such as cooperation probability, are similar to credits both in usage and principle. Thanks to the extended strategy space in quantum game theory, it is possible to simply consider increasing bounds as rewards and decreasing bounds as costs.

An environment involving movement path, link quality, and others can affect communications as a main domain. However, these do not affect incentive schemes directly, which is shown simply by a scenario argument without detailed considerations in the following analysis.

To sum up, for an analytical model, the cost and reward function can represent an incentive scheme. Other parameters such as buffer and energy can be considered as a part of cost, as they may reduce the cooperation degree. A simple argument briefly summarizes the impact of the environment. Furthermore, interactions among nodes must be considered, which can be solved by a quantum game model and Nash equilibrium.

2.2. Quantum Game Model

The proposed model has two parts, as shown in Figure 1. The quantum game model depicts how an incentive scheme motivates cooperation, and the utility formula $\$_m$ describes how a node is served. The utility formula $\$_m$, including the energy usage, buffer usage, and degree of security, accepts an incentive policy IP and entanglement λ as arguments. In both the quantum game model and the utility formula, λ communicates with a quantum game model to describe incentive processes. Therefore, the optimal value $\$^*$ is computed by the quantum game model with Nash equilibrium.

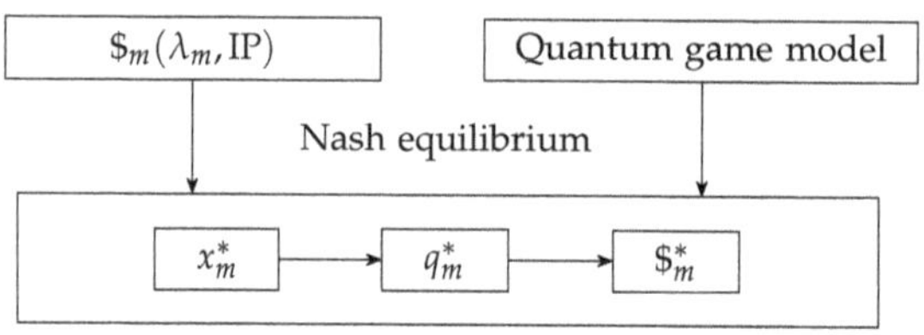

Figure 1. Model construction.

Quantum game theory is more suitable for analyzing incentive schemes because of the concept of entanglement and the extended strategy space, as discussed in Section 1. First, we briefly describe quantum game theory based on [11], shown in Figure 2. The actions of players in a quantum game ae based on a specific set of strategies. The state of node m is represented by $|vac_m\rangle$, a state vector in a Hilbert space. The whole system state is $|H\rangle = |H_1\rangle \otimes |H_2\rangle \cdots \otimes |H_n\rangle$.

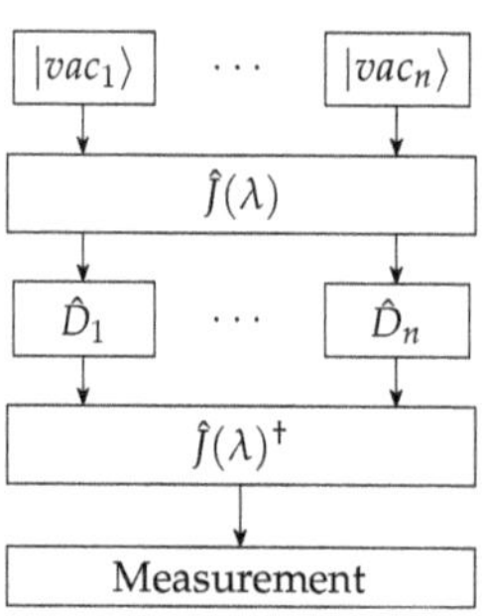

Figure 2. Quantum game model.

If players are dependent, entanglement must be used to express the system state in the above product. Entangling gate $\hat{J}(\lambda)$ with operation parameter x_m defines how to combine the strategies of each node. This expresses how players affect others, such as encounters due to moving paths and selfish policies, especially in opportunistic networks; in our study, entanglement is considered as the cooperation degree. The model includes the following three stages:

1. The initial quantum state of the system is $\psi_{init} = \hat{J}(\lambda)|vac\rangle$, achieved by entangling all nodes togather.
2. Node m operates on its state using unitary operator $\hat{D}_m$ with the feature $\hat{D}_m\hat{D}_m^{\dagger} = 1$.
3. With disentangling gate $\hat{J}(\lambda)^{\dagger}$ on the system, the final state is $\left|\psi_{fin}\right\rangle = |J\rangle(\lambda)^{\dagger}(\hat{D}_1 \otimes \hat{D}_2 \cdots \hat{D}_n)\hat{J}(\lambda)|vac\rangle$.

Each definition and proof are detailed in [11]. After forwarding $\left|\psi_{fin}\right\rangle$ to measurement, node m can quantize the relayed data to others as follows:

$$q_m = \frac{x_m}{n}\left[e^{(n-1)\lambda} + (n-1)e^{-\lambda}\right] + \sum_{j=1,j\neq m}^{n} \frac{x_j}{n}\left[e^{(n-1)\lambda} - e^{-\lambda}\right]. \tag{1}$$

The optimal value x_m^* can be computed by a utility formula with the Nash equilibrium for impacts on nodes. The Nash equilibrium is defined as follows:

$$\begin{cases} \dfrac{\partial \$_1}{\partial x_1} = \dfrac{\partial \$_2}{\partial x_2} = \cdots = \dfrac{\partial \$_n}{\partial x_n} = 0 \\ \dfrac{\partial^2 \$_m}{\partial x_m^2} < 0,\, m \in \{1, 2, \cdots, n\}. \end{cases} \tag{2}$$

Then, the optimal strategy q_m^* and optimal utility $\$_m^*$ can be calculated with x_m^* according to (1) and (3). The calculation depends on the utility formula, with the input incentive scheme to be discussed later. Then, we obtain a mathematical equation between an incentive scheme and $\$^*$, which can be used for evaluatation and reveal the insight if the utility formula depicts factors correctly. Moreover, $\$^*$ is called the incentive degree in this study.

2.3. Utility Formula

The utility formula describes the incentive degree of a node with entanglement λ_m, reward function $r(q_m)$, cost function $c(q_m)$, and constant cost c', as follows:

$$\$_m = \lambda_m q_m - r(q_m) - c(q_m) - c', \tag{3}$$

where $r(q_m)$ is represented as the incentive scheme. According to the introduction of incentive schemes in Section 2.1, an incentive scheme consists of a reward function $R(q_m)$ and a price function $P(q_m)$. The incentive degree is more dependent on their relation than each individual calculation, and the calculation makes the incentive process more rational and balanced. Therefore, in the utility formula, only the relationship between $R(q_m)$ and $P(q_m)$ is considered, as follows:

$$r(q_m) = q_m \frac{R(q_m)}{P(q_m)}. \tag{4}$$

We show the correction with simulations in Section 3. Entanglement λ_m means the cooperation degree determined by the scenario, such as selfish degrees, social groups, movement paths, etc., which can be avoided by fixing the entanglement in the model during analyses.

The cost constant c' is for each transmission, such as the fixed energy consumption [16]. The cost function $c(q_m)$ describes what relay nodes have to pay for when transmissions succeed, which is made up of an energy cost $E(q_m)$, buffer cost $B(q_m)$, and security cost $S(q_m)$, as follows:

$$c(q_m) = E(q_m) + B(q_m) + S(q_m)\,. \tag{5}$$

The security cost is simple:

$$S(q_m) = sq_m^2\,, \tag{6}$$

where s is the willingness to forward messages, determined by reasons such as security. Unlike entanglement λ, s is a node's own attribute and is not affected by other nodes.

The incentive degree does not continue to increase with larger buffer size and energy capacity, as it is limited by the cooperation between nodes. For instance, a node can only forward messages to parts of others, even if it has infinite buffer and energy. Therefore, they should have asymptotic values. Furthermore, the slope change for the buffer is slower than that for energy, as the energy capacity grants a node the basic ability to participate in forwarding, instead of enough resources, such as buffer and the cooperation degree. To sum up, they are defined as follows:

$$E(q_m) = \left(\frac{E_c}{E_s}\right)^5 r(q_m) \text{ and } B(q_m) = \left(\frac{B_c}{B_s}\right)^3 r(q_m)\,, \tag{7}$$

where E_c is the energy cost per message, E_s is the energy capacity, B_c is the buffer cost per message, and B_s is the buffer size.

2.4. Model With Incentive Scheme

To obtain a model for a special incentive scheme, $P(q_m)$ and $R(q_m)$ should be defined in the utility formula, and the Nash equilibrium then applied to the qunatum game model with the utility formula.

Most incentive schemes are made up of a price function Z and reward function R. The two functions are usually independent on q_m and λ_m, especially in credit-based schemes. In these cases, $P(q_m)$ and $R(q_m)$ are linear functions of q_m, which can be defined as

$$P(q_m) = Zq_m \text{ and } R(q_m) = Rq_m\,. \tag{8}$$

Now, we can calculate the analytical model. According to (8), the utility formula is expanded as follows:

$$\$_m = \lambda_m q_m - \frac{R}{Z}q_m - \left(\frac{E_s}{E_s}\right)^5 \frac{R}{Z}q_m - \left(\frac{B_c}{B_s}\right)^3 \frac{R}{Z}q_m - sq_m^2 - c'\,. \tag{9}$$

Then, we apply (2) to obtain

$$\begin{aligned}\frac{\partial \$_m}{\partial x_m} &= \frac{\partial \$_1}{\partial x_1} = \frac{\partial \$_2}{\partial x_2} = \cdots = \frac{\partial \$_n}{\partial x_n} \\ &= \frac{\partial q_m}{\partial x_m}\left\{\lambda - 2sq_m - \frac{R}{Z}\left[\left(\frac{E_c}{E_s}\right)^5 + \left(\frac{B_c}{B_s}\right)^3 + 1\right]\right\} \\ &= 0\,,\end{aligned} \tag{10}$$

where

$$\frac{\partial q_m}{\partial x_m} = \frac{e^{(n-1)\lambda} + (n-1)e^{-\lambda}}{n}\,. \tag{11}$$

Because optimal strategies are the same under Nash equilibrium, represented as $x_1^* = x_2^* = \cdots = x_n^*$, we can obtaun x^* with the above results as follows:

$$\begin{aligned}\frac{\partial \$_m}{\partial x_m} &= \lambda - 2sx^* e^{(n-1)\lambda} - \left[\left(\frac{E_c}{E_s}\right)^5 + \left(\frac{B_c}{B_s}\right)^3 + 1\right] \\ &= 0\end{aligned} \tag{12}$$

and

$$x^* = \frac{\lambda - \left[\left(\frac{E_c}{E_s}\right)^5 + \left(\frac{B_c}{B_s}\right)^3 + 1\right]\frac{R}{Z}}{2se^{(n-1)\lambda}}. \tag{13}$$

Substituting x^* in q^* yields

$$\begin{aligned}q^* &= x^* e^{(n-1)\lambda} \\ &= \frac{\lambda - \frac{R}{Z}\frac{E_c}{E_s} - \frac{R}{Z}\frac{B_c}{B_s} - \frac{R}{Z}}{2s}.\end{aligned} \tag{14}$$

Then, $\$^*$ can be computed by (3), which shows the incentive degree's relation with a special incentive scheme.

3. Verification of the Model with Simulations

The verification compares incentive degrees and delivery ratios with two typical incentive schemes, namely, the incentive and privacy-aware data dissemination scheme (IPAD) [12] and the probabilistic routing scheme based on game theory (PRGT) [13]. The two credit-based schemes use different calculation parameters. IPAD uses time-to-live for prices and degree of selfishness for rewards. PRGT evalutes social degrees and transmission history with a Markov chain to calculate prices and message importance for rewards. The following simulations are conducted by OMNet++ (https://omnetpp.org (accessed on 20 August 2022)) with the OPS framework (https://github.com/ComNets-Bremen/OPS (accessed on 20 August 2022)). The parameter setting is listed in Table 1 for maritime environments where communications are over long ranges, which is associated with sparse density of node points. Random movement is selected because it is the part of mobile model in the latter network capacity analysis used to show the average theoretically. Nodes' selfish behavior is partly cooperative [17] in our simulations. Certain parameters, such as movement and the type of selfish behavior, are reflected by λ_m indirectly.

Table 1. Simulation parameters.

Parameter	Value
Simulation time	8 hours
Area size	5×5 km^2
Transmission range	300 m
Transmission speed	1 Mbps
Node count	100
Message interval	15 s
Message size	truncnormal $(1.5, 1)$ MiB
Buffer size	56 MiB
Node selfish	bernoulli (0.7)
Movement change interval	truncnormal $(30, 15)$ min
Movement angle delta	normal $(0, 30)^\circ$
Movement velocity	normal $(30, 10)$ mps

Because successful transmissions in opportunistic networks are highly dependent on mobility and node forwarding, selfish nodes reduce delivery ratios [13] remarkably. Utility

$\$^*$ represents the incentive degree, which can be reflected by delivery ratios in the network with selfish nodes. In comparison, the co-domain of $\$^*$ is rescaled with the same co-domain of delivery ratios due to the different scalars. The rescaled values are analyzed by mean square errors (MSE), as shown in Table 2 and Figures 3 and 4. We analyze each of them in the following subsections. Another important metric for opportunistic networks is the contact time of the nodes. Because selfish nodes decrease contact times due to the lower forwarding rate and incentive schemes try to increase the rate shown by delivery ratios, it is implied in delivery ratios that there is a higher delivery ratio with a lower contact time if there are no extra factors, such as attackers.

Table 2. MSE between delivery ratios and $\$^*$.

Parameter	IPAD	PRGT
scheme parameter	5.18×10^{-5}	1.76×10^{-4}
buffer size	5.00×10^{-4}	7.04×10^{-4}
energy capacity	2.48×10^{-5}	4.14×10^{-5}

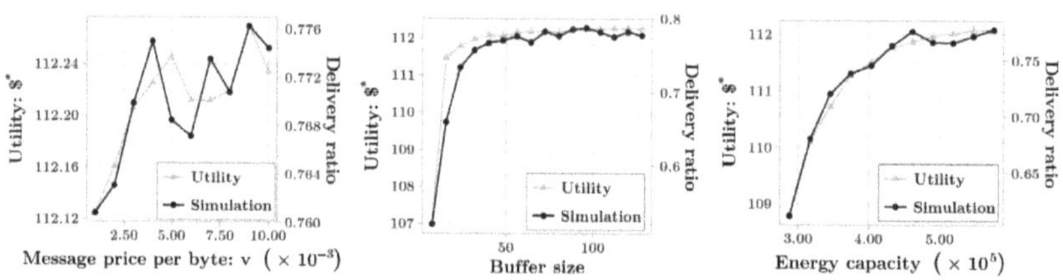

(**a**) comparison of v (**b**) comparison of buffer size (**c**) comparison of energy capacity

Figure 3. Verification with PRGT.

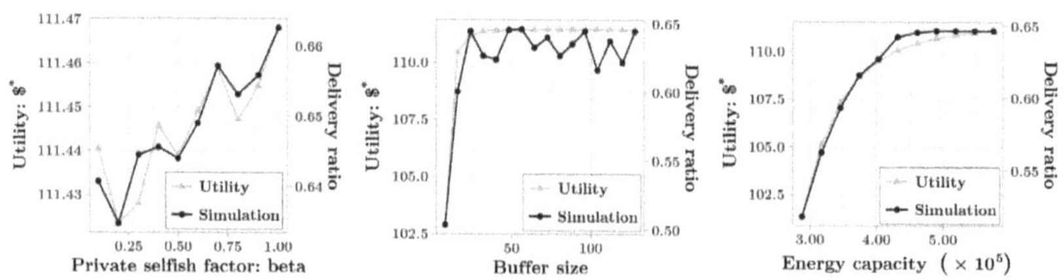

(**a**) comparison of β (**b**) comparison of buffer size (**c**) comparison of energy capacity

Figure 4. Verification with IPAD.

3.1. *Incentive Schemes*

The message price per byte v in PRGT affects the message price and reward directly; thus, it is selected as an argument in this verification. For PRGT, $\$^*$ and delivery ratios from simulations are shown in Figure 3a.

In IPAD, the private selfish factor β is the parameter for analysis, and is used in the payoff function for incentive. With a larger β, a larger price is needed and nodes have to forward more messages to obtain credit for their own messages, as shown in Figure 4a. As a whole, the similar relations show the correctness and generality of $r(q_m)$.

Furthermore, from $\$^*$ and delivery ratios, it can be found that the impact of parameters in incentive schemes is low. The main factor is the incentive strategy, represented by $r(q_m)$. The $\$^*$ of PRGT is larger than that of IPAD, which means that the degree of incentive of PRGT is greater than that of IPAD in the same scenario.

3.2. Buffer Size

Similar to the previous verification, delivery ratios and $\$^*$ with buffer sizes are shown in Figures 3b and 4b. The two similar relations suggest that buffer sizes can affect performance, and the proposed model reflects the realistic relationship with buffer size. Intuitively, message sizes have similar results as buffer sizes, because they are inversely proportional in the model and work together to store messages in simulations.

An increasing buffer size means that a node can forward more messages or a larger message at a time, however, the increase is slowed down because of the restriction of nodes' movement paths. Therefore, too large a buffer size is useless for network performance unless used with a larger message size.

3.3. Energy Capacity

The selfish behavior caused by energy in simulations is that if the energy is lower than the safety value, forwarding is stopped unless nodes need credits. The energy capacity is set to simple values here instead of considering a real scenario for easier calculations and understanding. Delivery ratios and $\$^*$ with energy capacities are shown in Figures 3c and 4c. These two similar relations suggest that this parameter can affect performance and the proposed model can reflect the realistic relationship.

Similar to messages size, the energy consumption per action has similar results to energy capacity, as they influence each other in the model and the simulations.

4. Capacity Analysis for Incentive Schemes

Network capacity can be analyzed with the proposed model by abstracting incentive schemes. For simplification, this analysis is based on an existing efficient mobile model of a sparse clustering regime [2], in which nodes' behaviour resembles the opportunistic networks with social groups discussed in Section 1. The network model is introduced at first. then the capacity and factors are discussed.

4.1. Traffic Model

We assume all nodes communicate with others at the same rate λ.

Definition 1 (The throughput capacity of opportunistic networks). *The throughput capacity λ is that nodes can sustain $\Theta(f(n))$ bits per second if there are deterministic constants $c > 0$ and $c' < \infty$ such that*

$$\begin{cases} \lim_{n\to\infty} \Pr[\lambda(n) = cf(n) \textit{ is feasible}] = 1 \\ \lim_{n\to\infty} \Pr[\lambda(n) = cf(n) \textit{ is feasible}] < 1\,. \end{cases} \tag{15}$$

4.2. Network Topolopy

The sparse clustering regime considers n nodes moving over a square with area n, and nodes are divided into $m = \Theta(n^v)(0 \le v < 1)$ groups. Each group covers a disk with radius $R = \Theta(n^\beta)$. The average number of nodes per cluster is assumed as $q = n/m = \Theta(n^{1-v})$. When $v + 2\beta < 1$ (i.e., $mR^2 = o(n)$), the strong node cooperation in which communications are mainly dependent on mobility is considered as a sparse clustering regime.

The time is divided into slots. Cluster centers and nodes are independent and identically distributed (IID) and uniformly chosen among the whole network and their own cluster, respectively, at the beginning of each time slot. That is, each cluster center randomly selects a point in the network area as its new center and nodes randomly select positions in the contained cluster. Nodes transmit messages without movement.

4.3. Scheduling Policy

Nodes move in clusters and clusters move in the whole area such that different usage of high and low quality communications is exploited. The tradition models can be considered as all nodes under the same cluster with different node density. The scheduling policy defines how clusters work under the regime, which is shown in Figure 5.

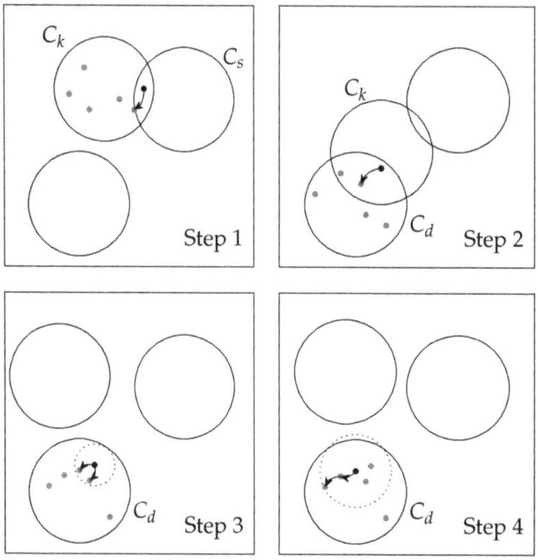

Figure 5. Scheduling policy.

1. A source creates relays R_s in source cluster C_s via multicast. The message is sent to a cluster $C_k(k = 1, \cdots, R_c{}^s)$ by one-hop unicast when the cluster meet others.
2. Relays $R_k{}^s$ are created in C_k via broadcast. After clusters repeat these actions some number of times, the message reaches the destination cluster C_d.
3. Relays $R_d{}^s$ are created by broadcast in the destination cluster until the message is near to the destination node.
4. If the distance between the destination node and the current node is within a range l^s, the message is transmitted via h^s-hop unicast to reduce resource usage.

4.4. Transmission Protocol

If sources i and j want to send messages to destinations k and j, respectively, for an arbitrary other node k, the following equation should be satisfied [18]:

$$|X_j - X_i| \geq \frac{\Delta}{2}(|X_i - X_k| + |X_j - X_l|), \tag{16}$$

where X_i is the position of i and $|X_i - X_j|$ is the distance between them. The basic inequation is often used in research directly, and is associated with incentive schemes similar to the physical model in [18].

To have transmissions, the incentive degree should have a lower bound, that is,

$$\$(\mu) \geq \$. \tag{17}$$

According to inequality properties, we have

$$\frac{\$_k}{\$_i + 1} \leq \frac{\$}{\$ + 1}. \tag{18}$$

If we assume successful transmission probabilities between nodes are the same, the closest node has the highest cooperation degree. Moreover, +1 in the left formula has little impact on the inequation. Therefore, from the previous formula we can infer that

$$\frac{\$\left(\frac{\mu_k}{|X_k - X_j|^\alpha}\right)}{\$\left(\frac{\mu_i}{|X_i - X_j|^\alpha}\right)} \leq \frac{\$}{\$ + 1}, \tag{19}$$

where α is the distance decay and μ replaces entanglement λ in (3) to avoid conflicts with the throughput λ. After expanding $\$(\mu)$, we have

$$\frac{|X_i - X_j|^\alpha}{|X_k - X_j|^\alpha}\frac{\mu_k}{\mu_i} \leq \frac{\$}{\$+1}. \tag{20}$$

In the worst situation, $\frac{\mu_k}{\mu_i}$ is equal to $\frac{\mu_\top}{\mu_\perp}$ where $\mu_\top$ is the maximum feasible entanglement in the scene and $\mu_\perp$ is the minimum feasible entanglement to lead to the emergence of communications in the scene which meets that $\forall \mu > \mu_\perp$ implies $q^*(\mu) > 0$ and $\$^*(\mu) > 0$. Then, we have the inequation with the distance and incentive schemes:

$$\begin{aligned} |X_k - X_j| &\geq \left(\frac{\mu_\top}{\mu_\perp}\frac{\$+1}{\$}\right)^{\frac{1}{\alpha}} |X_i - X_j| \\ &= (1+\Delta)|X_i - X_j|, \end{aligned} \tag{21}$$

where $\Delta := \left(\frac{\beta P_{\min}}{P_{\max}}\right)^{\frac{1}{\alpha}} - 1$.

4.5. Upper Bound of Network Capacity

Different from the other studies, because Δ is deemed as a variable associated with incentive schemes by (21), it is not simplified during derivations. According to the interference restriction and the meeting probability among nodes and clusters, the amount of messages that can be transmitted in a time slot between clusters is $O\left(\frac{mR^2}{\Delta^2}\right)$ if the transmission range is $\Theta\left(R\sqrt{\frac{m}{n}}\right)$ and the amount of messages in a cluster is the similar. Therefore, we can propose the following theorem:

Theorem 1 (Throughput capacity of incentive schemes). *In a sparse clustering regime with incentive schemes, the throughput capacity can achieve $\Theta\left(\frac{mR^2}{\Delta^2}\right)$ and the upper bound can be reached if and only if the transmission range satisfies $l_i = \Theta\left(R\sqrt{\frac{m}{n}}\right)$, where i is a duplication in a cluster, and $l_i = \Theta(1)$, where i is a duplication between clusters. Each node can achieve $\Theta\left(\frac{mR^2}{n\Delta^2}\right)$.*

We dismiss the proof here for simplification, as it is similar to the proofs of Theorem 1 and Theorem 2 in [19].

The above conclusion does not consider the limited ratio resources that messages cannot be sent out due to congested channels when many nodes send messages simultaneously in the same area. If the channel capacity can transmit W bits per second during the whole network alive time T, each bit b at hth-hop satisfies

$$\sum_{b=1}^{\lambda n T}\sum_{h=1}^{h(b)} 1 \leq \frac{WnT}{2} \leq WnT. \tag{22}$$

According to (16), there are no other nodes in the disk radius $\frac{\Delta}{2}$ times transmission range. Each node consumes $\frac{1}{n}$ in area n averagely, thus, the summing area that source nodes consumes meets

$$\sum_{b=1}^{\lambda n T}\sum_{h=1}^{h(b)} S_b \leq WT, \tag{23}$$

where S_b is the area of bit b and $h(b)$ is the hop of bit b. The following lemma can be achieved with the above idea:

Lemma 1 (Throughput inequation considering ratio resources)**.** *Considering ratio resources, the throughput of bit b under sparse cluster regime should meet the following inequation:*

$$\sum_{b=1}^{\lambda^s nT} \frac{\Delta^2}{4} \frac{\mathbb{E}[R_{d_b^s}] - 1}{n} + \mathbb{E}\left[\sum_{b=1}^{\lambda^s nT} \sum_{h=1}^{h_b^s + \frac{nRc_b^s}{mR^2}} \frac{\pi\Delta^2}{4} \frac{{l_b^h}^2}{mR^2} \right] \leq c_1^s WT \log n \,, \tag{24}$$

where c_1^s is a positive constant.

The proof is omitted for simplification as it is similar to Proposition 3 in [20] and Lemma 3.3 in [2]. With the above lemma and theorem, we can deduce the following theorem:

Theorem 2 (throughput capacity under sparse clustering regime with incentive schemes)**.** *In a sparse clustering regime, the throughput capacity with incentive schemes is*

$$O\left(\min\left\{ \sqrt{\frac{n \log n}{mR^2\Delta^2}}, \frac{mR^2}{n\Delta^2} \right\} \right) \tag{25}$$

and the upper bound can be achieved when

$$l = \Theta\left(R\sqrt{\frac{m}{n}} \right).$$

The proof is shown in Appendix A.

5. Factors Discussion

To design efficient incentive schemes with the formula, it is necessary to take a detailed discussion for each factor. The cooperation degree is introduced as an important factor for the incentive degree that provides a base of a network, as discussed in Section 2.1, so we discuss it first. Then, the mobility mode is analyzed because the capacity formula without incentive schemes should have the same result as the sparse clustering regime. Finally, each factor in utility $\$^*$ reveals impacts.

5.1. Cooperation Degree

As $\mu_\top$ and $\mu_\perp$ in (21) are defined by the network environment, intuitively, Theorem 1 should meet $\lim_{\Delta \to \infty} \lambda^s = 0$ and $\lim_{\Delta \to 1} \lambda^s = O\left(\frac{mR^2}{n}\right)$. The first proposition is easily derived because $\lim_{\mu \to \infty} \$ = \infty$ theoretically. When $\lim_{\$ \to \infty} \Delta = \left(\frac{\mu_\top}{\mu_\perp}\right)^{\frac{1}{\alpha}} - 1 = 1$, the second proposition is derived. It must be noted that μ has explicit upper and lower bounds in reality because nodes cannot communicate to others without restriction. Therefore, the real upper bound should consider the cooperation between nodes defined by network topology and mobility.

5.2. Mobility Mode

Without incentive schemes, m, R, and n, represented as mobility mode, are the main factors. The capacity defined in Theorem 2 is a minmum value from two formulas. To find the impactful range of each formula, the variation of network capacity versus the number of nodes is shown with results in Figure 6. With an increasing number of nodes, network capacity grows quickly at the beginning. It converges at last due to the limited ratio resources. The red vertical line is the break point represented as the ideal number of nodes in a special network. Here, m and R have corresponding results as above because they are inversely proportional to n.

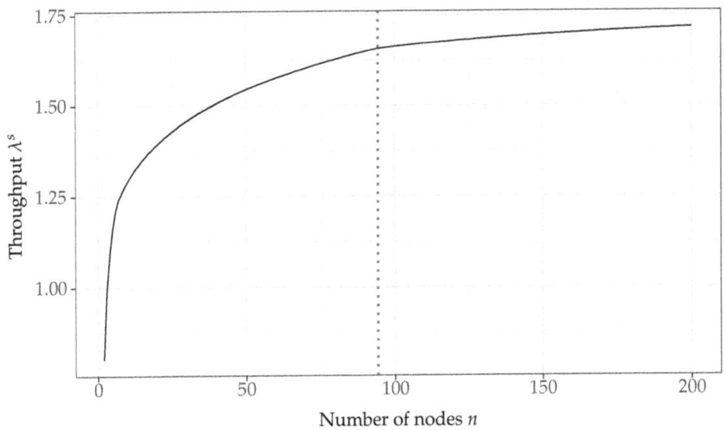

Figure 6. Number of nodes versus throughput.

5.3. Incentive Factors

In the following discussion under the two different formulas, the part from Theorem 1 in Theorem 2 is discarded for simplicity. The parameters used in analysis are listed in Table 3, and are selected according to simulations in Section 3 and cluster sparse regime [2].

Table 3. Value of parameters.

Parameter	Description	Value
n	number of node	100
v	Grow exponent of m	4/9
β	Growth exponent of R	1/3
$\mu_{\top}$	Maximum achievable entanglement	10
$\mu_{\perp}$	Minimum achievable entanglement	5
α	Decreasing coefficient of cooperation	1
μ	Entanglement	6
c	Constant cost	0
s	Safty coefficient	2
E_c	Energy cost per message	1
B_c	Buffer cost per message	1
E_s	Energy capacity	5
B_s	Message buffer	5
Z	Message price	100
R	Message reward	30

How incentive schemes affect network capacity is shown in Figure 7. An incentive scheme usually tries its best to increase its incentive degree to reach a higher capacity. However, the decreasing rate of change suggests that throughput cannot increase infinitely. The limitation is restricted by the cooperation degree discussed in Section 5.1. Moreover, the incentive degree has its own extremum, as shown in Section 3. Therefore, thoughput can reach its limitation earlier.

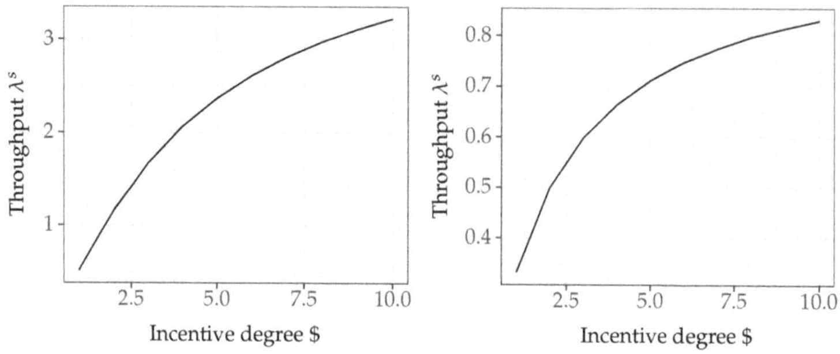

(**a**) Without ratio resources (**b**) With ratio resources

Figure 7. Incentive schemes versus throughput.

The limitation can be observed obviously through the relation with buffer usage and energy usage. Similar to the analysis in Section 3, the message size and energy cost per message work with buffer size and energy capacity together. As such, only the latter are discussed here; the relations are shown in Figures 8 and 9. The fact that they reach the limitation quickly suggests that both of them are the basic condition to ensure that nodes communicate. If a higher network capacity is needed, other factors should be adjusted to increase the achievable upper bound. According to this analysis, a suitable buffer and energy are more important; their value can be taken from our theorem.

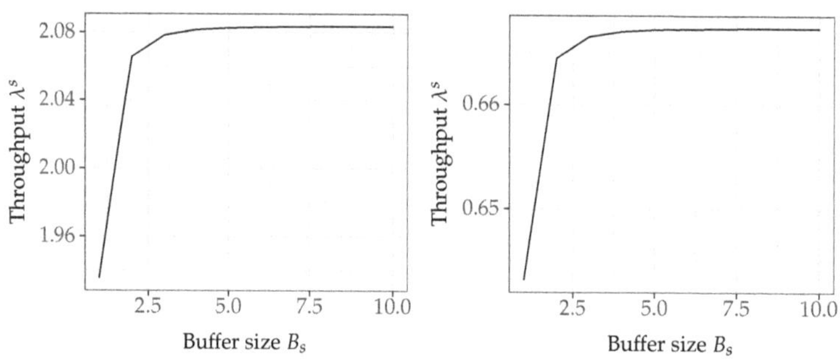

(**a**) Without ratio resources (**b**) With ratio resources

Figure 8. Buffer size schemes versus throughput.

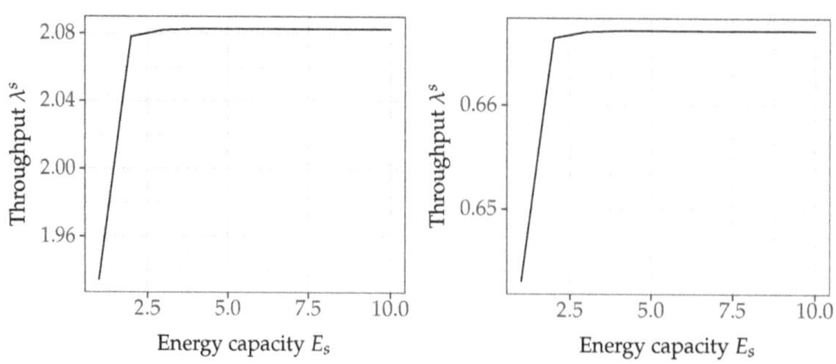

(**a**) Without ratio resources (**b**) With ratio resources

Figure 9. Energy capacity versus throughput.

The secure and incentive scheme can affect the achievable upper bound shown in Figures 10 and 11. Both of them have negative effects on network capacity. However, we cannot reduce them discretionarily. A too low reward or a too high price may block communications between nodes due to insufficient credits. A too low security coefficient may bring attacks to nodes. With these conditions, the relations can help us to design incentive schemes to select rewards and evaluate the security coefficient for both a node and the network.

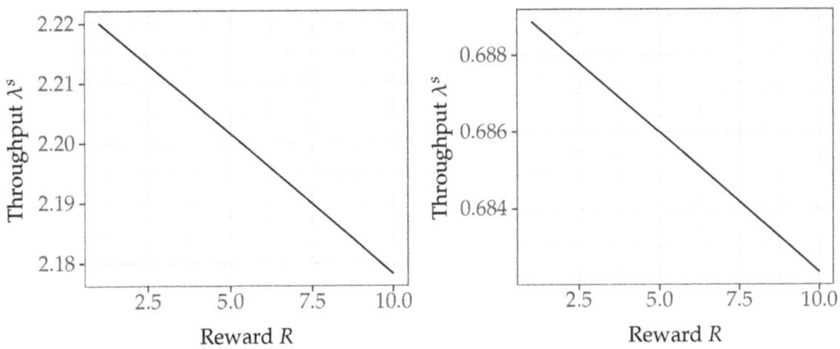

(**a**) Without ratio resources (**b**) With ratio resources

Figure 10. Reward versus throughput.

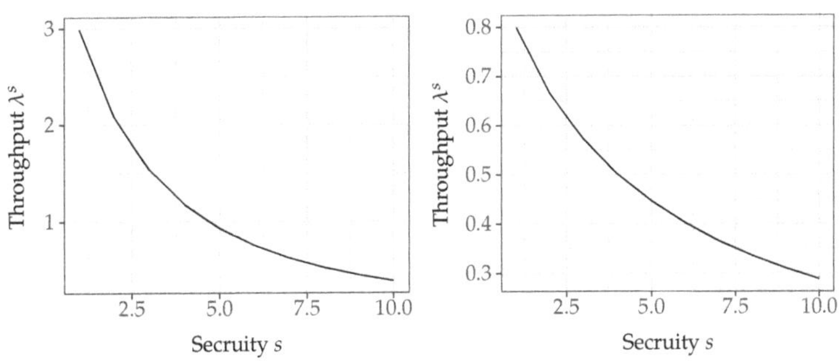

(**a**) Without ratio resources (**b**) With ratio resources

Figure 11. Security versus throughput.

6. Conclusions

In this paper, we design a mathematical model based on a quantum game model involving node parameters and use this model with a sparse clustering regime to analyze network capacity with incentive schemes. The mathematical model reflects the incentive degree for an incentive scheme in opportunistic networks with selfish nodes, finding that the character of incentive schemes is mainly determined by the relationship between reward and cost. According to the capacity analysis, how the factors affects on capacity are aware. The formula can be a guideline for new efficient incentive schemes. However, only energy usage, buffer usage, and security are considered. Other parameters, such as messages alive time and how the environment affects entanglement, can be taken into account as well. The capacity analysis associates the transmission model with incentive schemes simply. In the future, a more powerful model should be raised to reveal more details. In addition to sparse clustering regimes, other mobility models can be analyzed with incentive schemes.

Author Contributions: Conceptualization, S.J. and R.F.; methodology, R.F.; software, R.F.; validation, R.F.; formal analysis, R.F.; investigation, R.F.; resources, R.F. and S.J.; data curation, R.F.; writing—original draft preparation, R.F.; writing—review and editing, S.J. and Z.Z.; visualization,

R.F.; supervision, Shengming Jiang; project administration, S.J.; funding acquisition, S.J. All authors have read and agreed to the published version of the manuscript.

Funding: This research recevied was funded by The Innovation Program of Shanghai Municipal Education Commission of China, grant number 2021-01-07-00-10-E00121.

Institutional Review Board Statement: Not applicable.

Informed Consent Statement: Not applicable.

Data Availability Statement: Not applicable.

Conflicts of Interest: The authors declare no conflict of interest.

Appendix A. Proof of Theorem 2

Applying Cauchy–Schwarz inequality to the second monomial in (24) obtains

$$\left(\sum_{b=1}^{\lambda^s nT}\sum_{h=1}^{h_b^s+\frac{nRc_b^s}{mR^2}} l_b^h\right)^2 \le \left(\sum_{b=1}^{\lambda^s nT}\sum_{h=1}^{h_b^s+\frac{nRc_b^s}{mR^2}} 1\right)\left(\sum_{b=1}^{\lambda^s nT}\sum_{h=1}^{h_b^s+\frac{nRc_b^s}{mR^2}} l_b^h\right)$$
$$\le \frac{WTn}{2}\sum_{b=1}^{\lambda^s nT}\sum_{h=1}^{h_b^s+\frac{nRc_b^s}{mR^2}} (l_b^h)^2,$$

where the last step is from (22). Then, we have

$$\mathbb{E}\left[\sum_{b=1}^{\lambda^s nT}\sum_{h=1}^{h_b^s+\frac{nRc_b^s}{mR^2}} (l_b^h)^2\right] \ge \frac{2}{WTn}\mathbb{E}\left[\left(\sum_{b=1}^{\lambda^s nT}\sum_{h=1}^{h_b^s+\frac{nRc_b^s}{mR^2}} (l_b^h)\right)^2\right].$$

Applying Jensen inequality to the above inequation obtains

$$\frac{2}{WTn}\mathbb{E}\left[\left(\sum_{b=1}^{\lambda^s nT}\sum_{h=1}^{h_b^s+\frac{nRc_b^s}{mR^2}} (l_b^h)\right)^2\right] \ge \frac{2}{WTn}\left(\mathbb{E}\left[\sum_{b=1}^{\lambda^s nT}\sum_{h=1}^{h_b^s+\frac{nRc_b^s}{mR^2}} (l_b^h)\right]\right)^2,$$

where l_b^s is the transimssion range of bit b from source s and r_b^s is the distance of each transmission between two nodes. For multi-hops, the following inequation is obvious:

$$\sum_{h=1}^{h_b^s} l_b^h \ge l_b^s.$$

Appling Jensen inequality yields

$$\frac{2}{WTn}\left(\mathbb{E}\left[\sum_{b=1}^{\lambda^s nT}\sum_{h=1}^{h_b^s+\frac{nRc_b^s}{mR^2}} (l_b^h)\right]\right)^2 \ge \frac{2}{WTn}\left(\sum_{b=1}^{\lambda^s nT}\mathbb{E}[l_b^s]\right)^2$$
$$\mathbb{E}\left[\sum_{b=1}^{\lambda^s nT}\sum_{h=1}^{h_b^s+\frac{nRc_b^s}{mR^2}} (l_b^h)^2\right] \ge \frac{2}{WTn}\left(\sum_{b=1}^{\lambda^s nT}\mathbb{E}[l_b^s]\right)^2.$$

Substituting the above formula into (24) yields

$$\sum_{b=1}^{\lambda^s nT} \frac{\mathbb{E}[R_{d_b^s}] - 1}{n} + \frac{2\pi}{WTnmR^2} \left(\sum_{b=1}^{\lambda^s nT} \mathbb{E}[l_b^s] \right)^2 \leq \frac{4c_1^s WT \log n}{\Delta^2} . \tag{A1}$$

To continue the calculation, we should discuss $R_{d_b^s}$. Let $\hat{l_b^s(t)}$ be the minimum distance between the current cluster to C_d in slot t. If the value is negative, there is an overlap between the current cluster and C_d. For convenience, we define $\hat{L_b^s(t)} = \max(0, \hat{l_b^s(t)})$. For bit b, considering the network area and the transmission area of a cluster yields

$$\mathbb{E}\left[\frac{n}{[R + \hat{L_b^s(t)}]^2}\right] = \mathbb{E}\left[\frac{n}{R^2}\mathbb{I}_{L_b^s(t)\leq 0}\right] + \mathbb{E}\left[\frac{n}{[R + \hat{l_b^s(t)}]^2}\mathbb{I}_{L_b^s(t)>0}\right] . \tag{A2}$$

For the second monomial, we can obtain

$$\begin{aligned}
\mathbb{E}\left[\frac{n}{[R + \hat{l_b^s(t)}]^2}\mathbb{I}_{L_b^s(t)>0}\right] &= \int_0^{\sqrt{n}} \frac{n}{(R+u)^2}\, d\Pr\left[\hat{l_b^s(t)} \leq u\right] \\
&= 1 - \frac{n}{R^2}\Pr\left[\hat{l_b^s(t)} \leq 0\right] \\
&+ \int_0^{\sqrt{n}} \frac{2n}{(R+u)^3}\Pr\left[\hat{l_b^s(t)} \leq u\right] du .
\end{aligned} \tag{A3}$$

Therefore, the original inequation is

$$\begin{aligned}
\mathbb{E}\left[\frac{n}{[R + \hat{L_b^s(t)}]^2}\right] &= 1 + \int_R^{\sqrt{n}} \frac{2n}{u'^3}\Pr\left[R + \hat{l_b^s(t)} \leq u\right] du' \\
&= 1 + \int_R^{\sqrt{n}} 2\pi R_{c_b^s}(t)\frac{(R+u')^2}{u'^3}\, du' \\
&\leq 1 + 6\pi R_{c_b^s}(t) \int_R^{\sqrt{n}} \frac{1}{u'}\, du \\
&= 1 + 6\pi R_{c_b^s}(t) \log\frac{\sqrt{n}}{R} \\
&\leq 1 + 6\pi R_{c_b^s}(t) \log n \\
&\leq 6\pi R_{c_b^s}(t) \log n .
\end{aligned} \tag{A4}$$

For a message under all slots, we have

$$\begin{aligned}
\mathbb{E}[(R + \hat{L_b^s})^2] &\leq (R + l_b^s)^2 \\
\mathbb{E}\left[\frac{n}{[R + \hat{L_b^s}]^2}\right] &\geq \frac{n}{(R + l_b^s)^2} \\
\frac{n}{(R + l_b^s)^2} &\leq 6\pi R_{c_b^s} \log n \\
R_{c_b^s} &\geq \frac{n}{6\pi \log n (R + l_b^s)^2} \\
&\geq \frac{n}{24\pi \log n R l_b^s} .
\end{aligned} \tag{A5}$$

Substituting the above inequation into (A1) yields

$$\frac{1}{24\pi \log n R \sum_{b=1}^{\lambda^s nT} \mathbb{E}[l_b^s]} + \frac{2\pi}{WTnmR^2}\left(\sum_{b=1}^{\lambda^s nT} \mathbb{E}[l_b^s] \right)^2 \leq \frac{4c_1^s WT \log n}{\Delta^2} . \tag{A6}$$

If $h_b^s < \frac{nR_{c_b^s}}{mR^2}$, the ratio resources consumption does not decrease with a decreasing number of hops due to the unchanging transmission range. In such a situation, we deem $h_b^s = \Theta(\frac{nR_{c_b^s}}{mR^2})$, which is contained in the other situation and has the same result.

If $h_b^s \geq \frac{nR_{c_b^s}}{mR^2}$, we can calculate (A6) directly. If $\sum_{b=1}^{\lambda^s nT} \mathbb{E}[l_b^s] < \lambda^s TmR^2$, the first monomial has little impact on the result in (A6), and it is duplication in clusters which is restricted by duplication in clusters according to Theorem 1. We ignore this term and obtain

$$\begin{aligned} \frac{2\pi}{WTnmR^2}\left(\lambda^s TmR^2\right)^2 &\leq \frac{4c_1^s WT\log n}{\Delta^2} \\ \lambda^{s2} &\leq \frac{c_1^s nW^2 \log n}{mR^2\Delta^2} \\ \lambda^s &\leq O\left(\sqrt{\frac{n\log n}{mR^2\Delta^2}}\right) \end{aligned} \tag{A7}$$

If $\sum_{b=1}^{\lambda^s nT} \mathbb{E}[l_b^s] \geq \lambda^s TmR^2$, according to Theorem 1, the same upper bound can be achieved when $l_i = \Theta\left(R\sqrt{\frac{m}{n}}\right)$.

References

1. Li, P. Capacity of wireless ad hoc networks. In *Encyclopedia of Wireless Networks*; Springer: Berlin/Heidelberg, Germany, 2020; pp. 160–164.
2. Yao, S.; Wang, X.; Tian, X.; Zhang, Q. Delay-throughput tradeoff with correlated mobility of ad-hoc networks. In *Proceedings of the IEEE Infocom 2014—IEEE Conference on Computer Communications, Toronto, ON, Canada, 27 April–2 May 2014*; IEEE: Piscataway, NJ, USA, 2014; pp. 2589–2597.
3. Yoo, Y.; Agrawal, D.P. Why does it pay to be selfish in a MANET? *IEEE Wirel. Commun.* **2006**, *13*, 87–97. [CrossRef]
4. Ge, L.; Jiang, S. An Efficient Opportunistic Routing Based on Prediction for Nautical Wireless Ad Hoc Networks. *J. Mar. Sci. Eng.* **2022**, *10*, 789. [CrossRef]
5. Akestoridis, D.G.; Papapetrou, E. A framework for the evaluation of routing protocols in opportunistic networks. *Comput. Commun.* **2019**, *145*, 14–28. [CrossRef]
6. Aditya, A.; Srinivasan, S.; Richard, K.; Scott, S.; Christos, P. Selfish Behavior and Stability of the Internet: A Game-Theoretic Analysis of TCP. *SIGCOMM Comput. Commun. Rev.* **2002**, *32*, 117–130.
7. Li, W.; Chen, J.; Zhou, B. Game Theory Analysis for Graded Punishment Mechanism Restraining Free-Riding in P2P Networks. In Proceedings of the 2011 International Symposium on Computer Science and Society, Kota Kinabalu, Malaysia, 16–17 July 2011; pp. 262–266.
8. Park, J.; van der Schaar, M. A Game Theoretic Analysis of Incentives in Content Production and Sharing Over Peer-to-Peer Networks. *IEEE J. Sel. Top. Signal Process.* **2010**, *4*, 704–717. [CrossRef]
9. Sun, W.; Wang, S. General Analysis of Incentive Mechanisms for Peer-to-Peer Transmissions: A Quantum Game Perspective. In Proceedings of the 2017 IEEE 37th International Conference on Distributed Computing Systems (ICDCS), Atlanta, GA, USA, 5–8 June 2017; pp. 517–526.
10. Zhai, Y.; Bai, X.; Liu, Q. Incentive mechanisms in mobile delay tolerant network. In Proceedings of the 2017 IEEE 7th International Conference on Electronics Information and Emergency Communication (ICEIEC), Macau, China, 21–23 July 2017; pp. 184–188.
11. Wang, S.; Sun, W.; Ma, L.; Lv, W.; Cheng, X. Quantum Game Analysis on Extrinsic Incentive Mechanisms for P2P Services. *IEEE Trans. Parallel Distrib. Syst.* **2020**, *31*, 159–170. [CrossRef]
12. Lu, R.; Lin, X.; Shi, Z.; Cao, B.; Sherman Shen, X. IPAD: An incentive and privacy-aware data dissemination scheme in opportunistic networks. In Proceedings of the 2013 IEEE INFOCOM, Turin, Italy, 14–19 April 2013; pp. 445–449.
13. Qin, X.; Wang, X.; Wang, L.; Lin, Y.; Wang, X. An efficient probabilistic routing scheme based on game theory in opportunistic networks. *Comput. Netw.* **2019**, *149*, 144–153. [CrossRef]
14. Lu, R.; Lin, X.; Zhu, H.; Shen, X.; Preiss, B. Pi: A practical incentive protocol for delay tolerant networks. *IEEE Trans. Wirel. Commun.* **2010**, *9*, 1483–1493. [CrossRef]
15. Mingchu, L.; Xing, J.; Cheng, G.; Jia, L.; Guanghai, C.; Tie, Q. RIMNet: Recommendation Incentive Mechanism based on evolutionary game dynamics in peer-to-peer service networks. *Knowl.-Based Syst.* **2019**, *166*, 156–169.
16. Evjola, S. Energy consumption analysis of different routing protocols in a Delay Tolerant Network. *J. Ambient. Intell. Humaniz. Comput.* **2020**, *11*, 3833–3839.
17. Ghani-Ur, R.; Anwar, G.; Shad, M.; Madhusudan, S.; Dhananjay, S. Selfishness in Vehicular Delay-Tolerant Networks: A Review. *Sensors* **2020**, *20*, 3000.
18. Gupta, P.; Kumar, P. The capacity of wireless networks. *IEEE Trans. Inf. Theory* **2000**, *46*, 388–404. [CrossRef]

19. Ciullo, D.; Martina, V.; Garetto, M.; Leonardi, E. Impact of correlated mobility on delay–throughput performance in mobile ad hoc networks. *IEEE/ACM Trans. Netw.* **2011**, *19*, 1745–1758. [CrossRef]
20. Lin, X.; Shroff, N.B. The fundamental capacity-delay tradeoff in large mobile ad hoc networks. In Proceedings of the Third Annual Mediterranean Ad Hoc Networking Workshop, Bodrum, Turkey, 27–30 June 2004. Available online: https://citeseerx.ist.psu.edu/viewdoc/download;jsessionid=0F24A437D3E894EAA2445A8CF8D1AE33?doi=10.1.1.86.2431&rep=rep1&type=pdf (accessed on 20 August 2022).

MDPI AG
Grosspeteranlage 5
4052 Basel
Switzerland
Tel.: +41 61 683 77 34

Journal of Marine Science and Engineering Editorial Office
E-mail: jmse@mdpi.com
www.mdpi.com/journal/jmse

www.ingramcontent.com/pod-product-compliance
Lightning Source LLC
Chambersburg PA
CBHW061443210926
51850CB00008B/1594
* 9 7 8 3 7 2 5 8 5 5 4 3 8 *